W0043969

Stress and Tension Control 2

Stress and Tension Control 2

Edited by

F. J. McGuigan

United States International University
San Diego, California

Wesley E. Sime

University of Nebraska
Lincoln, Nebraska

and

J. Macdonald Wallace

West London Institute of Higher Education
London, England

PLENUM PRESS · NEW YORK AND LONDON

Library of Congress Cataloging in Publication Data

International Interdisciplinary Conference on Stress and Tension Control (2nd: 1983: University of Sussex)
Stress and tension control 2.

"Proceedings of the Second International Interdisciplinary Conference on Stress and Tension Control, held August 30–September 3, 1983, at the University of Sussex, Brighton, England"—T.p. verso.
Includes bibliographies and index.
1. Stress (Psychology)—Congresses. 2. Stress (Physiology)—Congresses. 3. Relaxation—Congresses. 4. Medicine, Psychosomatic—Congresses. 5. Job stress—Congresses. I. McGuigan, F. J. (Frank J.), 1924–   . II. Sime, Wesley E. III. Wallace, J. Macdonald. IV. Title. V. Title: Stress and tension control two. [DNLM: 1. Relaxation—congresses. 2. Stress, Psychological—prevention & control—congresses. W3 IN11887 2nd 1983s / WM 172 I603 1983s]
BF575.S75I57 1983   616'.001'9   84-18104

ISBN-13: 978-1-4612-9726-0   e-ISBN-13: 978-1-4613-2803-2
DOI: 10.1007/978-1-4613-2803-2

Proceedings of the Second International Interdisciplinary Conference on Stress and Tension Control, held August 30–September 3, 1983, at the University of Sussex, Brighton, England

©1984 Plenum Press, New York
Softcover reprint of the hardcover 1st edition 1984
A Division of Plenum Publishing Corporation
233 Spring Street, New York, N.Y. 10013

All rights reserved

No part of this book may be reproduced, stored in a retrieval system,
or transmitted in any form or by any means, electronic, mechanical,
photocopying, microfilming, recording, or otherwise,
without written permission from the Publisher

DEDICATION

To Edmund Jacobson (April 22, 1888 – January 7, 1983)

    This volume is affectionately dedicated to Edmund Jacobson, the father of the contemporary field of relaxation. His classic "Progressive Relaxation" published in 1929 established the basic principles of scientific and clinical relaxation. While his contributions are internationally recognized, they will be increasingly appreciated by future generations. His last laboratory research on profiles of tension states has been recognized as medicine practiced 50 years ahead of its time.

PREFACE

The Second International Interdisciplinary Conference on Stress and Tension Control, sponsored by the International Stress and Tension Control Society, was held at The University of Sussex, Brighton, England during the period August 30 – September 3, 1983. The Society has evolved from the American Association for the Advancement of Tension-Control, which met each year for five years in Chicago commencing in 1974, and for which proceedings such as these were published annually. Because of an international flavor which the association gradually acquired the name was changed to that of The International Stress and Tension-Control Association. That organization met in London in 1979, and then in Louisville, Kentucky in 1981 in conjunction with The Biofeedback Society of America. The proceedings of that first international conference in London were also published by Plenum Publishing Company. (<u>Stress and Tension Control</u>, McGuigan, Sime and Wallace, 1981). Because the results of that first conference were so gratifying, this second conference was scheduled, with similar consequences.

These proceedings are offered for the purpose of advancing our methods of coping with stress through tension control, for excessive bodily tension can indeed result from failure to adapt to the many stresses of life that we all continually face. As we are well aware, the consequences of chronic overtension can be disastrous in many ways for the human body.

There was minimal editing of the papers so that they retain the character created by the authors, though substantial modification was necessary in some cases to improve English comprehension. It was not possible to carefully edit such matters as citations, quotations and statistical analyses, so that they must remain the responsibility of the authors. The papers should be a fair representation of contemporary international work in the field. We do hope that the reader will profit from these presentations as much as the attendees at the conference did. Due to illness of one of the authors, there was a delay of several months in collating the papers, for which we apologize as we well recognize the importance of timely reporting of conference proceedings such as these.

We want to express our great appreciation to Elaine and David Wright* for the excellent editorial and typing assistance in the production of this book. Hiroshima Shudo University is also appreciated for the support that was furnished during the conclusion of this volume, as are United States International University and the University of Nebraska for earlier support.

F. J. McGuigan

* WordWrights, 405 Santa Dominga, Solana Beach, California, USA 92075

CONTENTS

CLINICAL APPLICATIONS IN MEDICINE, PSYCHOLOGY-PSYCHIATRY, SPEECH PATHOLOGY AND DENTISTRY

STUDIES OF STRESS AND TENSION IN OCCUPATIONAL SETTINGS

STRESS AND TENSION CONTROL IN EDUCATION

METHODS OF COPING WITH STRESS

FOREWORD

AN OVERVIEW OF CONTEMPORARY WORK IN THE FIELD

OF STRESS AND TENSION CONTROL

F. J. McGuigan

Executive Vice President
International Stress & Tension Control-Society

"Stress" and "Tension" are internationally recognized words. Their omnipresence in our public media - in our newspapers, on TV, in magazines, and on radio - as well as in our everyday conversations indicate that we are well aware of the problems of over-tenseness. Pulp newspapers and magazines increase their sales with promises of quick relief for stress and tension problems. Business executives complain at the end of the day of being "uptight", and often accept a hotel chain's invitation to "unwind" at their bar. Soap operas attract large audiences, in part by capitalizing on tension problems - irritable arguments between husband and wife seem interminable. Indeed, the entire world is aware of the need for stress management. Such widespread needs invite varied "solutions", with the most attractive appearing ones offering promises of easy relief, with minimal cost. Those who suffer headaches, spastic colon, "essential hypertension", back pains, phobias and anxiety are especial sensitized to tension disorders and potential cures. Organizations such as The International Stress and Tension-Control Society are dedicated to the elimination of such tension problems. Our first priority is the immediate application of tension control principles that now exist, and secondly to encourage scientific research to further develop our methods. Eventually the public can practice only those technics of tension control that have sound scientific and clinical validation; this can be accomplished in part by the objective measurement through electromyography of the tension that is the reaction to stress.

The classical, standard definition of tension is contraction of the skeletal muscles; and relaxation is the elongation of skeletal muscle fibers. The consequences of these scientific definitions of tension and relaxation for prophylactic and for therapeutic purposes are enormous, as these proceedings illustrate. Our long term goal is

3

to bring effective stress and tension control methods to the peoples
of the world. By working vigorously together, interdisciplinary efforts such as those in this volume should bring us increasingly close
to success.

Our objective in this volume is to present a fairly representative sample of work being conducted in the western world. A perspective thus emerges as to what scientists and technologists in stress
and tension control are seeking to accomplish and what they are, in
fact, accomplishing. It was possible to include the large majority of
presentations at the conference. Consequently the reader may note
some variability in the quality of the papers, both in terms of
methodological and content criteria. Each paper thus must stand on
its own merit.

The papers were categorized into six groups to facilitate comprehension. While other equally defensible categories could obviously
have also been formed, these have to do with:

(1) A dedication to Edmund Jacobson in recognition of his
    pioneering of our field over seven decades ago, and in
    view of his recent death on January 7, 1983 at the age
    of 93;

(2) Research strategies and approaches that offer some
    stimulating ideas for further advancement;

(3) Clinical applications in diverse health related fields
    that continue to emphasize the value of relaxation for
    both prophylactic and therapeutic purposes;

(4) Stress and tension control in the everyday world of
    occupational settings wherein so many of our tension
    maladies are generated;

(5) Applications in the critically important field of
    education where, if relaxation were more widely taught
    to the children of the world, the benefits for society
    would be countless;

(6) Some methods of coping with stress.

For an enlarged perspective of these six categories the reader can
consult McGuigan (1978, 1981).

The first section, that devoted to Edmund Jacobson, Ph.D., M.D.,
could well have comprised the entire volume had we but called on the
countless international devotees of Progressive Relaxation to offer
their widely varying contributions in diverse health related applica-

tions. Joseph Wolpe's presidential presentation, however, is sufficient to illustrate the importance of Jacobson's work. In discussing
how we can cope and change our undesirable tension habits through
Progressive Relaxation, he emphasizes the importance of differential
relaxation and his own important application in Systematic Desensitization. By directly controlling our muscles, we can produce autonomic effects that constitute "a remarkable therapeutic resource".
Progressive Relaxation, he points out, is effective for relieving
anxiety, and for reducing blood pressure. It can also change a Type A
coronary prone behavior pattern so as to overcome excessive habits of
anger and over-reactions to frustration in much the same way that we
can overcome neurotic anxiety. Wolpe recognizes the great importance
of Edmund Jacobson's work and states that his "discovery about a half-
a-century ago that Progressive Relaxation produces a state of emotional calmness was one of the most important developments in the
history of psychiatry".

The second category of papers is highly recommended as being most
thought provoking and should reward the reader with novel notions,
ideas, principles and strategies. Eysenck offers us another of his
intellectually challenging theses, this one having to do with the
possibility that stress and personality factors are causally linked to
cancer. He notes that there is a correlation between measures of
cancer and extraversion which he attempts to understand through conditioning mechanisms. He astutely recognizes the complexity of these
relationships and the difficulties we have in dissecting them out when
we are compelled to study normal humans living in their everyday
worlds. For instance, stress and personality factors must interact
with such behavioral characteristics as smoking and drinking.
Reminiscent of the speculations of R. A. Fisher, Eysenck suggests that
genetic factors are important for personality, for smoking and drinking habits and they thus may contribute to the development of
carcinogenic illnesses.

Seyffarth presents a scholarly analysis of how the human body
reacts to stress. The extensive involvement of the skeletal musculature to stress is discussed and illustrated in some detail. Excessive
tension, which he refers to as long-lasting static muscular work,
provokes ischemia and thence ischemic impulses in the nervous systems
that have pathologic consequences. However, he points out that the
body's reaction to stress depends on whether or not the person can
solve the stress problem. If the problem is solved, the reaction of
the body is healthy in that the potential for action increases. But
if the problem is insoluble, the potential for action decreases.
Seyffarth takes issue with Selye's position since Selye failed to make
this important distinction. For this and other reasons he mentions,
Seyffarth is highly critical of Selye's work and considers that it had
quite unfortunate consequences for Scandinavian medicine. This paper
is certainly a distinguished contribution to this volume.

Sainsbury discusses the nature of muscle tension, and states that the most conspicuous of the bodily accompaniments of psychological states are muscular contractions. He and Aubrey Lewis measured muscle tension through electromyography and concluded that all patients had excessively high tension scores and that there are high correlations between level of anxiety and tension symptoms. Other findings indicate a sophisticated dealing with the problem of tension and anxiety. In discussing objective measurements of motility depressives with high fidgit counts had significantly higher scores on measures of anxiety and neuroticism. While these superfluous movements do not convey information, these patients have something to communicate, but they are restrained from doing so in some way. Sainsbury is thus caused to ask "what is communicated by nonverbal movements?"

Jacobs' paper is a model for the clinician to be greatly admired. One of the misfortunes of clinical practice, in my opinion, is that there is relatively little effort to systematically collect data on the efficacy of various procedures. Jacobs not only addresses the important problem of discrimination of visceral stimuli, but offers a sensible methodological proposal whereby the clinician can engage in therapeutic efforts, while simultaneously evaluating them without retarding the patient's presumed progress. The clinician who wishes to validate her or his methods for the benefit not only of the clinician's own future patients/clients but also to furnish sound therapeutic knowledge for others would do well to apply Jacob's valuable research strategy.

Wickramasekera presents data on 103 of his patients which support the conclusion that patients at greatest risk are those who are positive for certain personality features and who are deficient in support systems and coping skills. The person at lowest risk is the patient who has none of those personality features and who has effective use of multiple support systems and coping skills. For those who are high in hypnotic ability and high in neuroticism there is an increased probability that psychophysiological factors cause their stress-tension disorder. He concludes that daily hassles can accumulate to affect somatic health, even after the effects of major life events are statistically eliminated, as in Holmes and Rahe's 1967 work.

Cork and Cox critically review research in the field of relaxation and correctly point out that it is typical that insufficient relaxation training is given participants in experimental evaluations. However, they indicate that relaxation training and sitting quietly have similar psychophysiological effects, a statement in direct conflict with the findings of Jacobson (1929, 1957). Jacobson found that rest was decidedly inferior to Progressive Relaxation when PR is properly learned and when tension is measured through electromyography. This difference possible reflects a frequent error in the stress and tension control literature, viz., the tacit assumption that all relaxation training methods are similar, which of course is not

the case -- few experimenters use Progressive Relaxation training as
specified by Jacobson. Cork and Cox also support a social psychophysiological method of relaxation training in which the expectations
and perceptions of the subjects influence their covert behavior, as
psychophysiologically measured. "Control conditions" they point out
are typically inadequate in that they are not control conditions at
all. To elaborate on this point we can contrast a control as against
a comparison condition wherein subjects are randomly assigned to experimental and control conditions; in contrast a comparison group is
not randomly composed of members of a sample, but is one that is
already formed, as in second grade school class (McGuigan, 1983).

Karnitschnig suggests that the endorphins, the opiate within us,
constitute a pleasurable chemical produced by stress. The endorphins
can prevent one from feeling pain during stress, they produce a state
of calmness accompanied by lowering blood pressure, respiration and
heart rate and reduced intestinal motility. Endorphins therefore
normalize the stress response of adrenalin. Naloxone blocks endorphins and opiates. Substance P, a neurotransmitter like ACTH, GABA
and dopamine, occurs throughout the body and there is a critical
balance between endorphins and substance P which is thought to be a
factor in the gate control theory of pain. Noxious stimuli release
substance P in the dorsal root ganglion. If enough messages come
through, one will feel pain. Consequently, humans need sufficient
substance P and not too many endorphins to feel pain. If one is
deficient in substance P, pain is not experienced nor is there ability
to heal injury. In human amputees, the spinal cord shows a depletion
of substance P on the side of the missing limb. It is speculated that
massage and other forms of manipulation that are a little painful
produce substance P, suggesting that substance P may be necessary for
the healing process. Substance P seems to normalize sleep patterns in
stressed rats, seems to help avoid overeating under stress, seems to
heal our wounds (as when bee stings increase substance P) and it is a
counterirritant that helps to heal rheumatism. The reader should find
that this paper is most interesting, if speculative, and has
intriguing and innovative ideas.

The section grouping various clinical applications well illustrate the diversity of therapeutic approachs in our field, as well as
the interdisciplinary character of our organization. When one considers that tension control technics are applied to the numerous
disorders in dentistry, medicine, psychology, psychiatry, education,
physiotherapy, and speech pathology, the therapeutic implications
become enormous. Add to that the prophylactic consequences of the
early educational application of tension control methods and the
benefits for society can be multiplied.

For instance, an interesting application of Progressive Relaxation by
Robert Rinehart, a rheumatologist, is with the treatment of arthritis,
bursitis, glandular disturbances, neuritis, rheumatism, neuralgia,

stiff neck, headache, sciatica, backache, and chest pains. Primary attention is placed on the relaxation of those muscles used in "the instinct to fight", the muscles of the arms, shoulders, neck, and chest, and those for "the instinct to flee" (the hips, back, and legs). Rinehart's theory is that overtension in these muscles leads to muscular fatigue and spasm (cramp). Since muscles are relatively insensitive, some cramping may persist for months or years without significant symptoms. When symptoms do occur they are apt to appear first in the neck and shoulder muscles, a pain in the neck, or in hip and lower back muscles. Eventually tightened joints, tendons, and bursae evidence signs of excessive wear and tear (inflammation). When mild and chronic, this is called osteoarthritis or "wear and tear" arthritis. When acute and severe, it is called rheumatoid arthritis and can become aggravated by an autoimmune allergic inflammation. Rinehart's clinical work suggests that most antirheumatic drugs relieve symptoms by suppressing inflammation (repair) and by relieving muscle spasm. Unfortunately this results in further weakening of muscle leading to more difficulties. His approach has been to use the classical rehabilitative measures of rest, exercise, and physical therapy. To those he has added intensive training in recognition and correction of maladaptive muscular tightening. Though a great deal of time and effort is required on the part of both patient and therapist, it is the only satisfactory way to achieve long-term gains. Rinehart has also discovered that muscle fatigue spasm is invariably the cause of pain and joint inflammation. In patients who had chronic pain, it is often too difficult to begin relaxation training immediately because the pain is so unbearable. Thus he prescribes physical therapy treatment such as heat, cold, massage, and vibration to loosen up the muscles and force circulation into the afflicted area. Once the pain has been temporarily alleviated in this manner, relaxation training can be used to control the tension and the pain more permanently.

Gessel, in a most stimulating paper, starts his thesis with the observation that among the medications that affect the biochemical system of the body, anti-depressants are universally recognized as effective for pain. Musculospasms increase the intensity of pain. It is thus clear that anti-depressants reduce pain, at least to some extent, because they are muscle relaxants. Tricyclics can help depression because they are muscle relaxants. Cardiologists who treat hypertension with diuretics have reported depression in some of their patients which is alleviated when sodium restrictions are reduced. There is thus an implied relationship between sodium metabolisim and depression. The hypothesis that chemically produced relaxation alleviates depression has sizeable therapeutic consequences and is consistent with clinical findings that natural relaxation sometimes reduces depression (McGuigan, 1981).

Mennell summarizes his extensive clinical experience in treating and in isolating causes of headaches. A major factor is the involvement of the musculoskelatal system. He offers us a rather precise mapping of muscular regions of the head that directly produce and control those sensitive regions. In particular, Mennell holds that myofascial trigger points in the muscles of the head and neck cause tension headaches. Trigger points can be diagnosed by palpation of the area of the muscle in which it is situated. The trigger point is locally tender but it is not near the place where the patient experiences pain. The pattern of pain can specify the location of a trigger point that is, furthermore, common among people. By palpating the trigger point the pain is reproduced and the muscle twitches with a "jump response". The treatment is to ensure that the muscle resumes its normal maximum length when at rest. This may be produced by fluori-methane spray and muscle stretch, or by injecting procaine. The muscle can also be strengthened by gentle stretching exercises followed by gentle moist heat. (Starling's Law holds that a long muscle is a strong muscle and a shortened muscle is a weak muscle). Furthermore, a general principle of therapy is that the structure in which a pathological condition occurs has to be rested from function while healing. This is a remarkable summary of information for diagnosing, treating and classifying various kinds of tension headaches.

Sedlacek, in his article on primary Raynaud's, summarizes three approaches to treatment: 1) surgery; 2) pharmacological agents; and 3) behavioral treatment with the latter being the most effective for primary Raynaud's at this time. In a thoughtful discussion, he examines the pros and cons of this approach, and there are some contra-indications.

Sedlacek and Cohen present some interesting data on an extremely complex problem, that of attempting to lower blood pressure in a "real-life appearing situation". The problem is complex and any conclusions within this very difficult research area must be carried lightly, but they conclude that a combination of EMG and thermal biofeedback was clinically useful. It is interesting to note that a Relaxation Response group (Benson) and a waiting list control group did not show significant decreases in blood pressure. Their work also reveals some interesting findings about motivational variables in field studies such as this.

The paper by Hill and Cox represents a valuable approach to the study of factors that lead to or contribute to myocardial infarction. Unfortunately, the results did not definitively ascertain any treatment effects, but they did develop a new MI scale based on a post hoc analysis of the questionnaire, which they are now using to further advance towards their goal.

Casteel in his work on speech pathology has given us an everyday analysis of the expression of being overly tense with such phrases as "I only need to get a grip on myself", "I'm scared stiff", "You give me a pain in the neck", "Keep a stiff upper lip", all indicating states of overtension. His analysis of speech pathologies indicates that there is excessive tension in the speech musculature which can often be relaxed away through the application of Progressive Relaxation.

Hodgson and Oatley discuss the nature of agoraphobia and the function of self-help groups in their role of coping. Interestingly they note that, contrary to popular belief and common definitions in medical references, agoraphobia means a fear of places where people gather ("agora" coming from the Greek meaning "meeting place"). It originally, then, did not mean a fear of open spaces. Hodgson and Oatley report that a primary ingredient in self-help procedures is the teaching of relaxation. They present some interesting survey data about characteristics of agoraphobics. Among their findings are that merely offering fellowship and a measure of hope without putting the person under any obligation to venture out, seems to reduce their depression. We look forward to more extensive data on this interesting approach and problem.

Lindsay presents a most interesting set of findings on the perception of fear by both dentists and patients. He sampled 100 adults, asking them to describe what they usually expect to experience during such dental procedures as drilling and injections. Forty dentists completed a similar questionnaire to estimate sensations, apprehensions and discomfort which they expected the average patient would experience during dental treatment. For the more intensely felt procedures, such as extractions and drilling, the patients attributed stronger sensations than did the dentists. Such a difference was not evident for the milder dental operations. A second sample of thirty patients rated, immediately before treatment, the sensations they expected from their treatments. They then immediatley afterwards rated what they had actually experienced on identical scales. Almost all subjects experienced less intense sensations than they had anticipated. Lindsay concludes that dentists have a more realistic assessment of the intensity of sensations than do patients, but both overestimate the fear and discomfort expected. The anticipation of treatment is more distressing than the treatment itself. Several years ago it was reported that dentists assumed the number one ranking for frequency of suicide, overtaking psychiatrists with this unenviable distinction. This may be partially explained by Lindsay's conclusion that dentists anticipate (empathize?) along with their patients that treatment will be more alarming and uncomfortable than it turns out to be. Both dentists and dental patients will find this paper most interesting.

The fourth section deals with the stresses that we experience in our work-world. Here, as in all phases of life, our ability to control our bodies "on the job" as we are harassed by bosses, are fatigued from overwork and are irritated by our colleagues, is critical for our survival as healthy human beings. Sime, Mayes, Witte, Ganster and Tharp engaged in a rather dangerous study of occupational stress on the job by going into and performing jobs in a nuclear power plant. They employed multiple physiological and psychological measures on large numbers of employees, and found a relationship between job strain, personality characteristics and personal health and/or organizational performance outcomes. While such field research is difficult to conduct, it has a clearly established important place in our study of stress and tension control.

Spillane reviews the Australian research literature on stress at work in a selective, comprehensive manner. He points out that stress research has a long history dating from 1916 in Australia. In addition to specific research findings that he summarizes, he presents an interesting view into the social fabric of Australia.

Hawkins reviews the stressors that affect those involved in aviation, including especially pilots and flight attendants. One of their prime methods of dealing with stress is alcohol. The effects of alcohol as well as the problems of other drugs form the basis of a valuable discussion. Finally, he discusses a non-pharmacological method for dealing with stress, Autogenic Training, and recommends that such techniques of controlling stress be used by management as well.

Haward undertakes the study of controlling stress for pilots. He first develops a short history of the term "stress", "stressor" and "strain" and indicates the considerable inconsistencies in their uses, with a criticism of Selye's work in this regard. Haward's main effort was to use Autogenic Training for this purpose. It should also be noted that the classic work of Jacobson in which Progressive Relaxation was applied by the United States Navy in training some 15,000 aviators during World War II to differentially relax is relevant to Haward's problem.

Leppanen employed questionnaires and group discussions to obtain data to assess the kinds and amounts of stress experienced in institutions by professional personnel who house and care for children and the mentally handicapped. These personnel proved to be a risk group from the point of view of occupational health, having more stress reactions than other Finnish groups who also have intensive social contacts. The subjects produced a plan to solve their problems by way of reduction of their stresses and an informal followup a year later indicated that some of their aims had been achieved. This is certainly a constructive, basically empirical, approach to help solve the problems of stress among workers by reducing the environmental factors and the working conditions that they themselves perceive as producing the increased stress.

In introducing the section on education we can do no better than
to repeat the well known principle that, by teaching our young to
differentially relax in the face of stress, the benefits for society
could be enormous. Educational applications of tension control
principles should become a, if not the, cardinal principle
of preventive psychiatry.

The paper by Setterlind and Patriksson addresses a most important
problem, viz, how to teach children at a young age to relax. The
obvious benefit is to allow them to carry the habit of relaxation
throughout their entire lives, rather than starting at an adult age
with less benefit. Furthermore, there are prophylactic measures possible so as to prevent the development of tension disorders. Various
details of the research methodology and of the statistical analysis in
this study are not included, but they are available directly from the
authors. The obvious methodological shortcomings in this research are
in part because of its preliminary nature and in any event are out-
weighed by the importance of the problem addressed.

Cowell discusses some varied methods for controlling stress primarily through a variation of Progressive Relaxation. By discussing a
number of individual cases, he makes application principally to pupils
in school and discusses some of the unique problems that an educator
(teacher) faces in such cases. For instance, some individuals actively resist the initiation of relaxation therapy. Teachers, especially,
may find this paper of some interest.

Germeroth presents an approach for the management of stress
through tension-control, essentially through Jacobson's Progressive
Relaxation, for a course in the community college. Through a number
of years of experience, Germeroth has developed a number of interesting insights and applications of Progressive Relaxation that he holds
are uniquely effective for the student. An important ingredient in
this apparently successful program is the actual measurement of
tension, electromyographically.

Runsdorf summarized some of the literature having to do with
stresses experienced by university and college faculty. He then
sampled faculty members at the University of Pittsburgh to ascertain
their reactions on a number of stress related issues. Based on the
responses of 164 out of a possible 546 faculty, she summarized their
reactions, though no statistical data are presented. University
faculty may find this study of particular interest.

Masson gave the most engaging presentation at the conference by
demonstrating her techniques that she developed for physical activity
of the elderly, generally ranging from 60 to 80 years. The techniques
were inspired by the works of Jacobson, Feldenkrais and of Tai Chi.
Though only verbal summaries of her medical data were presented, the

results seem to indicate lower blood pressure and pulse rates from
before to after the course.

The last section of papers in this volume, appropriately enough,
consists of technics for coping with stress, a topic that in today's
world is almost endless. These are not necessarily meant to be the
primary, well established methods widely used throughout the world,
but are those interesting methods that were discussed at our
conference. The first is by a reknowned authority in the world of
physical education, Harris, who presents a neuromuscular circuit view
of how cognitive processes are generated, and uses this introduction
for an interesting application in teaching youngsters self-control in
competitive sports situations. Exercise and competitive sports is
thus a laboratory for teaching and awareness of mind/body responses so
that the youngsters can become "tuned in" to their bodies. In the
growing field of sports psychology, perhaps one of the major contributions of psychologists and physiologists is to teach athletes Progressive Relaxation. Harris' paper makes a distinct contribution to
that aspect of sports psychology. In traditional Progressive Relaxation terms, a differentially relaxed athlete is an effective athlete,
one who does not become "up tight". She offers an effective discussion of the mental aspects of athletics, as in her statement that the
difference between a best and a worst performance has to do with
"self-thoughts and self-talk".

Edwards reviews the literature relevant to the effects of exercise on controlling stress. In summarizing a relatively large number
of selected studies, she reaches the conclusion, which is the con-
census in the field, that exercise is a straightforward manner to feel
better and to calm down. At the very minimum, "the human organism
must move or it will deteriorate", she says. As Spirduso elsewhere
has put it, the maintainence of physical fitness through daily exercise is an inexpensive and safe method to prevent motor and mental
deterioration during aging.

Lake shows how, on the basis of her clinical experience, various
kinds of poor posture exert physical effects on the body. As a result, certain nerves can become irritated and hyperactive. Further
effects on the skeletal muscular system can thus result in unique
patterns of pathophysiological disease when there is chronic or sudden
excessive stress overload.

Mitchell summarizes a method for learning to control the body
through the skeletal musculature to thereby prevent various kinds of
pathological conditions. The method bears much resemblance to the
original classic method of Progressive Relaxation developed by
Jacobson in 1929. Since Progressive Relaxation has been so thoroughly
validated, one should expect that Mitchell's method would be similarly
effective for coping with stress.

REFERENCES

Jacobson, E., 1938, "Progressive Relaxation", (Revised Edition),
        University of Chicago Press, Chicago.
Jacobson, E., 1957, "You must relax", 4th ed., McGraw-Hill, New York.
McGuigan, F. J., 1978, "Cognitive psychophysiology: Principles
        of covert behavior", Lawrence Erlbaum Assoc., Hillsdale,
        New Jersey.
McGuigan, F. J., 1981, "Calm Down: A guide for stress and tension
        control", Prentice-Hall, Englewood Cliffs, New Jersey.
McGuigan, F. J., 1983, "Experimental psychology: Methods of Research",
        4th Ed, Prentice-Hall, Englewood Cliffs, New Jersey.

EDMUND JACOBSON, FATHER OF THE FIELD OF RELAXATION

TENSION CONTROL FOR COPING AND FOR HABIT CHANGE[*]

Joseph Wolpe

Director, Department of Psychiatry/Behavior Therapy Section
School of Medicine, Temple University
Philadelphia, Pennsylvania

PERSPECTIVE ON PROGRESSIVE RELAXATION

Edmund Jacobson's discovery about half-a-century ago that progressive relaxation produces a state of emotional calmness was one of the most important developments in the history of psychiatry. Anxiety and its consequences are the commonest source of suffering and disability in the whole field of mental health. Since we can directly control our muscles, but not our autonomic nervous systems, the ability to use our muscles to bring about autonomic effects of a desirable kind is a remarkable therapeutic resource.

The contents of the program of this conference reflect the fact that most widespread use of progressive relaxation is to diminish ongoing anxiety. The subject is taught the skill of muscle relaxation so that he will eventually be able to "switch on" relaxation whenever he experiences anxiety or some consequence of it. By thus diminishing the anxiety when it arises, he copes with it.

The most practical manner of coping is by Jacobson's technique of differential relaxation. A person who has acquired the ability to relax a wide range of his muscles can be taught to obtain and to maintain emotional calm through relaxing whatever muscles are not actively being used. For example, the muscles of the arms and legs are usually free for relaxing while a person is talking, and the facial muscles while he is driving his car. With practice, such relaxations of limited muscle groups produce significant and sometimes profound calming.

[*] Presidential Address of the International Stress and Tension Control Society, University of Sussex, Brighton, England, August 1983.

As is apparent from the data provided in Jacobson's seminal first book, Progressive Relaxation, while patients consulted him for a large variety of complaints, most of which were anxiety-related, he dealt with all of them in essentially the same fashion. He taught each patient his detailed and exacting method of undoing tensions, muscle by muscle. He imposed a heavy requirement of practice to enable the person to attain calmness through profound relaxation of limited parts of the musculature. A great many people acquired, through this training, the power to reduce, often greatly, the anxieties that troubled them.

But what has not been widely noted is that among the successful relaxers there were great differences in the ultimate status of the illness. There were in effect two groups. In one group a basic change progressively developed: the target anxieties and related complaints became weaker and weaker over time, so that the relaxation was less and less needed to counteract them and finally not at all. It seemed that the consistent use of relaxation had led to the ex- tinction of the anxiety problem. The environmental conditions that had previously triggered undesirable anxiety had been lastingly de- prived of their ability to do so. To say this in another way, the undesirable anxiety habits had been eliminated.

By contrast, there were other subjects who did not lose their anxiety habits. They needed to continue, often for many years, to use their relaxation skills to counteract anxiety, because it continued to be elicited sufficiently to interfere in their lives. Clearly their vulnerability to distress had persisted. A survey of the eight de- tailed case histories that Jacobson gives in his 1964 volume, Anxiety and Tension Control, provides a sampling of the division be- tween cases in whom the relaxation treatment obtained recovery, and those who went on needing it as a coping device. Only one of the eight cases became independent of relaxation after a modest amount of treatment - 20 hours over 7 months. The other seven were much im- proved in different ways, displaying fundamental recovery in some directions after between 26 and 300 hours of treatment over periods ranging between 8 months and 3 years. But with one exception, all continued to have significant symptoms, and continued to need progressive relaxation.

Jacobson did not perceive that he was getting two different kinds of outcomes - on the one hand lasting recovery from the illness, on the other amelioration by indefinitely prolonged use of coping. He did not realize that the anxiety problems are based on previously formed emotional habits; and that in his completely successful cases the use of relaxation had had the effect of breaking such habits. He was a physiologist, and was never interested in psychology, in behav- ioral change processes, in the making and breaking of habits. His use of relaxation was not directed against the emotional habits; and when habit change occurred it depended on the chance of appropriate coincidences of anxiety arousal and counteracting calmness. The chance element is greatly reduced if one identifies the triggers to the anxiety responses and then methodically opposes them by the calming effects of muscle relaxation.

RELAXATION FOR HABIT CHANGE

When I had the good fortune to encounter Progressive Relaxation in 1950, I saw it from the viewpoint of a psychologist. I was already aware that neurotic anxiety is a matter of learned maladaptive habits, and I had found from experiments on neuroses that I had produced in cats that the anxiety habit could be overcome by counterposing to small measures of it the incompatible response of feeding. I also knew that children's fears had been overcome by Mary Cover Jones in the 1920's by the competitive action of feeding. In addition, rein-forced by the work of Andrew Salter (1949), I had repeatedly enabled patients to overcome interpersonal timidity by teaching them to ex-press legitimate anger and at times other emotions in relevant situa-tions. This, of course, is now known as assertiveness training. But until 1950 I had been unable to find a response that could be used to combat anxiety in cases in which assertive expressiveness was off the mark. I needed a response that I could bring to bear against anxie-ties aroused by nonpersonal stimuli and by personal stimuli in rela-tion to which assertiveness was inappropriate - for example, the anxiety and hurt feelings a person might experience when his remarks are ignored in a group conversation. Progressive relaxation was exactly what I wanted and I embarked upon its use with delight. At first I had the patient counterpose it to disturbing real-life stim-uli. But since this was cumbersome and often awkward, I later tried the experiment of counterposing the relaxation to weak anxiety-arous-ing imagined scenes.

This was how systematic desensitization originated. Since that time it has been increasingly recognized as an effective method that gives the therapist precise control of the therapeutic process. In the standard technique progressively more disturbing scenes within the problem area are presented to the imagination of the previously re-laxed and calm subject. Each scene is repeated until it ceases to elicit anxiety. Eventually the patient can imagine even the strongest scene of that problem area's anxiety, and thereafter he is in real life completely and permanently imperturbable by any scene whatever in that area. In the desensitization of 68 anxiety response habits that I reported some years ago, 45 of the fear systems were completely eliminated and 17 others markedly and lastingly ameliorated in an average of 11 desensitization sessions per fear. Of course, as Borkovec & Sides (1979) have shown, it is imperative for the prelim-inary training in relaxation to be adequate. Otherwise, poor results such as reported by Benjamin, Marks & Huson (1972) are to be expected.

In the systematic deconditioning of anxiety on the basis of the competition of relaxation we can if necessary elicit the anxiety in real-life situations instead of using imagination. There is no alternative to this with patients who are unable to imagine realistically. Real objects (or pictures) can be brought into the perceptual field of the deeply relaxed subject in ascending order of their power to disturb. The decline of the anxiety level at zero is the signal for the introduction of the next stronger anxiety source. This variant of desensitization has been successfully used with a large number of common phobias. It has even been successful in certain cases of agoraphobia. For example, a 38-year-old man whose life had been severely constricted by agoraphobia for 15 years had been treated with limited success first by imaginal desensitization and then by a program of assignments of actual excursions of increasing length. When progress appeared to have stalled, he was given more intensive training in muscle relaxation which he mastered extremely well. He was then instructed to employ differential relaxation on journeys of increasing length both on foot and by car. In the course of 3 weeks, he was able to go any distance without anxiety, and then found less and less need to use the relaxation. At a 1-year follow-up, he was free from agoraphobia. There is no doubt that this method should be much more widely used in the treatment of this frequently challenging syndrome. But I must emphasize the necessity for preliminary behavior analysis, of which I shall say more shortly, for the accurate identification of the stimulus-response chains is an absolute prerequisite to any rational habit change program. The core of many cases of agoraphobics is not in anxiety about spatial separation as such, but in social or hypochondriacal anxiety. In such cases, space-separation desensitization of any kind is of secondary relevance at most.

With regard to life situation desensitization of this kind, a further comment must be made. It is well-known that many patients recover from agoraphobia and other fears by systematic exposures to real stimulus situations either in the small-step manner of desensitization or else in the large-step manner of flooding, without the use of relaxation. Clinical experience seems to indicate that relaxation greatly facilitates change, which the case that I have just given exemplifies; but systematic research is very much needed.

It is important, however, to be aware that relaxation is not the only anxiety-competitive response, nor is it always the most suitable. For example, there are patients in whom relaxation increases anxiety because an underlying fear of abandoning control is activated when they attempt to relax, and for them another anxiety competitor must be used - for example emotional excitement of a pleasant kind that certain images may arouse. Another possibility is to resort to one of the other ways of inducing emotional calmness - such as autogenic training, biofeedback or transcendental meditation. Many cases of sexual anxiety are appropriately treated by sexual arousal in either real or imaginary contexts. Under certain circumstances, it is

desirable to use the flooding technique, in which the patient is continuously exposed for half-an-hour or longer to stimuli to relatively
strong anxiety. By a mechanism that is not fully understood the
anxiety may go down to a low level, and if it does the anxiety will
afterwards be found to have been to some extent deconditioned. In
cases of interpersonal timidity, assertiveness training, which I have
already mentioned, is still the treatment of first choice. And when
fear exists because of misconceptions, the necessary task is to correct these. For example, if a patient is fearful because he erroneously believes that his spells of dizziness are indications of "going
crazy" that belief needs to be corrected. No amount of relaxation or
desensitization can remove such a fear: it must be treated by providing corrective information.

The decisions on what method to use depend on the behavior
analysis of the case. This is the process of determining the stimuli
or the circumstances that trigger the anxieties that are the cause of
the patient's suffering. Behavior analysis consists of a routine of
questioning that is easy to describe but whose effective use only
comes with constant practice under expert supervision.

RELAXATION FOR COPING

While systematic behavior therapy programs are our most powerful
way of procuring lasting recovery from maladaptive anxiety, there is
nevertheless a wide and important role for relaxation as an isolated
mode of therapy. First of all, in the usual range of neurotic conditions, any skilled relaxation therapist can achieve the kinds of
success that Jacobson reported. Second, relaxation is a valuable
instrument for controlling and perhaps lastingly diminishing the emotional stress of the Type A personality who is seemingly especially
susceptible to coronary thrombosis. Third, relaxation can significantly alleviate the daily stresses of normal people.

As far as the neuroses are concerned, it is inescapable fact that
there are at present insufficient behavior therapists to deal with the
large numbers of neurotic patients who require treatment so that
relaxation therapists will for a long time to come be called on for
their treatment. A great many of these patients benefit from relaxation alone as did Jacobson's patients, even though falling short as a
rule of lasting freedeom from their anxieties. They cope by using
relaxation to diminish anxiety whenever necessary. Expecially when
relaxation training can be administerd to groups, this is an eco-
nomical way of benefitting substantial numbers of people.

The isolated use of relaxation training has to date been particularly important in certain of the syndromes that have traditionally been called "psychosomatic" - notably migraine, tension headache,
insomnia and essential hypertension. Migraine and tension headaches

have been estimated to affect about 30% of the population of the United States. Recent studies have shown that about 65% of patients with tension headache and 50% of those with migraine can, on the basis of a reasonable amount of training and practice of progressive muscle relaxation, experience a decrease in incidence and severity of headache of at least 50%. Follow-up studies suggest that most who have not responded to this treatment have been those who failed to <u>practice</u> the relaxation. In the case of migraine, however, it is worth mentioning that superior results can be obtained from the use of biofeedback techniques designed to reduce cerebral blood flow – which makes good sense in the light of the fact that the pain of migraine seems to be due as a rule to dilatation of cerebral arteries.

Relaxation training per se has also been found outstandingly effective in the treatment of those cases of insomnia that are not due to medical or psychiatric conditions (Turner & Ascher, 1979). The patients complain of difficulty in falling asleep and often of not sleeping well enough to feel refreshed the next day. The evidence is that relaxation training is the most effective available treatment for these patients, diminishing, on an average, the time to fall asleep by about 60%.

The power of relaxation to lower blood pressure was noted by Edmund Jacobson in the 1930's. He reported that careful training of a small group of hypertensive patients resulted in significant lowering of their blood pressure. Jacobson gave renewed attention to this problem in the years before his death and obtained significant and lasting diminution of diastolic blood pressure in 31 out of 52 patients. Agras (1983) has recently provided impressive support for Jacobson's observations. He demonstrated that relaxation training conducted in 1/2-hour sessions once a week for 8 weeks resulted in a mean decrease of 12 mg in diastolic pressure, in contrast to a 2 mg decrease in a control group. The important finding was that the improvement transferred to the work situation of the patient and persisted at follow-up of 15 months. Agras concluded, "We have a procedure that works – that can significantly lower blood pressure more effectively than drugs alone and does not require the elaborate apparatus required for biofeedback".

In their book <u>Type A Behavior And Your Heart</u>, Friedman and Rosenman (1974) presented evidence, largely based on their own research, showing that coronary heart disease occurs predominantly in people displaying a certain constellation of behavioral responses, that has come to be called the "Type A coronary prone behavior pattern". The cardinal features of this pattern are time urgency, chronic impatience, intense striving for achievement, over-commitment to work, excessive aggressive behavior, competitiveness and abruptness of gesture and speech. While in some individuals the existence of this combination of features is obvious, in others it is obscured by their having learned to control, conceal or disguise characteristics like impatience or aggressiveness.

The emotional core of the Type A problem appears to be an excessive arousability to anger. There is a proneness to hostility, manifested in part by a liability to be irritated by trivial things. Despite the sharp difference between this kind of emotionality and the anxiety of the neurotic, it is interesting to note that behavioral techniques, such as relaxation training and systematic desensitization, can be successfully used to overcome excessive habits of anger and other over-reactions to frustration (Roskies, 1979) - in much the same way as they are used for overcoming neurotic anxiety. It seems highly likely that change in autonomic reactivity will have to be achieved in Type A subjects if their susceptibility to heart attacks is to be decreased. Yet the centrality of the emotional factor has been almost universally missed by those who are currently striving for this goal. For example, a recent paper (Levi, 1983) advocates such measures as decreasing stresses, improving the fit between person and environment, promoting realistic expectations, fostering social support, teaching good eating habits, and insuring that the workplace promotes health and well-being. The environment is to be made gentler, but nothing is to be done about the subject's oversensitivity. Among those who do give the emotional factor its due, there is a strong tendency to emphasize increasing the subjects' awareness of their tension states and the nature of the stimuli leading to them. Others, e.g. Sigg (1974), have proposed that the emotional problem be tackled by the use of B-adrenergic blocking drugs like propanolol.

As far as tension control methods are concerned, Benson, Marzetta and Rosner (1974) showed that regular practice of a version of transcendental meditation reduced blood pressure and heart rate but provided no evidence of alteration in Type A behavior. Rosenman and Friedman (1977) have proposed the use of relaxation training and biofeedback within a multifaceted program that includes philosophic reorientation, guided practice for changing habits and the daily routine. I know of no data regarding the success of their program, but encouraging results of biofeedback and relaxation have been reported by Patel (1983).

Outside of specific pathologies of one kind or another, there has recently been, especially in the United States, a great upsurge of interest in relaxation methods for the purpose of diminishing the effects of daily stresses. There has been great impact in the large corporations. According to an article in Time magazine (June 6, 1983), about 1 of 5 of America's largest companies now have some sort of stress management program. What lies behind these programs is not solicitude for employees, but the idea that by encouraging workers to reduce the strains on their hearts, backs and psyches, corporations can begin to reduce the $125 billion dollars or more they spend annually on health care for employees.

Whatever the motivation, much benefit could accrue from such programs. Unfortunately, there is no cause for jubilation. The demand for stress management programs has hatched a small army of entrepreneurs eager to fill the vacuum. Most of them are poorly trained and therefore most of the programs are of poor quality. As Time charitably puts it, "Not all of the relaxation programs are bargains." But what is more important is that the prestige of genuine stress management professionals is undermined by the shoddy work of these entrepreneurs.

The generally high quality of the contributions to this conference is very largely a product of the enterprise and judgement of the chairman, Dr. J. Macdonald Wallace. The conference is a glowing manifestation of the scientific rigor and clinical dedication that characterizes the field of tension control. In this year of the death of the discoverer of progressive relaxation, Edmund Jacobson, it is appropriate that a multitude of torch bearers have assembled to display the brightness of his light.

REFERENCES

Agras, W. S., 1983, Relaxation therapy and hypertension, Hosp. Pract., 129-137.
Benjamin, S., Marks, I. M., and Huson, J., 1972, Active muscular relaxation in desensitization of phobic patients, Psychol. Med., 2:381.
Benson, H., Rosner, B. A., Marzetta, B. R., and Klemchuk, H. P., 1974, Decreased blood pressure in pharmacologically treated hypertensive patients who regularly elicited relaxation response, Lancet, Vol. 1, (7852) 289-291.
Borkovec, T. D., and Sides, J. K., 1979, Critical procedural variables related to the physiological effects of progressive relaxations: A review, Behav. Res. Ther., 17:119.
Friedman, M., and Rosenman, R. H., 1974, "Type A Behavior And Your Heart," Aldred A. Knopf, New York.
Jacobson, E., 1938, "Progressive Relaxation", University of Chicago Press, Chicago.
Jacobson, E., 1964, "Anxiety And Tension Control", Lippincott, Philadelphia.
Jacobson, E., 1978, Relaxation technology applied to hypertensives, Archiv fur Arzneitherapie, 2:152-159.
Jones, M. C., 1924, A laboratory study of fear, The Case of Peter, J. Genet. Psychol., 31:308.
Levi, L., 1983, Stress and coronary heart disease – causes, mechanisms, and prevention, Activ. Nerv. Sup., 25:122.

Patel, C., 1983, Effects of biofeedback, relaxation and stress
     management in reducing coronary heart desease risk, <u>presented
     at:</u> International Conference on Stress and Tension Control,
     University of Sussex, Brighton, England.
Rosenman, R. H., and Friedman, M., 1977, Modifying Type A behavior
     pattern, <u>J. Psycho. Res.</u>, 21:323-331.
Roskies, E., 1979, Considerations in developing a treatment program
     for the coronary prone (Type A) behavior pattern, <u>in</u>:
     "Behavioral Medicine: Changing Health Lifestyles", P. Davidson,
     ed., Bruner/Mazel, New York.
Salter, A., 1949, "Conditioned Reflex Therapy," Creative Age Press,
     New York.
Taggart, P., 1983, Beta-blockade as prevention and therapy in coronary
     prone behavior, <u>Activ. Nerv. Sup.</u>, 25:137.
Turner, R. M., and Ascher, L. M., 1979, Controlled comparison of
     progressive relaxation, stimulus control, and paradoxical
     intention therapies for insomnia, <u>J. Consul. & Clin. Psychol.</u>,
     47:500-508.

ON EDMUND JACOBSON'S WAY OF LIFE

Yves Chesni

Neuropsychiatrist
Saint Philbert de Grand Lieu, France

During his long, happy and useful life, Edmund Jacobson became precociously able to place the right stress on the right point, so to speak, and he continued to do so until his death at the age of ninety four.

By relaxing scientifically and systematically both his body and his mind, he got rid of extraneous, illusory efforts, emotions and thoughts. But, instead of merely providing himself with the comfort of relaxation, with 'ataraxia' or 'apathy', if not with 'nirvana', he made room within himself for a true and full life: research, service, love and friendship, with all their struggles, duties and pleasures.

At one and the same time he discovered, enjoyed and taught us a way of life which includes purgation as a condition of fulfilment and which, from that point of view, is comparable with the way of life of the greatest spirituals. But his method of progressive relaxation is completely original.

ON THE PROFESSIONAL ACCOMPLISHMENTS OF EDMUND JACOBSON

F. J. McGuigan

Executive Vice President
International Stress & Tension-Control Society

I can think of no greater professional privilege than to pay my personal tribute to Edmund Jacobson, my revered friend, colleague and teacher. As a scientist/clinician he is without peer. We are all aware of his pioneering direct electrical measurements of mental activities through quantitative electromyography as a scientist and of his development of the field of relaxation as a clinician.

More than any person I have known, Jacobson has marched to the beat of his own drummer. As a graduate student at Harvard in 1908, he tested a theory for Münsterberg, but the results that Jacobson obtained conflicted with Münsterberg's expectations. A dissastisfied Münsterberg discharged him to work on his own. Fortunately for all of us, Jacobson's genius was applied to experimentation on the startle reflex which formed the empirical grounding of Progressive Relaxation. In that research, he presented a startle stimulus and found that some subjects over-reacted while others remained relatively tranquil. Jacobson reasoned that the prior state of skeletal muscle tonus accounted for these individual differences such that overly-tense individuals exhibited an exaggerated startle reflex with reduced latency. Conversely, relatively relaxed individuals emitted less intense startle reactions with longer latency. The question was how to make the former kind of person into the latter type? Jacobson thus launched a lifetime of empirical research in the psychophysiology of relaxation and indeed of mind itself.

Another reason for the development of Progressive Relaxation was his observation of the nervous behavior of another of his professors, the great William James. James had written the "bible" on relaxation, but according to Jacobson's reminiscences, James was an extremely tense individual, constantly fidgeting, with his beard going in and

out of his mouth as he spoke. Jacobson was distressed that all three
of his great professors at Harvard — Münsterberg, James and Josiah
Royce — died of tension maladies.

Another major impetus for the development of Progressive Relaxation was Jacobson's highly developed ability to precisely introspect
on internal psychophysiological states of the body. This facility was
furthered when he moved to Cornell to work with Titchener and to teach
his brand of introspection to such graduate students as E. G. Boring.

Jacobson then left Cornell to take his medical degree in Chicago.
Among his accomplishments in the field of medicine was the salt-free
diet he developed in 1917. At the 1921 meetings of the American
Medical Association he introduced concepts and definitions of a double
approach to maladies, "mental" and "physical", which now is widely
known as Psychosomatic Medicine. Among the results of these investigations were his pioneering measurements of nervous states in man and
in animals as conditions of muscular tension, and the development of
new methods of treatment of nervous and mental maladies through Progressive Relaxation. Though the name didn't take, he later called
Progressive Relaxation "Self-Operations Control", which today is a
major movement referred to as "Stress Management", "Self Regulation"
and by other similar terms.

While in the Department of Physiology at the University of
Chicago (1926-1936), he conducted now classical research on the knee
jerk reflex with the world-famed physiologist A. J. Carlson. Jacobson
had been engaged in his clinical practice throughout this period of
time but, with his scientific bent, required an objective measure of
progress of his patients. With Carlson, he found that the amplitude
of the knee jerk reflex decreased when patients became more relaxed.
In fact, the reflex is not even elicited in a totally relaxed person.
It can be observed that at this point he had accomplished his early
Harvard goal of eliminating the "nervous start". I routinely observe
this effect in my own students — as part of their "final exams" in my
Progressive Relaxation courses, I produce a loud "BANG" when they are
relaxing and have yet to observe even the "bat of an eye". The characteristic of using objective measures of degree of relaxation and
tension was critical in guiding him in his life-long efforts to
develop the most effective method of relaxation that was reasonably
possible.

While the knee jerk reflex served him well, it is obviously a
cumbersome measure so that Jacobson attempted to replace it with a
more feasible and sensitive method, that of quantitative electromyography. We need not stress the critical importance today of the
direct measurement of tension states through electromyography. It was
his refined development of this measure, which in the '20's and '30's
was considered impossible, that also allowed him to successfully
measure mental activities in a direct manner as small scale covert
responses.

Another of his technological achievements came during World War II when he was called on to collaborate with the United States Navy by training flight officers to teach Progressive Relaxation to air cadets. The Navy had observed that flyers who were excessively tense had an elevated casualty rate. On requesting Dr. Jacobson's help to relax pilots, a program was instituted in which some 15,000 naval flight officers were eventually trained. Every once in a while I still come across an ex-naval pilot who recalls the benefits of that program.

In the early 1950's Jacobson also developed a technique of transducing electromyographic signals so that they could be visually displayed for his patients, this in a futile effort to further facilitate the ability of one to learn to relax. Some years later the technique came to be known as muscular biofeedback.

At the age of 94, Dr. Jacobson had a long and distinguished career. Being the recipient of so many national and international recognitions, it would be hard to select out other accomplishments to highlight. Unfortunately, his last work remains unfinished, that on the nature of purpose and design in the universe.

I am confident that his scientific genius will be increasingly recognized: first, scientifically for his direct electrical measurement of the mental activities that constitute mind; and second, clinically for his development and widely applied methods of Progressive Relaxation for benefiting so many of those afflicted with tension maladies. While we are all so grateful to him, future generations will be more so.

RESEARCH STRATEGIES AND APPROACHES

STRESS AND PERSONALITY

AS CONTRIBUTORY FACTORS IN THE CAUSATION OF CANCER

H. J. Eysenck

Professor, Department of Psychology
University of London
London, England

In the past, the concept of psychosomatic disease was popular, but lacked good methodological support. So also, in recent years the link between stressful life experiences and disease had become a topic for much research, but some has been methodologically weak (Kasl, 1983). There has been a considerable change in the contents of two edited volumes on stressful life events, both edited by Dohrenwend and Dohrenwend (1974, 1981). The earlier volume was taken up largely with reports of empirical findings and reviews of evidence; it only had a small section on methodology. The more recent volume, however, is overwhelmingly methodological, with a secondary emphasis on theoretical issues. Empirical findings from new studies are almost non-existent. This seems to suggest that earlier studies were more enthusiastic than methodologically sound, and that recent interest in methodology has suggested a more critical outlook.

Kasl (1983), in an excellent critique of past and current research, suggests the following:

(1) Stressful life events must be conceptualised, to a greater extent, as characteristics of the person rather than as a separate (and separable) characteristic of the person's environment; this has serious consequences for research design strategies which seek to detect and define the etiological role of stressful life experiences.

(2) Past and current methodological criticisms and research have emphasised measurement issues rather than causal-etiological problems; this emphasis on measurement is a displacement to a secondary, albeit more manageable, problem.

35

(3) The current disquietude regarding methods and knowledge in
    the area of stressful life events and disease is in no
    small part attributable to the availability of a simple
    short instrument, mainly the Schedule of Recent Experience
    (Hawkins, 1957), and the many modifications and variations
    on this early instrument (e.g. Dohrenwend et al, 1978;
    Holmes and Rahe, 1967; Horowitz et al, 1977; Hurst et al,
    1978; Paykel et al, 1971; Rahe, 1978). Kasl is appropri-
    ately critical of this questionnaire and the assumptions
    underlying it.

In this paper I wish to take seriously his first criticism, and
try to bring personality into the general picture linking stress and
disease, in this case lung cancer. Even to ears attuned to the siren
songs of the psychosomatic disease proponents, the idea that personality might be correlated with major diseases like cancer must sound
strange. Yet one obvious link has already been the subject of a great
deal of research. There are theories linking cancer, cardiovascular
disease and other disorders with certain behaviors, such as smoking
cigarettes, or eating saturated fats, etc., which are certainly related to personality. If indeed there is a link, say, between cigarette smoking and cancer, a point on which the evidence is far from
conclusive (as Eysenck, 1980, has pointed out), and if extraverts
smoke more than introverts, then here we have a possible link between
extraversion and lung cancer. Different personalities have different
styles of life, and these life styles may be characterised by behaviors directly or indirectly contributing to disease. It is not,
however, with this particular relationship that we are here concerned.

Hippocrates already postulated certain personality-disease relationships, and more recently Walshe (1846), postulated a hypothesis
linking cancer with the "sanguinous" temperament, i.e. with extraversion. Gengerelli and Kirkner (1954) edited a symposium on "the psychological variables in human cancer", but the contents do not live up
to the promise of the title, methodologies being poor and conclusions
premature. The first serious effort to link personality and cancer
seems to be a study by Kissen and Eysenck (1962), in which 116 male
lung cancer patients and 123 non-cancer controls were tested on the
MPI, both groups being patients at surgical and medical chest units
tested <u>before</u> diagnosis. It was found that, as regards, neuroticism,
the control group had much higher N scores than the cancer group,
regardless of psychosomatic involvement. As regards extraversion,
there were no differences between cancer and control patients without
psychosomatic disorders, but in comparing the groups with psychosomatic disorder, it was found that the cancer group was considerably
more <u>extraverted</u> than the control group. The finding concerning
neuroticism could be interpreted in two ways. Kissen suggested that
the low N scorers were <u>suppressing</u> their emotional reactions, while
Eysenck would have preferred to interpret the finding of a low N score
as indicating the <u>absence</u> of emotional reactions. The study in the
nature of things could not provide direct evidence on this point.

In the same year, Hagnell (1962) reported an epidemiological survey of the 2,550 inhabitants of two adjacent rural parishes in the south of Sweden, constituting a 10 year follow up during which it was observed that a significantly high proportion of women who had developed cancer had been originally rated as <u>extraverted.</u> Hagnell of course was not concerned with lung cancer as such, but it is notable that he too found extraversion to be a crucial personality variable in relation to cancer. Coppen and Metcalfe (1963), using the MPI for their enquiry studied patients in two gynaecological and two surgical wards, and other patients attending the surgical clinic. Forty-seven patients had malignant tumours, and two control groups were used, one consisting of hospital control patients, and the second consisting of subjects representative of the general population of the London area. The cancer group had significantly higher <u>extraversion</u> scores than both control groups; the mean neuroticism score did not differ significantly.

The relationship between lung cancer and lack of neuroticism was taken up again by Kissen (1964a,b), reporting on the neuroticism scores on the MPI of lung cancer patients and other chest unit admissions. He again found that the lung cancer patients had very significantly lower N scores than did the other patients, and concluded that very low scorers on N have about a six-fold probability of developing lung cancer as compared with very high scorers.

Berndt et al (1980) compared control groups of patients who after completion of the questionnaire were found to suffer from breast cancer or bronchial carcinoma. The size of the female control group was 953; that of the breast cancer group was 231. The male control group numbered 195, and the male bronchial carcinoma group 123. There was also a small female bronchial carcinoma group. All three comparisons showed significant differences on the N scale, with cancer patients having lower scores than controls (Eysenck, 1981).

Other studies tending in the same direction have been summarised by Eysenck (1980), e.g. Ure (1969), Abse et al (1974), Greer and Morris (1975), Pettingale et al (1977); there seems to be little doubt that there are connections between lung cancer and personality, mostly in the direction of implicating <u>low neuroticism</u> and <u>high extraversion</u> in the causation of lung cancer. These relations appear to extend beyond lung cancer to mammary cancer, and possibly other types of cancers as well.

These results relate to correlational studies carried out within a given country; in a particularly interesting study Rae and McCall (1973) attempted to demonstrate that an association between cancer and personality holds across a series of countries. National extraversion and anxiety levels in eight advanced countries, and statistics of the number of cigarettes smoked per adult per annum in these countries, were compared with mortality rates per 100,000 of the population due to lung cancer (males and females separately) and cancer of the cervix (females). Rank order correlations were then calculated between national personality levels and cancer mortality rates. The findings were very clear cut and definite. Extraversion showed a high correlation with lung cancer in males (0.66) and in females (0.72). Corresponding correlations for cigarette consumption and lung cancer were quite insignificant. For cancer of the cervix, the correlation with extraversion was again significant (0.64), whereas for cigarette consumption it was insignificant. Correlations between anxiety and lung cancer were negative in both sexes (-0.52, and -0.71). This study thus suggests that the correlation between lung cancer and personality can be observed not only within, but also between countries.

It is usual to find a high degree of emotionality in psychosis, and hence the work of Kissen and Eysenck might suggest that psychotics, too, would be relatively protected from lung cancer. Bahnson and Bahnson (1964a) suggested in the title of their paper that: "cancer is an alternative to psychosis", and Rassidakis et al (1971, 1972, 1973a,b,c,d) have shown that mentally ill populations, especially patients with schizophrenia, seem to be at relatively low risk for cancer. They found that the percentage of mental patients that died from cancer was considerably lower than that of the general population, being only about one third of the general population average. Other randomly selected causes of death (cardiovascular, diabetes) showed no appreciable differences. In England and Wales, about 20% of deaths are caused by neoplasms compared with 7% of deaths in mentally ill populations; for Scotland the figures are 17% and 5% respectively (Eysenck, 1980).

Schizophrenic patients seem to be more resistant to neoplasms than patients with other forms of illness (Levi and Waxman, 1975). Other studies, and possible causal explanations of the phenomenon, are cited by Eysenck (1980); here we need merely note the fact that these data support the general conclusion of Kissen and Eysenck.

Given that personality appears to be fundamentally involved with the development of carcinomas, what is the position with respect to stress? Bammer (1981) reviewed evidence to show that stress can increase metastasis in animals, and there is also some evidence that this is true for humans. This can happen either by the physical action of the stressor or by stress induced impairment of immune function. Sklar and Anisman (1981), conclude from their review of the evidence that:

> "Although many animal tumor systems have been shown to be responsive to stress, the animal studies have used such a variety of procedures that drawing general conclusions is difficult. It is suggested that much of the disparity in animal research linking stress and cancer can be explained by differences between studies in the stressors and backgroud environmental conditions employed".

They then go on to draw a general conclusion which is of fundamental importance for the development of the theory here outlined. As they say:

> "Enhancement of tumor development has usually been reported in studies using acute, uncontrollable physical stress, chronic social stress, or stimulating housing conditions. Chronic uncontrollable physical stresses tend to be associated with tumor inhibition. There is considerable correspondence between brain neurochemical responses to stress and cancer development under stress. Stress increases the synthesis and utilization of compounds such as norepinephrene, and if synthesis does not keep pace with utilization, brain depletion is observed. Brain neurochemical activity in cancer development has been shown to respond similarly to the difference between acute and chronic stress, to the availability of coping responses, and in some cases to social conditions."

They also found that there is evidence for changes in immune functioning under stress which corresponds to neurochemical, hormonal, and tumor development effects. The neural systems affected by stress are involved in hormone increase and immune reactions.

In relation to the development of cancer in humans, retrospective studies have shown that life stress events frequently precede the appearance of forms of neoplasia, e.g. Bahnson and Bahnson, (1964b); Greene, (1966); Horne and Picard, (1979); Jacobs and Charles, (1980). Green and Swisher (1969) succeeded in eliminating genetic factors by looking at leukemia in monozygotic twins discordant for the illness, and found that psychological stress was an important feature in the origins of this disease. Reviews by Bloom et al (1978) and Fox (1978) give a good survey of the literature.

One of the stressors most frequently studied has been loss of spouse, and here again there are a number of studies (Bloom et al, 1978; Greene, 1966; LeShan, 1966; Lombard and Potter, 1950; Murphy, 1952; Peller, 1952; Ernster et al, 1979) showing that cancer appeared in higher than expected frequency among such individuals. Retrospective studies of course are exposed to many difficulties (Fox, 1978; Sklar and Anisman, 1981), but the findings are remarkably uniform in suggesting the importance of stress in the causation of cancer.

This is not a complete or detailed account of the relationship between stress and cancer. However, what does emerge is that (a) stress and cancer are related, in the sense that stress may facilitate the origin and spreading of cancer (possibly largely through inhibiting the immune reaction), and (b) that a crucial variable in this relationship is the fact that chronic stress may afford some protection against the growth and spreading of tumours, possibly through a stimulation of the immune system. This is a crucial point to which we will return soon.

We may now turn to the relationship between stress and personality. Denney and Frisch (1981) have argued that: "If Eysenck's construct of neuroticism corresponds to an inherent reactivity to stress,
then neuroticism scores should function as a moderator variable influencing the relationship between life stress and illness." They
report two studies in which the diathesis-stress model of illness is
evaluated, and found that in both studies neuroticism and life stress
emerge a significant independent predictors of self-reported health
problems, but the tests of neuroticism as a moderator variable were
not significant.

We are not here concerned so much with the specific aspect of
neuroticism as a moderator variable, but rather with the fact that
this study draws attention to a general paradoxical position in the
general set of theories here proposed, and illustrated in Figure 1.
Clearly neurotic and psychotic persons, and generally persons having
high neuroticism scores, are under greater stress than are persons
with low N scores, and psychiatrically "normal". If it is true that
stress is instrumental in promoting the development of neoplasms,
either directly or through inhibition of the immune reaction, then one
would expect high H scorers to be more prone to the development of
such neoplasms. Yet, as we have seen, correlations both within and
between countries are in the opposite direction; high neuroticism
seems to protect the individual, as does psychosis, with its attendant
high anxieties. Unless an explanation can be found for this paradox,
the whole theory here developed would seem to be endangered. It is
suggested that an explanation for this effect may be found in what the
author has called the "inoculation effect" (Eysenck, 1983).

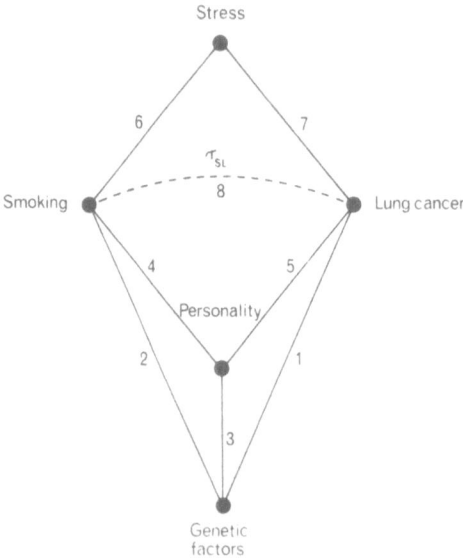

Figure 1. Hypothetical relations between stress, personality,
smoking and lung cancer.

By "inoculation" here is simply meant that <u>repeated stress will</u> <u>inoculate or desensitize the individual against future stress</u>, so that this later stress does not have the same kind of effect on cancer growth and immune reaction, as it would have done had it not occurred against the background of earlier stress. This inoculation effect is of course well-known in experimental psychology (Gray et al, 1981). To take but a few examples, Brown and Wagner (1964) trained two groups of rats to run an alley for food reward, giving both groups continuous reinforcement, but giving electric shocks to one group, initially of low intensity, but gradually increasing in intensity on successive trials. In the next phase of the experiment, both groups were given reward together wtih a high intensity shock on every trial. Comparing the two groups, it was found that the group which previously received shock showed much less reluctance to go to the good-plus-shock end than did the group that had not received shock earlier. Other examples come from the work of Seligman et al (1971) on "learned help-lessness"; Miller's (1976) concept of "toughening up", and studies by Weiss et al (1975), Goldman et al (1973), Lewis (1960), Brown and Wagner (1964), Chen and Amsel (1977), and other workers whose studies have been summarised by Gray et al (1981). This is not the place to summarise the evidence; let us merely note that the "inoculation" effect seems to be a definite reality, and that repeated exposure to stress may produce stress tolerance, which may make reactions to future stress weaker than they would have been without the prior stress experience.

If neuroticism and stress are linked in the manner described above, thus generating a model which can account for the observed facts, then a rather different model has to be found for the relationship between extraversion and cancer. Such a model has been proposed by Eysenck (1983) in terms of the well known fact that introverts tend to condition more quickly, more strongly and more lastingly than extraverts, under suitable conditions (Eysenck, 1981). This greater degree of conditionability of introverts may be relevant to the development of neoplasia because it is now apparent that the immune reaction can be conditioned (Ader, 1981; Ader and Cohen, 1975; Rogers et al, 1976; Wayner et al, 1978; Cohen et al, 1979; Bovbjerg et al, 1982). Conditioning the immune reaction might thus serve as a protection against the development of cancerous growth and metastases, thus accounting for the observed correlation between cancer and extraversion.

It should be noted that the postulated effects of conditioning might also affect the negative correlation between neuroticism and lung cancer. Under certain specified conditions, strong emotions also facilitate the conditioning process, and hence anxiety and neuroticism can be found positively correlated with conditioning (Spence and Spence, 1966). Thus the conditioning mechanism may also mediate a better conditioning of the immune reaction in high N scorers, particularly as a strong degree emotion necessary for producing the effects is more likely to be found under conditions of environmental stress.

I have in this paper concentrated on the relationships between
stress and personality, but of course the picture is very much more
complex due to the interaction of both factors with reactions to
stress, such as smoking or drinking, which in turn are correlated
with personality, and which may in turn again lead to increases or decreases in stress. A fuller picture of such interaction, and the
importance of genetic factors on both personality, smoking and drinking, and the devlopment of carcinogenic illnesses has been presented
elsewhere (Eysenck, in press). There is no space here to go into
these complex issues. What is clear, however, is that Kasl
(1983 p. 80) was right in stating that:

> "Stressful life events must be conceptualised, to a
> greater extent, as characteristics of the person rather
> than as a separate (and separable) characteristic of the
> person's environment."

I have always argued that individual differences are of fundamental
importance in psychology, whether experimental, social, clinical,
educational, industrial or social; they are equally important in
relation to medicine, although physicians are usually reluctant to
admit the influence of yet another complex and complicating factor
in the genesis of disease. Unfortunately this has led to a reluctance
to perform experiments involving personality, and hence the data base
on which this paper is grounded is less adequate than might be wished.
It is my hope that by reviewing the evidence, and suggesting tentatively some formal hypotheses regarding the relationships between
disease, personality and stress, I will persuade others to include
personality factors in their research, and thus enlarge the data
base on which any conceptualisation must rest.

REFERENCES

Abse, D. W., Wilkins, M. M., Castle, R., Buxton, W. D., Demars, J.,
     Brown, R. S., and Kirschner, L. G., 1974, Personality and
     behavioral characteristics of lung cancer patients,
     Journal of Psychosomatic Research,18:101-113.
Ader, R., ed., 1981, Psychoneuroimmunology, Academic Press, New York
     and London.
Ader, R., and Cohen, N., 1975, Behaviorally conditioned immuno-
     suppression,Psychosomatic Medicine, 37:333-340.
Bahnson, C. B., and Bahnson, M. B., 1964a, Cancer as an alternative to
     psychosis, in: "Psychosomatic Aspects of Neoplastic Disease,"
     D. M. Kissen, and I. I. Le Shan, eds., Lippincott, Philadelphia.
Bahnson, C. B., and Bahnson, M. B., 1964b, Denial and repression of
     primitive impulses and of disturbing emotions in patients with
     malignant neoplasms, in: "Psychosomatic Aspects of Neoplastic
     Disease," D. M. Kissen, and I. I. Le Shan, eds., Lippincott,
     Philadelphia.

Bammer, K., 1981, Stress, spread and cancer, in: "Stress and Cancer,"
    K. Bammer, and B. H. Newberry, eds., C. J. Hogrefe, Toronto,
    pp. 137–163.
Berndt, H., Günther, H., and Rothe, G., 1980, Persönlichkeitsstruktur
    nach Eysenck bei Kranken mit Brustdrüsen und Bronchialkrebs und
    Diagnoseverzögerung durch den Patienten, Archiv. für
    Geschwülstforschung, 50:359–368.
Bloom, B. L., Asher, J. J. and White, S. W., 1978, Marital disruption
    as a stressor, a review and analysis, Psychological Bulletin,
    85:867–894.
Bovbjerg, D., Ader, R., and Cohen, N., 1982, Behaviorally conditioned
    suppression of a graft–versus–host response, Proceedings of the
    National Academy of Science, 79:583–585.
Brown, R. T., and Wagner, A. R., 1964, Resistance to punishment and
    extinction following training with shock or non–reinforcement,
    Journal of Experimental Psychology, 68:503–507.
Chen, J. S., and Amsel, A., 1977, Prolonged, unsignaled, inescapable
    shocks increase persistence in subsequent appetitive
    instrumental learning, Animal Learning Behaviour, 5:377–385.
Cohen, N., Ader, R., Green, N., and Bovbjerg, D., 1979, Conditioned
    suppression of a thymus–independent antibody response,
    Psychosomatic Medicine, 41:487–491.
Coppen, A., and Metcalfe, M., 1963, Cancer and extraversion, British
    Medical Journal, July 6th, 1963, 18–19.
Denney, D. R., and Frisch, M. B., 1981, The role of neuroticism in
    relation to life stress and illness, Journal of Psychosomatic
    Research, 25:303–307.
Dohrenwend, B. S., Krasnoff, L., Askenasy, A. R., and
    Dohrenwend, B. P., 1978, Exemplification of a method for
    scaling life events: The PERI life events scale, Journal of
    Health and Social Behaviour, 19:205–229.
Dohrenwend, B. S., and Dohrenwend, B. P., eds., 1974, "Stressful Life
    Events," Wiley, New York.
Dohrenwend, B. P., and Dohrenwend, B. S., 1981, Socioenvironmental
    factors, stress and psychopathology, American Journal of
    Community Psychology, 9:128–159.
Ernster, V. L., Sacks, S. T., Selvin, S., and Petrakis, N. L., 1979,
    Cancer incidence by marital states, U. S. Third National Cancer
    Survey, Journal of the National Cancer Institute, 63:567–585.
Eysenck, H. J., 1980, "The Causes and Effects of Smoking,"
    Maurice Temple Smith, London.
Eysenck, H. J., 1981, "A Model for Personality," Springer Verlag,
    New York.
Eysenck, H. J., 1983, Stress, disease and personality: the
    'inoculation effect', in:"Stress Research," C. L. Cooper, ed.,
    John Wiley, New York, pp. 121–146.
Fox, B. H., 1978, Premorbid psychological factors as related to cancer
    incidence, Journal of Behavioral Medicine, 1:45–133.

Gengerelli, J. A., and Kirkner, F. J., eds., 1954, "The Psychological Variable in Human Cancer," University of California Press, Berkeley and Los Angeles.

Goldman, L., Coover, G. D., and Levine, S., 1973, Bidirectional effects of reinforcement shifts on pituitary adrenal activity, Physiological Behaviour, 10:209-14.

Gray, J. A., Davis, N., Owen, S., Feldon, J., and Boarder, M., 1981, Stress tolerance: possible neural mechanisms, in: "Foundations of Psychosomatics," M. J. Christie and P. G. Mellett, eds., Wiley and Sons, New York.

Greene, W. A., 1966, The Psychosocial setting of the development of leukemia and lymphoma, Annals of the New York Academy of Science, 125:794-801.

Greene, W. A., and Swisher, S. N., 1969, Psychological and somatic variables associated with the development and course of monozygotic twins discordant for leukemia, Annals of the New York Academy of Science, 164:394-408.

Greer, S., and Morris, T., 1975, Psychological attributes of women who develop breast cancer: A controlled study, Journal of Psychosomatic Research, 19:147-153.

Hagnell, O., 1962, Svenska Laki-Tibu, 58,4928.

Hawkins, N. G., Davies, R., and Holmes, T. H., 1957, Evidence of psychological factors in the development of pulmonary tuberculosis, Amer. Review Tuberc. Pulmon. Dis., 75:768-780.

Holmes, T. H., and Rahe, R. H., 1967, The social readjustment rating scale, Journal of Psychosomatic Research, 11:213-218.

Horne, R. L., and Picard, R. S., 1979, Psychosocial risk factors for lung cancer, Psychosomatic Medicine, 41:503-514.

Horowitz, M., Schaefer, C., Hiroto, D., Wilner, N., and Levin, B., 1977, Life event questionnaires for measuring presumptive stress, Psychosomatic Medicine, 39:413-431.

Hurst, M. W., Jenkins, C. D., and Rross, R. M., 1978, The assessment of life change stress: A comparative and methodological inquiry, Psychosomatic Medicine, 40:126-141.

Jacobs, T. J., and Charles, E., 1980, Life events and the occurrence of cancer in children, Psychosomatic Medicine, 42:11-24.

Kasl, S. V., 1983, Pursuing the link between stressful life experiences and disease: A time for reappraisal, in: "Stress Research," C. L. Cooper, ed., John Wiley and Sons, New York, p. 79-102.

Kissen, D. M., 1964a, Relationship between lung cancer, cigarette smoking, inhalation and personality, British Journal of Medical Psychology, 37:203-16.

Kissen, D. M., 1964b, Lung cancer, inhalation and personality, in: "Aspects of Neoplastic Disease," D. M. Kissen, and C. L. Le Shan, eds., Pitman, London.

Kissen, D. M., and Eysenck, H. J., 1962, Personality in male lung cancer patients, Journal of Psychosomatic Research, 6:123-137.

Le Shan, L. L., 1966, An emotional life history pattern associated with neoplastic disease, Annals of the New York Academy of Sciences, 125:780-793.

Levi, R. N., and Waxman, S., 1975, Schizophrenia, epilepsy, cancer, methionine, and folate metabolism, Pathogenesis of schizophrenia, The Lancet," July 5th, 11-13.

Lewis, D. J., Partial reinforcement: A selective review of the literature since 1950, Psychological Bulletin, 57:1-28.

Lombard, H. L., and Potter, E. A., 1950, Epidemiological aspects of cancer of the cervix: hereditary and environmental factors, Cancer, 3:960-968.

Miller, N. E., 1976, Learning stress and systematic symptoms, Acta Neurobiol. Exp., 36:141-156.

Murphy, D. P., 1952, "Heredity in uterine cancer," Harvard University Press, Cambridge, Mass.

Paykel, E. S., Prusoff, B. A., and Uhlenhuth, E. H., 1971, Scaling life events, Arch. Gen. Psychiat., 25:340-7.

Peller, S., 1952, "Cancer in Man," International University Press, New York.

Pettingale, K. W., Greer, S., and Tee, D. H., 1977, Serum IGA and emotional expressions in breast cancer patients, Journal of Psychosomatic Research, 21:395-399.

Rae, G., and McCall, J., 1973, Some international comparisons of cancer mortality rates and personality: a brief note, The Journal of Psychology, 85:87-88.

Rahe, R. H., 1978, Life change measurement clarification, Psychosomatic Medicine, 40:95-8.

Rassidakis, N. C., Kelepouris, M., and Fox, S., 1971, Malignant neoplasms as a cause of deaths among psychiatric patients (I), International Mental Health Research Newsletter, 13:3-6.

Rassidakis, N. C., Kelepouris, M., Goulis, K., and Karaiossefidis, K., 1972, Malignant neoplasms as a cause of death among psychiatric patients (II), International Mental Health Research Newsletter, 14:3-6.

Rassidakis, N. C., Erotokristow, A., Validou, M., and Collaron, T., 1973a, Anxiety, schizophrenia and carcinogenesis, International Mental Health Research Newsletter, 15:3-6.

Rassidakis, N. C., et al, 1973b, Schizophrenia, psychosomatic illnesses and malignancy, International Mental Health Research Newsletter, 15:3-6.

Rassidakis, N. C., et al, 1973c, Malignant neoplasms as a cause of death among psychiatric patients (III), International Mental Health Research Newsletter, 15:3-6.

Rassidakis, N. C., Grotoerton, A., and Validou, M., 1973b, An essay on the study of the etiology and pathogenesis of schizophrenia, the psychosomatic illnesses, diabetes melitus and cancer, International Mental Health Research Newsletter, 15:3-8.

Rogers, M. P., Reich, T. B., Strom, T. B., and Carpenter, C. B., 1976, Behaviorally conditioned immuno-suppression: replication of a recent study, Psychosomatic Medicine, 38:447-452.

Seligman, M. E. P., Maier, S. F., and Solomon, R. C., 1971, Unpredictable and uncontrolled aversive events, in: "Aversive Conditioning and Learning," F. R. Brush, ed., Academic Press, New York.

Sklar, L. S., and Anisman, H., 1981, Stress and Cancer, Psychological Bulletin, 89:369-406.

Spence, J. T., and Spence, K. W., 1966, The motivational components of manifest anxiety: drive and drive stimuli, in: "Anxiety and Behaviour," C. D. Spielberger, ed., Academic Press, London.

Ure, D. M., 1969, Negative association between allergy and cancer, Scottish Medical Journal, 14:51-54.

Walshe, W. H., 1846, "Nature and Treatment of Cancer," London.

Wayner, E. A., Flannery, G. R., and Singer, G., 1978, Effects of taste aversion conditioning on the primary antibody response to sheep red blood cells and Brucella abortus in the albino rat, Physiology and Behaviour, 21:995-1006.

Weiss, J. M., Glazer, H. I., Pohorecky, L. A., Brick, J., and Miller, N. E., 1975, Effects of chronic exposure to stressors on avoidance-escape behavior and on brain norepinephrine, Psychosomatic Medicine, 37:522-534.

RESEARCH ON STRESS OUGHT TO BE REVISED

Henrik Seyffarth

Specialist in Neurology
Oslo, Norway

INTRODUCTION

Research on stress is conducted too much in laboratories. Human beings, however, have stress reactions, particularly when they are nervous. Frequently they react to situations that do not induce stress in a healthy person.

Stressors can be divided into two major groups:

a) Pathologic anatomical changes in the skeletal muscular system and the internal organs which signal physical impulses to the nervous systems, making it oversensitive and creating pain and anxiety. In other words, the somatopsychic reactions.

b) Unsolved conflicts, that is psychic stimuli, which make the nervous system oversensitive. At the same time, reactions occur in the part of the brain which controls our emotions, i.e., the limbic system where anxiety is provoked. Simultaneously impulses are discharged to the motoric nervous system, producing muscular tension. When anxiety and tension combine, the term "psychosomatic reactions" is applied.

NORMAL PHYSICAL ACTIVITIES IN HEALTHY INDIVIDUALS

We know that a certain amount of physical (and psychic) activities are necessary for the maintenance of body and mind, and/or to increase our action potential (See Table 1). Among these activities are daily working functions and that activity which is provoked when we have to solve conflicts.

47

Table 1. Normal activities lead to increased action potential.

A. Normal healthy activities	No Stress	Action Potential increases

Physical stimuli

Voluntary muscular functions	Outlet good circulation	Action (fight, flight) no pat. anat. changes

Psychic stimuli

Soluble conflicts no problems	Outlet	Action
	concentration just before action	"healthy" tension
	good circulation	no pat. anat. changes
Short lasting static muscular work	Insufficient circulation	a) feeling of fatigue + b) diminished strength
	ischemia	= fatigue

Normally, dynamic muscular functions are involved, but we also notice "awareness reactions" which mainly consist of static muscular functions when a person "binds" or "tightens". This reaction can be noticed in a cat, lurching on his prey. A similar reaction is found in an athlete who concentrates immediately before a competition, or in a person about to make a speech. The awareness reaction is a healthy tension which increases our action ability.

Tension is static muscular work, and we know that static muscle functions are part of muscular activity which most rapidly leads to fatigue, the reason being that they cause insufficiency of circulation, i.e., ischemia. Fatigue decreases the potential for action, but this is merely passing. After a rest, the potential for action is restored to its previous level, even increased. Thus it is a common experience that by increasing muscular strength through training, the exertion will lead to fatigue. Static muscular work will then be of a particularly beneficial effect.

All in all, we may conclude that normal fatigue is a healthy, reversible reaction to muscular activity. Normal fatigue does not produce pathologic anatomical changes and the releasing activity should consequently not be termed as stress. If, on the other hand, the static muscular functions continue for too long, the provoked ischemia will produce pathologic anatomical changes.

WHAT IS STRESS?

Stress simply means that activity produces reactions which, in duration or strength, exceed the tolerance level of an individual and thus contribute to the formation of pathologic anatomical changes of the skeletal muscular system. These reactions are provoked by what we call "stressors" and the reactions are named "stressor reactions". The pathologic anatomical changes produced in this manner are termed "diseases due to strain".

Much of the disagreement on the conception "stress" is caused by not limiting the notion of stress as described here - but by maintaining that all sorts of physical activities constitute "stress".

NORMAL PHYSICAL STRAIN

In the last 40 years I have been studying the reactions of the skeletal-muscular and the nervous systems during physical strain (Seyfartth, 1982). Major results of this work include:

1. Electromyographic (EMG) experiments have shown that muscular fatigue is provoked by increasing ischemia in the muscles.

2. EMG also shows that muscular fatigue consists of stimuli simultaneously invading two different parts of the nervous system (Seyffarth, 1941) (Figure 1).

   a) Ischemia impulses due to inhibition produces weakness (paresis), i.e., a somatic reaction.

   b) At the same time ischemic impulses induce the feeling of fatigue, a reaction in the limbic system in cortex cerebri, i.e. a psychic reaction (Table 1).

According to this, we may assume that the effect of ischemic impulses provoke paresis and the feeling of fatigue, because the various parts of the nervous system are sensitive to the ischemic impulses. Also a certain degree of increased irritability of the nervous system may develop. From this it may be concluded that ischemic impulses in general have their effects on the areas in the nervous system which are oversensitive to ischemia.

To summarize: Normal physical strain produce ischemia. Ischemic impulses produce weakness and the feeling of fatigue, i.e., a psychic and somatic reaction. This process is reversible, and does not provoke pathologic anatomical changes.

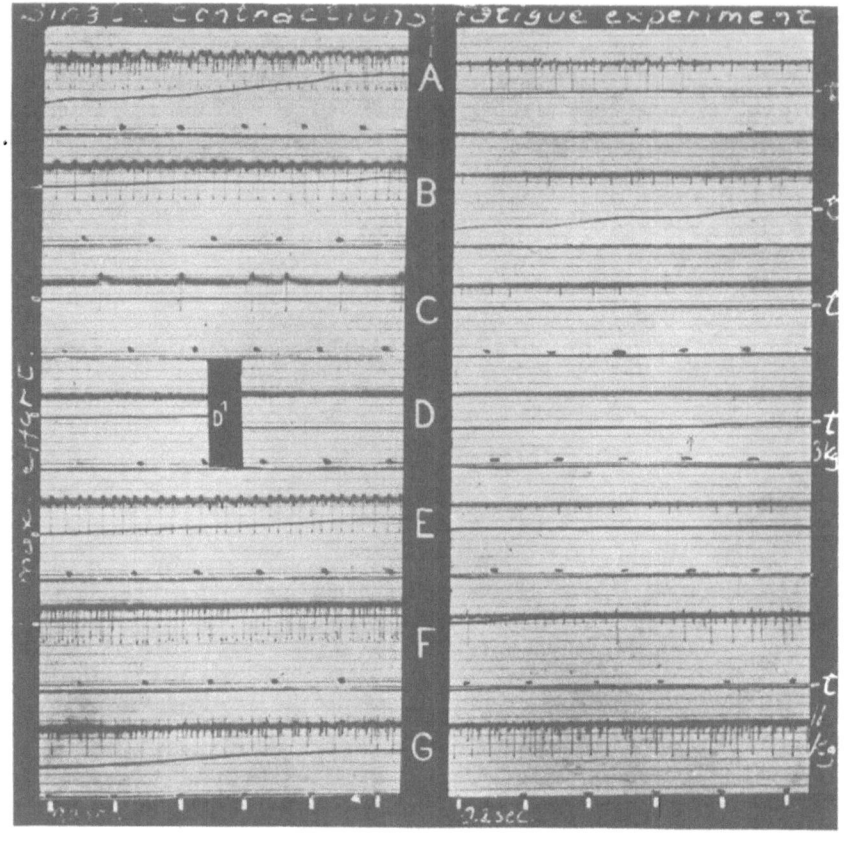

Figure 1. The effect of ischemia on unit frequency.
    1a. Maximum contractions.   1b. Fatigue experiment.
(0.2 sec/mark & 1 Kg/division – zero tension is on time mark line)
(A blood pressure cuff (330 mm Hg) on the upper arm, produces ischemia
in the forearm of a healthy subject. The response is recorded from
m. brachio-radialis during ischemia and after the cuff is relaxed.
The single thick line indicates the dynamometer tension recorded during elbow flexion. For maximum contractions (1a), the consensual
behaviour of motor units as ischemia causes a decline in frequency is
the same as in an ordinary single contraction when slow relaxation
occurs. Thus the last unit to act is the first to drop out. As far
fewer contractions are made than would produce fatigue in a healthy
arm, it is assumed that the decline in frequency is not due to fatigue
of the single motor unit. We may assume that the decline in frequency
of motor units may be brought about by inhibiting impulses which arise
in the muscle during ischemia. The same conclusion is made when studying the behaviour of motor units during the fatigue experiment (1b).
Despite the subject's maximum efforts, the frequency of units drops to
zero: the consensual behaviour of motor units is always the same.)

PHYSICAL STRESSOR REACTIONS IN THE HUMAN BODY

Long lasting ischemia induces exudation and growth of newly formed fibrosis tissue in the muscle, thus the local physical stressor reaction of ischemia produces the palpable fibrosis that is sore and is a firm part of the muscles. Also in the internal organs and suprarenal glands pathological anatomical changes are produced.

Fibrositis is microscopically detected (Rais, 1961) (Figure 2). In the case of lidocaine injections in fibrositis, the strength with which we have to press the piston, i.e., the injection resistance, is considerably increased.

We conclude that the physical stressor-reactions (muscular activity) produce pathologic anatomial changes in the human body.

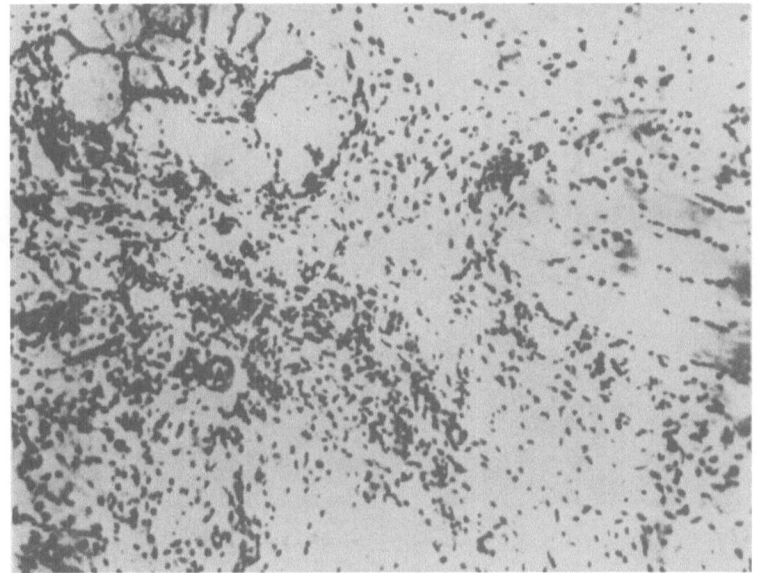

Figure 2. Section of the abductor pollicis longus muscle from a patient in whom the lesion was of 25 days duration (unaccustomed work). Note ingrowth of newly formed connective tissue rich in capillaries into the muscle (A). Moderate numbers of inflammatory cells appear in the connective tissue.

PHYSICAL STRESSOR REACTIONS IN THE NERVOUS SYSTEM

Fibrositis and other local pathological anatomical changes are
responsible for the greater part of the ischemic impulses invading
various parts of the nervous system. Thus the paresis, chronic pain
and the feeling of fatigue decrease when ischemia is diminished by
means of lidocaine injections in the fibrositis (Figure 3). (The improvement of ischemia is clearly observed in case of injections of
keloids.)

The ischemic impulses provoke increased sensitivity of the
nervous systems. They are the real stressors provoking stressor-
reactions (Table 2).

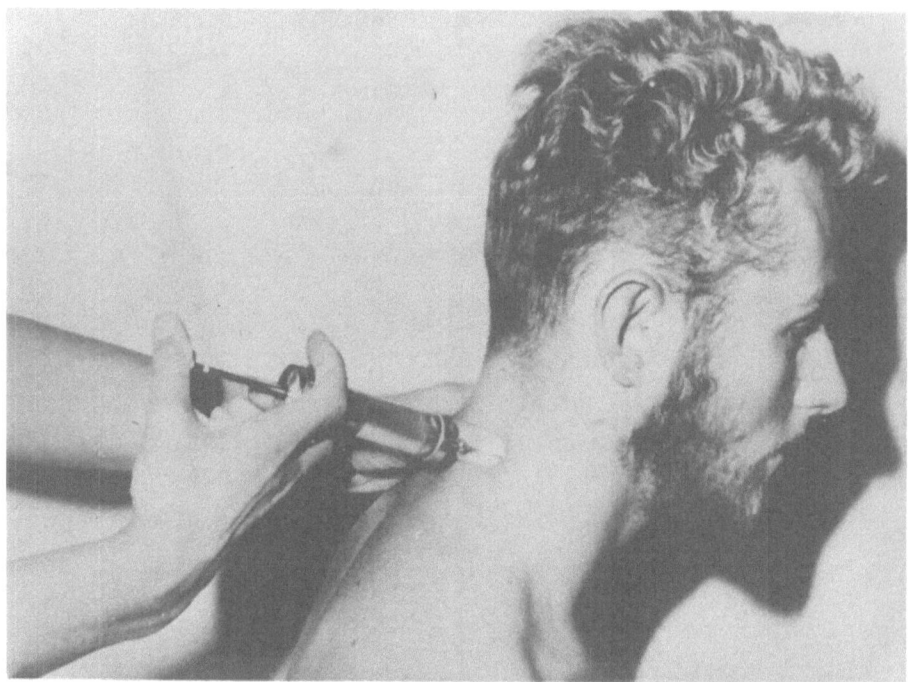

Figure 3. The posterior Scalenus Syndrom.

(Injection of a quarter percent lidocaine in the fibrositis in
m.m. scalenus post. and medius (increased resistence), is immediately
followed by an increase in the strength of the paretic shoulder. We
assume the reason is the improvement of the circulation around the
thoracic nerve, innervating serrat. ant. muscle.)

Table 2. Stressor-reactions lead to pathological anatomical change.

Stressors	Stressor reactions no outlet(=stress)	Pat. anat. changes action poten. decreases
Physical		
Longlasting static m. work or loading	a) Ischemia in skeletal m. syst.	fibrositis, arthrosis with ischemia
	b) Ischemic impuls increased irritab. of the nervous system (3 parts)	ischemic tissues in intestini
	1. limbic syst.: pain, anxiety	pat. anat. changes in the nervous system
	2. motoric syst.: inhibition	chron. fatigue in muscles - "paresis"
	3. sympat. syst.: ischemia	in intestini and suprarenal glands
	Circulus vitiosus	somatopsych. disease psychosomat. disease
Psychic stimuli	Increased irritab. of the nervous system (3 parts):	
Insoluble conflicts uncertainly	1. limbic syst.: pain, anxiety	pat. anat. changes in the nervous syst.
	2. motoric syst.: nervous tension (static m. work)	in skeletal m. syst.
	3. sympaticus: vasomotor reflex, ischemia	in intestini and suprarenal glands sec. fibrositis

There are three categories of reactions to ischemic impulses in the various parts of the nervous system:

1. Remote reactions of ischemic impulses to the limbic system induce chronic pain and anxiety that hampers normal mental activity. They are in other words somatopsychic reactions.

   (There are several indications that ischemic impulses can produce anxiety. Thus anxiety caused by angina pectoris is well known. Lidocaine injections reveal that ischemic impulses from the tender tissues in the front of sternum particularly induce anxiety. Also, ischemic pains that frequently precede anxiety are often observed. In many patients it is observed that headaches and anxiety are decreased after lidocaine injections in fibrositis of the neck.)

2. Reaction in the motoric system induces nervous tension that hampers all movements, you "tense yourself", i.e., an unsolved conflict between muscles.

3. Remote reactions in the sympathetic nervous system produce ischemia in the muscles and intestini.

Reactions 1 + 2 + 3 reduce the ability to act from the very beginning.

In short, pathologic anatomic changes in the skeletal-muscular system produce ischemic impulses invading the nervous system, working as a somatopsychic stressor.

PSYCHOSOMATIC STRESSOR REACTIONS

Stressors are unsolved conflicts, including frustrations and situations provoking uncertainty. These psychic impulses, the psychosomatic stressor-reaction in various parts of the nervous system, include the following:

1. In the limbic system: Stressor-reactions produce chronic pains and anxiety and/or depression.

2. In the motoric nervous system: Nervous tension results. The nervous tension is recognized by asking the patient whether the notices that he holds his breath, tightens his muscles and binds himself while working. We observe him when he speaks and/or writes, and check his daily functions such as sitting, standing, walking, breathing and working.

3. In the sympathetic nervous system: The vasomotoric nervous system is affected, resulting in circulatory disorders in the skeletal-muscle system and the internal organs.

It is the psychosomatic stressor-reactions here mentioned that, after a while, may induce pathologic anatomical changes. They are then called dysfunctions. These are consequently the real strain in the case of stress, the pathological anatomical changes being only symptoms of these dysfunctions.

The view here mentioned can be designated "functional medicine". As opposed to this, we have the ordinary "anatomic medicine", which takes the pathologic anatomical changes as its starting point.

The conclusion is that psychic stimuli provoke psychic and psychosomatic stressor-reactions which in time produce anatomical changes in the nervous systems and the human body.

PATHOLOGIC ANATOMICAL CHANGES PROVOKED BY
ISCHEMIC AND PSYCHOSOMATIC IMPULSES

The corresponding pathological changes that are observed in these systems are as follows:

1. By way of a working hypothesis we suppose that <u>anxiety may induce lasting changes in the limbic system</u>. Otherwise, it would be difficult to explain that throughout our entire life we preserve the ability to swim once this has been learned.

2. <u>In the muscles the lasting changes are induced by the nervous tension</u>, which means static muscular work rapidly results in circulatory insufficiency, i.e., ischemia. Ischemia in turn causes a proliferation of connective tissue along with the secretion of exudate from palpable infiltrates in the chronically tensed muscles, in other words fibrositis. By palpating the muscles, we can therefore attain a good indication of the patient's level of nervous tension, something which some patients may otherwise be able to conceal during the examination itself.

3. <u>Reactions in the sympathetic nervous system induce vascular reflexes</u> giving remote secondary fibrositis and pathologic anatomical changes in the internal organs. At the same time <u>a sympaticus irritation of the endocrine system occurs</u>, with increased adrenaline secretion etc.

Impulses from the pathologic anatomical factors mentioned above, especially the chronic ischemic pain, increases the irritability of the nervous system. A vicious circle, **psyche - soma - psyche**, materializes, contributing to a chronic psychosomatic disease that

corresponds to what we call "chronic nervousness". One talks about
psychic fatigue developed by psychic stimuli (conflicts), and somatic
fatigue developed by somatic impulses. Weariness produced by psychic
stimuli has less somatic reaction.

However, there is a difference in degree between the two forms of
fatigue, as the reactions of the nervous system seem to be about the
same whether they are triggered by a conflict or by ischemic impulses.

The greater part of the patients' complaints are due to the
anatomical changes produced by ischemia. Only a minor part of the
symptoms is provoked by the stressor-reaction. The patients blame
it on the external strains and do not realize that, but for their own
inner reaction, there would have been no stress.

TREATMENT

It is necessary to break this vicious circle and therefore we
need both physiotherapy and psychotherapy.

Treatment of the stressor-reaction consists in teaching the patient to relax, thus relieving the nervous tension. At the same time,
the pathologic anatomical changes in the skeletal-muscular system are
treated by physiotherapy, thus to stop pain and other afferent impulses that increases the irritability of the nervous system. Then,
a rehabilitation programme is started with regard to the natural functions such as sitting, standing, walking, breathing and working
properly.

The somatopsychic and psychosomatic reactions can be considerably
reduced or made to disappear by treating the pathologic anatomical
changes such as neck fibrosis in the case of headache or brachialgia
due to the posterior scalenus syndrome (Figure 2). This can be
achieved by physiotherapy or, still better, by adding 0.25% lidocaine
injections in the fibrosis. As noted the local anesthesia induces
better circulation. This may explain the fact that lidocaine injections in fibrotic, sore, firm areas in back, neck, abdomen and soles
of the feet, neutralize paresis and pain, caused by inhibiting impulses from the ischemic fibrosis. The same result is achieved by
injecting lidocaine in fibrosis close to nerves (Figure 4).

Psychotherapy ought to make the patients aware of themselves and
try to make them understand their own reactions and the causes of
these. The lasting pathologic anatomical changes can explain that
patients with chronic nervousness are difficult to heal. But we can
help them function in daily life.

The three tender firm areas in the neck, around Th2 and L1 which
give remote reactions to the upper or the lower extremities. With

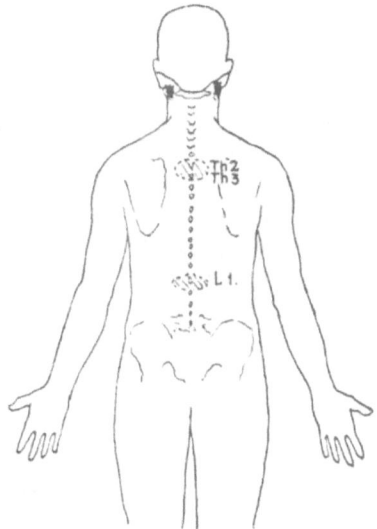

Figure 4. The Ischemic Areas

injection of local anaesthesia (20-60 ml 0.25% lidocaine) in the
fibrositis in one of the areas, all remote reactions in one of the
extremities are affected (fibrositis, pain, paresis, defense
musculaire).

PROPHYLAXIS

Relaxation plus correct breathing techniques are the best aids
in reducing stress reactions and are of utmost importance for
prophylaxis.

We have to learn the activities sitting, standing, walking,
breathing and working. The prophylaxis also should include exercise to
strengthen the organism.

To learn this and techniques for the avoidance of stressor
reactions should be a part of the physical education of children in
the schools.

"RESEARCH ON STRESS OUGHT TO BE REVISED" - WHY?

The Canadian Hans Selye calls <u>all</u> activities stress and concludes that it is not possible to exist without a certain degree of stress. This is right when stress means healthy stress, increasing the potential. However, I strongly believe that stressor-reactions can and ought to be avoided.

Selye's apprehension of stress is based on experiments with animals in laboratories. Yet it has been predominant in the medical literature also in respect to man stress reactions (Levi, 1972). He takes account of the psychic and physical functions only to a small degree, and he does not palpate the pathologic anatomical changes in the skeletal muscle system.

Hans Selye's concept of stress has had a bad influence on Scandinavian medicine.

When Hans Selye studied stress by restraining rats to a table, he set off two different reactions. First, the rat fights for its life, believing it will free itself. Secondly, the rat tires, loses courage, becomes frightened. It then reacts with nervous tension, which means anxiety, and static muscular work with ischemia, producing pathologic changes in the body. The potential for action decreases from the very beginning. These are the real stressor reactions.

Selye did not distingish between reactions to solvable and unsolvable conflicts. But the fact is that when you can solve a problem it is a challenge and your potential for action increases. If the problem is insoluble, the potential immediately decreases.

The dilemma remains as to how a conflict can be solved at all when the solution, which is essentially dependent on muscular activity, is regarded by Selye as stress.

REFERENCES

Levi, L., 1972, "Stress and Distress in Response to Psychosocial Stimuli," Pergamon Press, New York.

Rais, O., 1961, Acta Chir. Scand., 88:268.

Selye, H., 1979, "Guide to Stress Research," Van Nostrand Rainhold, New York.

Selye, H., 1974, "Stress without Distress," Hodder and Stoughton, London.

Seyffarth, H., 1982, Det perifere grunnlag for kroniske smerter, (The Peripheral Basis for Chronic Pains), Den norske Laegeforening, (The Norwegian Society of Physicians), 1466-71.

Seyffarth, H., 1941, The behavior of motor units in healthy and paretic muscles in man, Acta Psychial. Neurol., 17.

Wolff, H. G., 1953 and 1968, "Stress and Disease," C. C. Thomas, Springfield, USA.

SOME CLINICAL OBSERVATIONS ON MUSCLE TENSION AND EXPRESSIVE MOVEMENT

Peter Sainsbury

Director, MRC Clinical Psychiatry Unit
Graylingwell Hospital
Chichester, Sussex, England

"Any defect of psyche or soma is the occasion of the greatest disproportion in the other." I came to appreciate this Plato dictum early in my career when I worked as Aubrey Lewis's research assistant at the Maudsley Hospital. As my first assignment was "to consider nervous tics" I was not surprised to find these introductory passages to a WHO's Report on Psychosomatic Disorders which he edited (WHO 1964), "In all ages and places men have expressed their feelings in bodily happenings and, through verbal and other symbols, in bodily as well as mental terms", and he went on to quote from Tuke's "Illustrations of the Influence of the Mind upon the Body in Health and Disease" published in 1872: "Mind or brain influences, excites, perverts or depresses the sensory functions, muscular contraction, nutrition and secretion." I would add that the most conspicuous of these bodily accompaniments of psychological states are "muscular contractions."

Consider, for example, the taut expression of the examinee expectantly perched on the edge of his chair; the fidgeting of the bored or the impatient; the gesture and mime of the political demagogue; or, in the clinical setting, the sorry demeanour of the agitated melancholic, with raised brow, bent posture and wringing hands.

So, in psychomotor behaviour we reveal needs and attitudes, express feelings, communicate ideas and information, and convey something of our disposition and temperament. But at the neuromuscular level all this display boils down to changes in muscular tone and to "muscular contraction", to tension and motility.

My colleague John Gibson (1959) remarked "Psychiatry urgently needs objective criteria whereby the phenomena observed can be measured and whereby those measurements can be repeated by other observers. Only thus can we establish sound principles that may well be the key to some of the most stubborn problems of medicine."

59

It was to make a start in this direction that he and I developed an objective measure of "muscular tension", by which we meant that component of neuro-muscular activity which is distinct from reflex postural tone, but dependent on some central arousing mechanisms, and which prepares the individual to respond to the demands of his personal environment (Sainsbury and Gibson, 1954).

Apart from the questions of devising a method, our aims were:

1) to ask how muscle tension relates to anxiety and
2) to see whether certain common aches and pains
   can be attributed to muscular hypertension.

Aubrey Lewis suggested that a suitably modified electromyograph might be an appropriate aid to thse ends. Accordingly, an apparatus was designed which included an integrator, enabling us to count the signals produced by any active motor units when electrodes are placed on the skin overlying a muscle: a novel application of the EMG at that time.

We avoided the awkward problems of provoking tension by exposing patients to "stresses" such as loud noises which, in fact, they may not find in the least disturbing, by simply recording their muscle activity when they were lying comfortably. Any activity recorded under these conditions, we maintained, should provide a measure of tension or inability to relax. The next question was where to record from. Trial, error and the literature pointed to the muscles principally concerned with expression as being the most tension-prone. We therefore put electrodes over the frontalis muscles and the forearm extensors, but with the arms pronated to avoid recording reflex postural tone. As we were interested in muscular symptoms, we also recorded from the muscles at the back of the neck, because tension headaches are so often occipital. To do this we used bio-feedback, now a fashionable technique, but then unheard of so far as I was aware. The patient sat facing an oscilloscope incorporated in the EMG circuit, electrodes were placed over the muscles at the back of the head and the patient asked to relax and position his head so as to maintain the oscilloscope display of the muscle's activity at the lowest possible level. In this way we were able to obtain reliable and steady counts of the resting tension in a group of muscles never wholly devoid of reflex-tone.

We made two 7 minute recordings from 30 psychoneurotic patients and 30 healthy controls, each of whom also completed a questionnaire listing the somatic symptoms of anxiety and symptoms attributable to muscle tension - backache, headache, writer's cramp and so forth.

The results were clear cut: patients had higher muscle tension scores from all recording sites than had the controls; and their muscle tension counts correlated reasonably well with their anxiety and tension-symptoms scores (0.49).

Another of the findings threw some light on the mechanism of psychosomatic aches and pains. When the patients were divided into those with and without symptoms of pain or tightness in their arms, forehead and neck, those <u>with</u> symptoms had significantly more muscle tension at these sites than did those <u>without</u> symptoms: but in no instance were there any differences at the other sites. Muscle hypertension therefore appeared to account adequately for many common muscular complaints.

A third finding, one to which I will return presently, was the high correlation between the individuals' tension records on two occasions. From this we inferred that the method is reliable; but more interestingly, it also implies that muscle tension is a stable characteristic of the individual. While a significant concordance found between the tension counts from three distinct groups of muscle suggested that there is, in addition, an increased innervation of the body musculature as a whole in anxious or otherwise predisposed individuals.

Since then my colleagues and I have been able to confirm and add to these early observations. We have shown for example that forearm tension of normal subjects relates significantly to other personality factors, notably to Eysenck's neuroticism and introversion, Taylor's Manifest Anxiety, and a high symptom score on the Cornell Medical Index. So neurotic, highly strung people might be expected to be vulnerable to such complaints as backache, "fibrositis" and headaches. We tested this prediction in the following way. For six weeks every patient attending each of the out-patient clinics at a District General Hospital completed a neuroticism questionary (M.P.I.) the doctors having recorded the diagnoses. The 80 patients diagnosed "backache" or "prolapsed disc", for example, had much higher neuroticism scores than the controls. And those with the highest scores were differentiated by being female, having a history of recurrent back pain and of other psychosomatic disorders. Moreover, their replies to that item on the M.P.I. Scale which asks "would you rate yourself as a tense or highly strung individual" were answered affirmatively by the backache patients significantly more often than by patients in the control groups (Sainsbury 1960).

We next looked at the psychomotor behaviour of 29 patients with a primary depressive illness. Their level of muscle tension was also higher than that of matched controls; and on recovery their tension had decreased in direct proportion to their clinical improvement. We then compared the "agitated" (restless) depressives and "retarded" (underactive) ones. The agitated depressives were more anxious and had higher levels of muscle tension than the retarded, both when ill and when recovered; on recovery they were also more anxious and tense then their matched controls. Consequently, we inferred that if an <u>anxious</u> individual suffers a depressive illness, it is more likely to take on the agitated rather than the retarded form of the disorder.

The data also suggested that muscle tension to some extent determines
the kind of somatic symptoms the depressives have, because more of the
muscularly tense and agitated patients were preoccupied with pains in
the head, back and elsewhere: symptoms which abated when tension
decreased on recovery.

To summarise these remarks on muscular hypertension:

- Muscle tension can be reliably measured.

- It has definite clinical correlates since it is
  consistently associated with anxiety states,
  whether uncomplicated or occurring in a depressive
  setting.

- The sustained muscular contractions which
  accompany prolonged anxiety or heightened arousal
  are the psycho-psychological mechanisms of many
  common aches and pains.

And lastly an individual's level of muscle tension is stable and
can be considered an attribute of personality. Other personal quali-
ties also tend to be linked to muscle tension especially a predisposition to anxiety, neurotic tendencies, and expressive style as manifested in posture, hand-writing, countenance and gait.

From muscle tension which is one manifestation of psychomotor
behaviour I want to turn to the other and more obvious one – motility.
By which I mean the play of facial expression, gestures and other
spontaneous bodily movements of an apparently superfluous kind such as
pacing, picking, rubbing and tics.

I have already mentioned that my brief from Aubrey Lewis was to
find out "something about 'tics'".   I did not have much success in
this, but while experimenting with the EMG and a movie camera, I did
find a way of measuring 'tics', and of gestures too.

Since it is the lack of "objective criteria whereby the phenomena
observed can be measured" that limits the scope of psychiatric research, and since psychomotor behaviour presents such unique opportunities for objective assessment, I want to say something more about
the measurement of gesture and expressive movements.

Simply quantifying the amount of movement would provide one objective criterion by which to assess the course of a disorder, the
effects of treatment, and the factors affecting their manifestations.
Much might be inferred in this way about disordered movements such as
chorea or tics; about the psychomotor behaviour which is such a conspicuous feature of some mental illnesses, especially the affective

disorders and schizophrenia; and about factors that affect a person's use of expressive movements. If, moreover, one can then derive some elementary classification of expressive movements in terms of their supposed functions, a start in assessing their content becomes a feasible proposition as well.

Methods of measuring motility have suffered from two serious drawbacks; they required cumbersome attachments which inhibited the subjects spontaneous behaviour and precluded recording in a natural setting.

To avoid these snares Olson (1929) used a time-sampling technique to count the nervous habits of children and combined this with an elaborate anatomical code to describe their habits. I therefore adapted Olson's time-sampling method to measure tics and gestures. The tiqueur was filmed while a small light placed behind him flashed every five seconds to define the sample of time. When the film was projected, the observer scored "one" if a tic or tics occurred during the five seconds, and zero if there were none. The test-retest reliability was very high (0.93). So by filming over quite short periods changes in tic scores were readily discerned. Gestures could, we found, also be measured in the same way and very accurate scores obtained by repeatedly projecting the film. For this reason we have used it to validate other methods.

The EMG offered a more practical approach, though it could only be effectively applied to gestures of the arms and hands; and had the further disadvantage of constraining the subject. In the event these drawbacks were not a serious hindrance as the electrodes could be fixed over the patient's forearm muscles, beneath his clothes with the wires running up the sleeves to the back of his collar: out of sight, they were quickly forgotten.

We adapted the EMG to record movements independently of muscle tension. This is feasible because movements recruit many more motor units and so give rise to far larger signals than does tension. An integrator was therefore adjusted to respond only to signals above a predetermined level, and this system gave reliable and valid counts of motility.

Recently, however, we have developed a new method of a quite different order (Haines and Sainsbury 1972). It consists of a small ultrasonic transmitter and receiver which can be placed or hidden in a familiar room. Any movements arising within the ambit of the ultrasonic beam interferes with it such that the received wave is no longer in phase with the transmitted one. The electronics are able to measure these phase differences, translate them into pulses, and then count them. Calibration experiments show a nice linear relation between amount of movement and number of counts.

Before describing the results of some of our movement studies, I want to outline the elementary classification of expressive movements that we use; a scheme which is imposed by the limitations of our recording methods; though I have been encouraged to find that our categories are very similar to those Eckman (1972) derived from his detailed analyses of non-verbal behaviour.

We distinguish four categories: first, the apparently <u>superfluous movements</u> everyone makes such as finger-tapping, scratching, or chin-stroking, which in ourselves we speak of as "fidgetting", but which in our patients we are more likely to call "restlessness" or "nervous habits", or "autistic gestures" depending on how far we are prepared to indicate their supposed origin (these movements correspond to Eckman's "adaptors"). Their pertinence for us to-day is that we commonly recognise overactivity of this kind as a manifestation of stress. However designated they are best elicited by leaving the subject alone and unoccupied for 5-10 minutes. Secondly, "<u>descriptive gesture</u>" refers to movements used as an aid to speech, to convey an idea or meaning, or to depict an object. To elicit these gestures we either give the subject a list of illustrative words to define, such as "corkscrew", "chin", "knock", "elephant" and so on; or ask him to describe his living room at home in detail. The third category, we refer to as <u>emotional</u> or <u>communicative</u> gestures because they are best observed in an interview in which stressful topics are discussed. They serve to punctuate conversation, to convey feelings and emphasis; but in so far as descriptive content enters there is some overlap with descriptive gesture. (Eckman uses the term "illustrators" to include both categories). Lastly, for some purposes a measure of the subject's <u>total activity</u> may be the appropriate aspect of his psychomotor behaviour, when assessing, for example, the effects of drugs on restlessness. Total activity includes both expressive and purposive or goal directed movements, whether they be repetitive ones such as smoking and knitting or acts of a more obviously coping kind. They can be represented by the total movement of a patient when left alone for an hour with, for instance, just a jig-saw puzzle to occupy him.

In the daily exchanges between people each unwittingly takes account of the other's expressive behaviour. But what can we confidently glean from doing so? Since we have seen that muscle tension reveals something about the psychomotor expression of personality, it is likely that the more easily observed expressive movements might also disclose information about the individual's feelings and disposition, whether outgoing or reserved, energetic or listless, tense or tranquil, rigid or flexible, under stress or at ease.

In addition, expressive movements convey <u>meaning</u> to another, and this function is especially important when words are beyond our reach. But how accurately can we recognise and translate the cues? This problem is even more evident in the psychiatric interview, where again words often do not suffice, and it is here in this situation that

another connotation of gesture becomes apparent. We know from clinical experience that patterns of psychomotor behaviour relate to
diagnosis and impart information about the nature, origin and likely
outcome of the disorder. A more precise description of motility
should, therefore, be clinically valuable.

Before discussing how expressive movements can contribute to the
description of personality; what it is they communicate; and what
clinically useful information they render, we need to establish that
the four categories of expressive movement can be reliably measured.
Table I shows the product moment correlations between counts of expressive movements obtained electromyographically over 10 or 15 minutes on two occasions separated by quite short intervals. They were
recorded from healthy controls and from three groups of patients, and
include three of our categories of gestural movement. With the exception of the descriptive movements psychoneurotics make when defining
such words as giraffe, and spiral-staircase, the correlations are
high and the method therefore reliable. To obtain such results, both
the normal and psychologically stressed subjects must have been consistent in their use of expressive movements, at least over short
periods. This is an important finding similar to that for tension
measurements. But how stable are they over longer periods of time?

Table I. The reliability of three categories of expressive movements
when EMG counts are obtained for 5 - 15 minutes and repeated within a
short period.

Group	N	Category of Expressive Movement	Interval between Recordings	Correlation
Healthy controls	22	Alone	1 day	0.75
Psychonuerotics	30	Descriptive	3 days	0.56
Psychonuerotics	16	Interview	1 day	0.87
Manics[1]	21	Alone	15 mins	0.69
Manics[1] (recovering)	17	Interview	3 days	0.88

[1]Dr. Costain, unpublished MD thesis.

Table II. Constancy of three categories of expressive movements when
EMG counts are obtained on two occasions, four or more weeks apart.
(adapted from Sainsbury and Costain, 1971)

Group	N	Category of Expressive Movement	Interval between Recordings	Correlation
Healthy controls[1]	29	Alone	5 weeks	0.62
Healthy controls[1]	29	Descriptive	5 weeks	0.79
Endogenous depressives[1]	29	Descriptive	weeks till recovered	0.61
Agitated depressives[1]	15	Alone	weeks till recovered	0.77
Retarded depressives[1]	14	Alone	weeks till recovered	0.37
Manics[2]	22	Alone	4 weeks	0.52
Manics[2]	22	Interview	4 weeks	0.71
Anxiety states and neurotic depressives[3]	9	Alone	30 days	0.65

([1]Pierce (unpublished), [2]Costain (unpublished), [3]Pearce (1964).)

When the measures are repeated some weeks later the correlations show
very little decrease (Table II). This is remarkable as all the patients were being treated so their levels of activity will have
changed appreciably on that account. Nevertheless their rank order
with respect to gesture is little affected. Moreover, this holds for
doodling and fidgetting while sitting alone, for descriptive gestures,
and for communicative gestures during an interview.

Consistency in gesturing was examined in another way using the
ultrasonic method. Eighteen English and eighteen French students,
matched on age, sex and education were first asked to describe in
detail their living room at home for seven minutes. This task elicited descriptive gestures. The French students, however, described
their living room twice: once in French and again in English, the
order being randomised. Then, to obtain emotionally-toned gesture –

that is gesturing under stress - both nationalities described the
members of their families with whom they got on best and those with
whom they got on least well; again for seven minutes. Movement counts
were recorded every minute or every half minute. The amount they
talked was also measured electronically to give a ratio of gesture to
words.

A consistent level of gesturing is maintained from minute to
minute (Figure 1). The individual's stability of descriptive and of
emotionally-toned gesture was also clearly shown in an analysis of
variance of the gestures of both groups of eighteen students. In both
groups and for both tasks the between subject variance was high,
whereas the within subject variance was low. That is to say, a student's counts differed little from one minute to another, while the
difference between students was considerable. What is more, the
correlation between descriptive and emotional gestures of the French

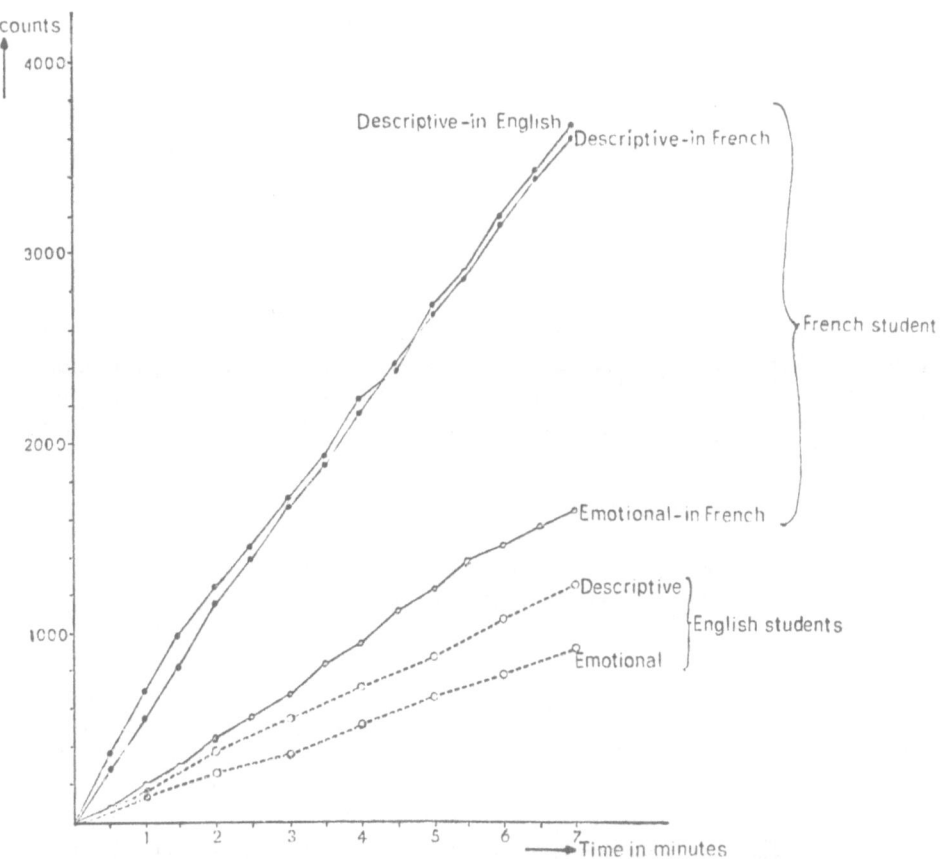

Figure 1. Mean counts per half minute of descriptive and
emotionally-toned gesture of French and English students
– using ultrasonic recording.

students was 0.86 and that of the English was 0.87. Both observations
strongly support the notion that the individual is consistent in the
extent to which he uses gesture to express himself. His gesture, in
other words, is characteristic of him and of his expressive style - it
is a kind of signature.

These considerations of a person's expressive style raise another
question, namely whether other attributes of personality are associated with the different kinds of gesturing? When gesture is related
to body build, for instance, the gestures of pyknics are more free and
smooth than those of asthenics (Brengelmann 1960). We reached a
similar conclusion from finding a significant correlation between EMG
counts of descriptive gesture and extraversion (scored on the EPI) in
both normal controls ($r = 0.41$, df 29) and recovered depressives
($r = 0.42$, df 29). Further support for a link between personality and
gesture was our finding that people who had high scores on extraversion and neuroticism (a combination some authorities equate with
hysterical trends,) used more descriptive gestures than those with
other permutations of these two traits. These extraverted neurotics
also had higher ratings on hypochondriasis and on a check list of
somatic symptoms; this may partly account for the conspicuous and
intractable nature of hypochondriasis in certain patients.

That out-going, sociable people should gesticulate more during an
interview than the introverted was to be expected; but what kind of
person is likely to fidget and make other superfluous movements that
do not apparently convey information? That such movements are often
referred to as "nervous habits" may not be altogether inappropriate,
because the depressives with high fidget counts also had significantly
higher scores on Taylor's Manifest Anxiety and on Eysenck's neuroticism, both of which to some extent assess susceptibility to tension
and stress. Moreover, these patients continued to be more fidgety
after recovery. Fidgeting in a group of psychoneurotics also correlated significantly with a measure of rigidity (Pearce), implicit in
which is the notion that these patients have something to communicate,
but are restrained from doing so in some way.

This brings me to the second question: What is communicated by
non-verbal movements? This is a matter about which I can only give
one or two hints.

The extent to which gesture is used as an aid to verbal communication is partly determined by cultural factors, and this was very
apparent when we compared the gesturing of the French and English
students. The differences between them were greater for the descriptive rather than the emotionally stressful tasks (see Table III), that
is, when the miming component of gesture is predominant or, as Eckman
puts it, when the act looks like what it means. It is this aspect of
gesture especially that is fostered by the social group in which we
first learn to communicate with others.

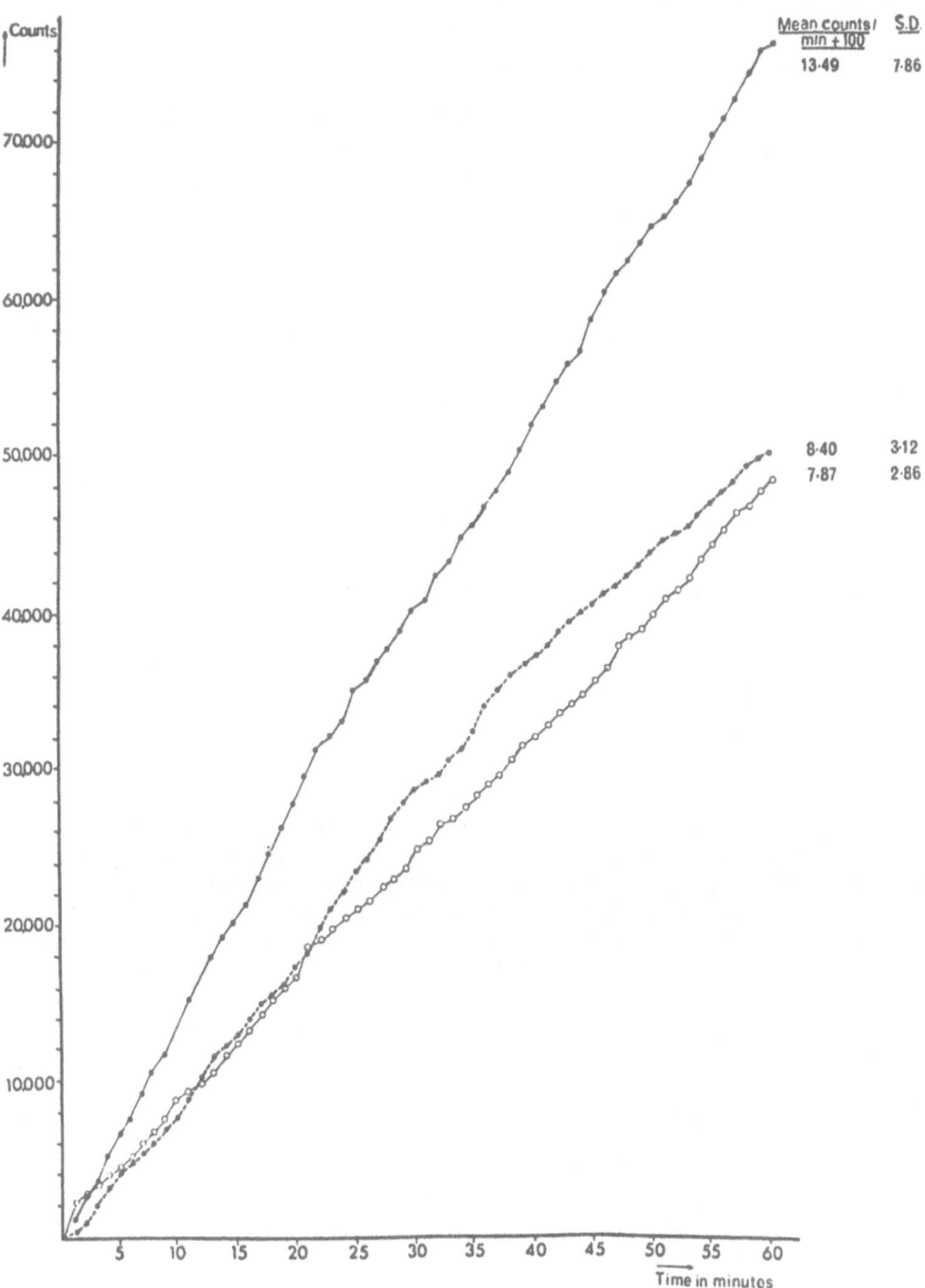

Figure 2. Counts per minute of three anxious and depressed patients
ultrasonically recorded while left alone for an hour
with a jig-saw puzzle.

Table III. Ultrasonic counts of descriptive and emotional gestures made by matched French and English students – N = 18 for each group. (Without allowing for amount of speech.)

Gesture type	French Students log scores		English Students log scores		t unpaired	p
	mean	dev	mean	dev		
Descriptive	3.46	0.34	2.82	0.60	3.48	<0.001
Emotional	3.14	0.30	2.72	0.53	2.80	<0.01
t paired	7.55		5.27			
p	<0.001		<0.001			

Besides the cultural predisposition to gesticulate, psychophysiological mechanisms also play a part in exciting gesture when an individual is under emotional stress. This was clearly shown in another experiment designed to see whether gestures during psychiatric interview would increase when stressful topics were discussed. The gestures of four patients were counted electromyographically during interviews in which neutral matters and topics believed to be disturbing to the patient were introduced; significantly more movements occurred when the stressful topics were discussed. In addition each "utterance" of the patient was rated for disturbance of affect, and again those given the highest ratings were accompanied by most gesture. Moreover, when the disturbing topics were reintroduced at a second interview, they were again those that increased most; and, incidentally, heart rate too (Sainsbury, 1955). The emotions were therefore consistently expressed in gestural movements. The effect of stressful topics in exciting gesture was also apparent in our experiment with the students; the Englishmen (who are less culturally prone to gesture) used more gesture when describing their feelings about their families than when describing their living room.

The communication of distress by gesture was further demonstrated by comparing the counts of the French students' movements when they described their living rooms in French and in English. The latter is obviously the more difficult task. When describing in English, and especially when speaking in English before they had done so in French, they gestured significantly more. These effects on gesturing of impeding verbal expression are relevant to many clinical problems in neuropsychiatry; the psychomotor behaviour of dysphasics, for example or of patients struggling to convey elusive ideas and feelings.

My third question, was whether improved measurement of gesture contributes something of clinical value. I will give two examples of it having done so.

First, we studied psychomotor behaviour in mania (Sainsbury and Costain 1971) to see if motility measures could differentiate types of mania and predict outcome. The movements of 22 consecutively admitted manics were measured by the EMG method soon after admission and again 4 weeks later after treatment. The measurements were made with the patient alone, and then during a standardised psychiatric interview. After 4 weeks their motility had decreased. This was not surprising because ten of the patients were clinically assessed as recovered, five as unrecovered from mania, and seven as having become depressed. What is of interest, however, is:

(1) These three categories of outcome are also differentiated by their scores on three measures of psychomotor activity (Table IV) and

(2) The motility scores recorded __before__ treatment began distinguished those manics who became depressed (Table V).

The motility score therefore effectively differentiated the outcome groups.

Table IV. Mania: Psychomotor scores.

Psychomotor Activity		Outcome of 22 manias: mean log scores			
		Manic (5)	Normal (10)	Depressed (7)	Significance
Speech	week 1	4.5	4.6	4.4	N.S.
	week 4	4.7	4.4	4.1	<0.01
"Interview" moves	week 1	3.8	3.8	3.3	N.S.
	week 4	3.6	3.1	2.7	<0.01
"Alone" moves	week 1	3.2	3.4	2.9	N.S.
	week 4	2.8	2.7	2.0	N.S.

Table V. Mania: First week motor score.

Outcome at 4 weeks	Speech	"Alone" movements	"Interview" movements
Depressed (7)	4.46	2.88	3.32
Manic and recovered (15)	4.60	3.36	3.82
	N.S.	$p < 0.025$	$p < 0.025$

Motility scores also provide a precise method of measuring the effects of psychotropic drugs. Chlorpromazine was prescribed for 11 of the patients and haloperidol for 7. All the psychomotor measures of the patients treated with chlorpromazine decreased significantly, but decrease was less with haloperidol (See Table VI). Although the difference between the two treatments was not significant, the data illustrate the sensitivity of the measures and the feasibility of using them to evaluate.

My second example deals with the question of the relation between expressive movements and the tendency to develop somatic complaints. We postulated that gesticulators would have fewer somatic complaints because their feelings find expression in motor activity. Though the results were all consistently in the predicted direction, they only approached but did not reach significance. The descriptive gestures of the depressed patients and the alone movement of controls, for example, related inversely to the number of their somatic symptoms at the 10% level of probability.

In summary, these observations suggest that we each have our own expressive style or psychomotor personality which, among other things, determines the way in which the motor system responds to stress and its accompanying emotions. The physiological effects of the emotion can be manifest on the one hand as increased muscle tension and as bodily symptoms which arise from this, such as aches and pains; and on the other as motility. This may be apparent in a number of ways, but notably as restless, stereotyped movements or more symbolically in gesture.

Table VI. Mania: Comparison of psychomotor scores
of patients treated with largactil and haloperidol.

Psychomotor Activity		Haloperidol N = 7 mean log score	Largactil N = 11 mean log score	t	Sig.
Speech	week 1	4.504	4.539	0.23	N.S.
	week 4	4.314	4.352	0.27	N.S.
		t = 1.52 p < 0.2	t = 2.586 p < 0.05		
Communicative Gesture	week 1	3.581	3.601	0.08	N.S.
	week 4	3.004	3.106	0.28	N.S.
		t = 2.034 p < 0.1	t = 3.369 p < 0.01		
Autistic Gesture	week 1	3.070	3.139	0.29	N.S.
	week 4	2.221	2.573	0.77	N.S.
		t = 1.965 p < 0.1	t = 4.095 p < 0.005		

REFERENCES

Allport, G. W., 1938, "Personality," Constable, London.
Allport, G. W., and Vernon, P. E., 1933, "Studies in Expressive
    Movement," Macmillan, New York.
Brengelmann, J. C., 1960, Expressive movements and abnormal behaviour,
    in: "Handbook of Abnormal Psychology," J. Eysenck, ed.,
    Pitman, London.
Eckman, P., and Friesen, W. V., 1972, Hand movements, The Journal
    of Communication, 22:353-374.
Effron, D., and Foley, J. P., 1947, Gestural behaviour and social
    setting, in: "Readings in Social Psychology,"
    T. M. Newcomb and E. L. Hartley, eds., H. Holt, New York.

Gibson, J. G., 1959, Psychophysical relationships, Ulster Medical Journal, 28:1-10.

Gibson, J. G., 1962, Emotions and the thyroid gland: a critical appraisal, Journal of Psychosomatic Research, 6:93-116.

Gibson, J. G., and Willcox, D. R. C., 1957, Observations on thyroid adrenocortical relationships, Journal of Psychosomatic Research, 2:225-235.

Grant, E. C., 1971, Facial expression and gesture, Journal of Psychosomatic Research, 15:391-394.

Haines, J., and Sainsbury, P., 1972, Ultrasound system for measuring patients' activity and disorders of movement, Lancet, ii:802-803.

Olson, W. C., 1929, "The Measurement of Nervous Habits in Normal Children," University of Minnesota Press, Minneapolis.

Pearce, K. I., 1964, "Measures of verbal and muscular activity as correlates of clinical states," (unpublished).

Pierce, D., "Psychomotor behaviour and depression," (unpublished).

Sainsbury, P., 1955, Gestural movements during psychiatric interview, Psychosomatic Medicine, 17:458-469.

Sainsbury, P., 1960, Neuroticism in unselected outpatients attending physical medicine and orthopaedic departments, Annals of Physical Medicine, 5:310-317.

Sainsbury, P., and Costain, W. R., 1971, The measurement of psychomotor activity: some clinical applications, Journal of Psychosomatic Research, 15:487-494.

Sainsbury, P., and Gibson, J. G., 1954, On symptoms of anxiety and tension and physiological changes in the muscular system accompanying them, Journal of Neurology, Neurosurgery and Psychiatry, 17:216-224.

Sainsbury, P., and Wood, E., "Gesture: its measurement and cultural relations," (unpublished).

W.H.O., 1964, Psychosomatic Disorders, in: "Technical Report Series No. 275," Geneva.

Wolff, W., 1943, "The Expression of Personality," Harper and Macmillan Bros., New York.

DISCRIMINATION OF VISCERAL STIMULI, BIOFEEDBACK AND SELF-REGULATION:

A METHODOLOGICAL PROPOSAL

David W. Jacobs

Department of Community Medicine, School of Medicine
University of California at San Diego
    and
Director, Biofeedback Institute of San Diego
San Diego, California

This paper has two purposes. The first is to persuade an audience, composed primarily of clinicians, of the relevance of a seemingly academic problem, discrimination of visceral stimuli, to their clinical work. The second is to persuade both clinicians and academicians that changes in their research methods are necessary if the problems of visceral perception are to be solved.

Concerns about the role of perception in self-regulation are usually addressed by clinicians in a somewhat different language than that used by researchers. One frequently hears discussions in clinical contexts about the 'awareness' a patient has concerning some particular body process. The implication is that the degree of awareness is positively related to the degree of self-control. A good example of this is in electromyographic (EMG) biofeedback, in which the therapist (often reflecting comments made by the patient) will say that the patient has become more aware of the tension in a given group of muscles and has correspondingly learned to keep them relaxed for longer periods of time. While such statements are heard less frequently in the context of self-regulation of visceral processes, it is often present by implication. Indeed, it is precisely because awareness plays such an unclear role in the self-regulation of visceral processes that research in the area is of such interest to practitioners.

My interest in this problem was initially stimulated by the publication of Blanchard and Young's (1974) classic review on the efficacy of biofeedback. The authors concluded that EMG biofeedback showed greater efficacy than applications of the technique to other

organ systems, although no compelling reasons for this difference were
given. The most obvious explanation is that EMG biofeedback training
is done on striate muscles, which are innervated by the central nervous system (CNS), while most other applications of biofeedback training involve visceral organs, which are innervated by the autonomic
nervous system (ANS). But there is no inherent reason why the CNS
should be more responsive to biofeedback training than the ANS, although it has been proposed that we are more aware of processes controlled by the CNS than the ANS. Clearly we are more aware of striate
muscle activity than of visceral activity. Such processes as dilation
in the peripheral vasculature, secretion of bile from the liver and
maintainence of core body temperature are all 'silent' processes of
which we are unaware, while we are exquisitely sensitive to movements
of limbs and digits. A mechanism which would supposedly account for
the differences in awareness was proposed by Sherrington (1906) who
argued that the lack of visceral awareness could be attributed to a
paucity of afferent fibers in the ANS. However, more recent work has
shown that there is abundant afferent activity at all levels of the
autonomic nervous system (Adam, 1980). Of course the distinction
between 'silent' and 'aware' processes is far too simple. What really
exists is a continuum between these two poles. There is, for example,
little awareness of activity in the lower insertions of the trapezius
muscle, although it is innervated by the CNS, and is regarded as part
of the voluntary musculature. On the other hand, a recent study indicates that a rather 'fine grained' capacity to discriminate blood
pressure can be found in at least some biofeedback subjects (Vidargar
et al, 1983).

Despite lack of information about its neurological substrate, the
importance of awareness in self-regulatory activity is assumed by many
writers working is this field. Indeed, its importance pervades much
of western thought. The distinction between voluntary and involuntary
behavior assumes that one cannot control actions of which one is not
aware. This same assumption also pervades the work of many of those
involved in the self-regulation of physiological functioning. An excellent example can be found in Gannon who, in discussing the relation
between interoception and learned visceral control, argued "when external feedback coincides with naturally occurring afferent information, persons can learn to recognize and direct their attention to
interoceptive cues and thus become 'aware' of their ongoing visceral
activity. This awareness can then be utilized to inform the person if
he/she is producing the instructioned . . . changes" (1977).

Logically, four positions may be developed with regard to the
relation between awareness and body control. One is the view that
perception of the bodily state is a necessary and sufficient condition
for self-control. Perhaps the most vociferous spokesman for this
position is Brener (1977). He argues that people with good self-
regulatory skills have a clearly defined capacity to discriminate the
sensory accompaniments of a visceral response. By implication, at
least, people with poor control have little or no such capacity.

These sensory accompaniments are integrated into a 'response image' which develops with increased physiological control. Once developed, the image, which is a perception of the self-regulated response, initiates the response as the image becomes activated. At the other extreme is the argument that awareness is neither necessary nor sufficient for bodily control. This view is espoused by Lacroix (1981). He argues that self-regulative learning is dependent on identification of already available responses that have previously provided success in body control. These responses should be distinguished from the identification of pertinent afferent stimuli in that it is the efferent processes that are the focus of selection and refinement during training. In the main, awareness of the sensory correlates simply have no affect on self-control. There are few writers taking either of the other two logical positions, although they are interesting to consider. If the view is taken that awareness is a necessary, but not sufficient condition for body control, one could begin to consider what else would be necessary to produce a reliable self-regulated behavior. For example, some form of response training by operant procedures might be necessary in addition to the development of the perception in order to achieve self-control. If the other view is considered, i.e., that awareness is sufficient but not necessary for self-control, alternative forms of training/learning for bodily self-control could be considered.

Choosing among these positions is difficult for there are fundamental conceptual and methodological problems involved in this line of inquiry. Powers (1973) has pointed out that the very distinction between perception and response is itself challengeable. All exteroceptive perceptual acts require behavioral components for their successful completion. At the least, adjustment of body position is necessary before perception can begin, and the perceptual act itself is rife with response elements. Something of the same sort must occur during interoceptive perception as well, since the internal responses made by subjects must form a prime source of stimulation for subsequent self-regulative responses. In such a situation, it is difficult to distinguish between the stimuli and the responses involved in physiological self-regulation.

The issues of identifying the stimulus and response involved in visceral self-regulation are not trivial from the clinical perspective. Patients may achieve self-control in many different ways and a treatment program which is to be maximally effective must be informed about the self-regulative techniques being used by a specific patient. Consider, for example, the problem of the nature of the stimulus. Chernigofskiy (1967) has pointed out that there are four classes of stimuli which may be involved in visceral self-regulation; mechanical, chemical, osmotic, and thermal. Receptors exist for each of these classes of stimuli in the visceral afferent system. When a patient is demonstrating discrimination of or control of a visceral response, it is often impossible to determine which of these categories of stimuli are involved in either the discrimination or the control.

Compounding this problem is the fact that subjects can learn to use exteroceptive responses which are inextricably involved in the discriminative act. An excellent example of this can be found in the paper by Mandler and Kahn (1960), in which they demonstrated the ability of a subject to reliably discriminate changes in acceleration and deceleration of heart rate. In a subsequent interview, it was determined that the subject had discovered the relation between respiration and heart rate change, and had utilized that information both to change heart rate and to report on the discrimination of that change. This capacity of subjects, whether wittingly or not, to use whatever information is available to them in achieving discrimination and control, presents a difficult problem. From the point of view of experimental rigor, it is important to demonstrate that an increase in awareness and/or control over visceral states does not reflect some easily controllable behavior such as breathing.

With this analysis in mind, it is now time to consider the demands on experimental designs which will yield information about these issues. I want to argue that there are four characteristics of an optimal design for experiments addressing the relation between awareness and visceral control.* The first is that the design should unambiguously distinguish between measures of awareness and measures of response. This means that the procedures that define and measure awareness must be operationally separable from the procedures that define and measure the response. Second, the experimenter should be able to manipulate these measures independently of each other, i.e., changes in awareness should be achievable without there being any necessary linkage to the response measures, and vice versa. Third, the measures should be relatable to each other over time. If the intention of the research is to seek out causal relations between awareness and acquired self-control, then it must be demonstrated that appropriate temporal and functional relations exist between the measures to establish that causality. Finally, the design should be such as to permit monitoring of other relevant physiological events simultaneously with the events that are under investigation. This will permit the experimenter to determine the physiological systems a given subject has used in the development of both awareness and acquired self-control.

As Brener (1982) has pointed out, in such a design it is also important the the experimenter be operationally clear what is meant by 'awareness' and 'self-control'. Brener has suggested that researchers already have operationalized the definition of 'awareness' to be the capacity to make accurate discriminations of physiological state. This definition does some violence to the initial meaning of the term 'awareness', which has more phenomenological implications. However, a very strong argument can be made that if a patient can reliably tell the state of a physiological system at any given point in time, he is

---

* The issues raised here follow concerns voiced by Brener (1982) in a most informative article.

in some sense perceiving that system, even if the phenomenological components of awareness are not present. Indeed, some very interesting epistomological issues arise out of such capacity for discrimination without conscious experience.

Brener (1982) also has pointed out that self-control has a meaning which can be taken directly from the literature on learning. The original definition of self-control is to be found in the notion of 'voluntary' behavior. In most learning studies, this has been operationalized as compliance with instruction. Again, most experimenters would be persuaded that acquired self-control of a physiological response is demonstrated when, upon appropriate instruction, a subject can make the appropriate physiological response.

An optimal design will meet these four criteria and use operationally defined terms. It seems to me that this requires the use of single case experimental designs as opposed to the group designs now more typically employed. A few comments about the differences between 'between group' and 'single case' designs may be appropriate. Interested readers may wish to consult Herson and Barlow (1976) for a more detailed examination. Between group designs employ many subjects, each of whom receives the same treatment. Experimenters using these designs seek regularities in behavior across subjects, treating the individual differences between them as unwanted 'noise', which interferes with understanding the processes which are common to all individuals. These designs are most useful when the differences between subjects are small and the processes being described are highly similar across individuals. Single case designs, on the other hand, do repeated measures on single cases and seek out the differences between individuals as the prime focus of inquiry. They are most useful when there are significant differences between individuals in the processes under study and there is much to be learned from the intense study of single individuals. Of course, the disadvantage of such designs is that generalizations to any population other than the single individuals being studied must be done with great caution.

These two types of design yield different results when applied to the same problems. Pennebaker (1982b) has pointed out that results of body perception studies are different when the experimenters use a group design or a single case design. Between group designs yield estimated correlation levels of about +.30 between given physiological states and the perception of those states. Single case designs, on the other hand, demonstrate an incredible diversity of individual differences in capacity to accurately perceive bodily states of various kinds. The point is that the group design obscures a process rather than illuminating it. It is simply not the case that about 9% of the subjects correctly perceive their physiological states. That statistic hides a complicated interaction between perception, judgment and response that is different from one individual to another.

Table 1. Within-subject correlations (in %) by subject between
selected self-reported symptoms and their autonomic referents.

SUBJECT	SEX	GSR WITH SWEATY HANDS	HR WITH FAST PULSE	FPV WITH WARM HANDS
1	M	37	58	48
2	M	82	−17	−40
3	M	−29	−21	63
4	M	21	67	21
5	M	12	34	68
6	M	00	35	23
7	M	−02	−07	46
8	M	33	−11	22
9	M	−36	42	18
10	M	46	59	−02
11	M	70	41	13
12	M	−31	01	40
13	M	−17	41	66
14	M	−16	16	07
15	M	20	57	−49
16	F	−04	16	14
17	F	−44	52	−45
18	F	−40	−11	−12
19	F	28	07	−14
20	F	22	45	−02
21	F	00	−11	02
22	F	66	02	−24
23	F	−08	00	17
24	F	16	14	51
25	F	−07	05	07
26	F	−22	34	65
27	F	−17	45	19
28	F	−12	07	11
29	F	−28	−14	38
30	F	13	06	25

Note: Accuracy results are indicated if the correlations are negative
for GSR (galvanic skin resistance) with self-reported sweaty hands,
and positive for HR (heart rate) with self-reported fast pulse and FPV
(finger pulse volume) with self-reported warm hands. For any of the
correlations to be significant (one-tailed test, 38 df, corrected for
experimentwise error rate), r must equal −31% for GSR and +31% for HR
and FPV. From Pennebaker et al (1982a) as reprinted in Pennebaker
(1982b p65).

This can be demonstrated by examining some of Pennebaker's data. Table 1 shows correlations between perceptions of three bodily states and objective measures of those same states for thirty subjects. Forty judgements were taken from each subject for sweaty hands, pulse rate and hand-skin temperature. These were correlated respectively with skin resistance level, heart rate and finger pulse volume. An ideal observer would show correlational values of -1.0 for skin resistance level versus sweaty hands, (since resistance varies inversely with sweat production) and +1.0 for the other two. It is instructive to consider this data as it would have been reported in a group design. Table 2 shows such an analysis. Overall perception of sweaty hands is close to chance, although males seem to be a bit worse at it then females. Males fare a bit better than females in perceiving pulse rate and skin temperature, although no entry in this table suggests ability to recognize bodily states in an accurate manner. The discerning observer of this data will become somewhat more uncomfortable with such conclusions, after noting the standard deviations and ranges of these scores. It is clear that the variability among the correlations is so large as to render measures of central tendency meaningless. Table 1 shows that the source of that variability lies

Table 2. Averaged correlations (in %) with measures of variability.[*]

		GSR	HR	FPV
Males (n = 15)	Mean	22.67	26.33	22.93
	Dev	±22.69	±20.89	±21.59
	Range	-36 to +82	-21 to +67	-49 to +68
Females (n = 15)	Mean	-1.6	13.80	10.13
	Dev	±17.50	±17.19	±18.67
	Range	-44 to +66	-14 to +52	-45 to +65
Total (n = 30)	Mean	5.53	10.06	15.53
	Dev	±20.35	±20.46	±20.75
	Range	-44 to +82	-44 to +67	-49 to +68

[*] Original data from Pennebaker (1982). See also Table 1.

Table 3. Comparison of acquired skin temperature response:
Acquisition by two migrainous patients.[*]

| Session | Patient | | | |
| | SAR | | DK | |
	Mean	Dev	Mean	Dev
1	74.5	±1.5	73.7	±0.4
2	76.6	±0.4	72.1	±0.9
3	80.1	±1.5	85.3	±1.6
4	85.6	±1.8	93.1	±0.7
5	73.3[**]	±0.3	92.9	±1.1
6	81.4	±1.4	93.1	±0.6
7	86.6	±1.9		
8	86.5	±1.8		
9	86.1	±2.0		
10	86.6	±1.7		

[*] Mean and standard deviation for each session.
[**] Patient in pre-menstrual period c̄ head pain.

in the individual differences among subjects. Subject 1 is a good
perceiver of his pulse rate and skin temperature but totally misper-
ceives his sweaty hands. Subject 17 is an excellent perceiver of
sweaty hands and pulse rate, but cannot accurately identify her skin
temperature. Subject 21 might as well not have an afferent component
to her ANS. Subject 12 does fine with sweaty hands and warm hands,
but cannot identify pulse rate, etc.

Similar variability can be demonstrated with regard to the behavioral side of visceral control studies. Table 3 displays data
taken on two patients at the Biofeedback Institute of San Diego within
the last six months. Both patients are migrainous and were referred
to the Institute for biofeedback training for treatment of their
disorder. They were trained on an SRS computerized biofeedback system. Training consisted of weekly sessions, using a simultaneous
video display of electromyographic and temperature data. Cassette
tapes with relaxation instructions were provided to each patient and
were used daily according to their logs. Session data were taken at
fourteen intervals, each ninety seconds long. Means and standard
deviations are given in °F for temperature.

Before considering the data, it should be noted that the cases
were not randomly chosen, but were selected to illustrate the argument. The differences are apparent. First, there is a difference in
asymptotic level of performance. SAR reached a level of 86.5°F and
did not improve beyond that during the last month of training. DK
reached a level of about 93°F during the last three weeks of training.

It should be noted, incidentally, that after six weeks of training, DK was experiencing no more headaches and treatment was discontinued. A second major difference concerns the rate of response acquisition. DK's performance during the first three weeks of training was relatively low, rapidly moved to asymptote and stayed there. SAR's performance was a more gradual increase during a ten week period, interrupted only by a pre-menstrual week. It is also interesting to note the difference in the standard deviations. SAR's standard deviations tended to stay fairly constant, whereas DK°s standard deviations were highly variable. This reflected a difference in within session acquisition curves that is not displayed here. DK typically had small variations in performance while SAR invariably started at a very low temperature level and moved to progressively higher levels within training sessions. In effect what happened to SAR was that the within-session acquisition rate became more rapid, although initial temperatures did not change by any great amount across sessions. DK's initial temperatures rose steadily from session to session, beginning with the third.

The point of this analysis is that looking at the data from the point of view of individual differences, one would be disposed to plan different treatment programs for the two patients, since the problems in control faced by them are somewhat different. SAR's problem is that of raising the initial temperature. She was placed on a program of frequent, short periods of meditation and relaxation to increase average hand temperature during the day. DK required no such training to achieve the same affect.

To reiterate, it is clear that both for perception and response, individual differences predominate. Individuals do not perceive their own bodies nor do their bodies respond in ways sufficiently similar to justify the use of between group designs. If that argument be accepted, it is appropriate to consider the kinds of experiments, using single case methodology, that might be most appropriate. The last section of this paper describes a design which is currently being developed at the Institute.

Table 4 presents a rather sketchy outline for such an experiment. It uses a modified A-B-A design in which the phases are repeated over sessions. In these designs, baselines (A) are followed by the implementation of some independent variable (B) which in the last phase (A) is removed. Designs of this type are useful to determine whether an effect of an independent variable can be demonstrated. Only five trials have been indicated in each phase, although the experiment being planned utilizes fifty trials in each phase. To illustrate the procedure, consider an experiment in blood pressure training. For any given trial in either the pre-training or post-training periods, a trial is initiated with the onset of a light, which signals the beginning of that trial (S(1)). The subject is instructed to "attend to blood pressure" and then attempt to reduce (raise) it (depending on

Table 4. Proposed single case design (A-B-A)
for visceral perception-visceral response training.*

PHASE	Trial #	VISCERAL RESPONSE			VERBAL RESPONSE	
		S(1)	S(2)	G	Judgement	KR
Pre Train ing	1 2 3 4 5					
Train ing	6 7 8 9 10					
Post Train ing	11 12 13 14 15					

* S(1) and S(2) are the start and end of the trial interval respectively. G is a reinforcement. KR is knowledge of the results. See text for details.

instruction) by the end of the intra-trial interval. At the end of the trial period (S(2)), the subject is then asked to judge whether his blood pressure was higher or lower at the end of the trial period than it was at the beginning. In the pre-training and post-training periods, no information and no reinforcement is provided. In the training phase reinforcement (G) is provided on those trials in which blood pressure is changed in the desired direction and knowledge of results (KR) is provided at the end of each trial advising the subject whether his judgment was correct or incorrect. The reinforcers will probably be praise, money or both.

Before turning to the types of analyses possible in an experiment like this, it may be worthwhile to consider this design in the light of the previously mentioned criteria for experiments in the relation between awareness and visceral control. The first criterion is that the design should distinguish between measures of awareness and measures of response. The design does this by utilizing a somatic measure of response and a verbal measure of awareness. The second criterion is that the design should permit the manipulation of these measures independently of each other. It is clearly possible to do this by introducing changes in knowledge of results without making any necessary changes in the somatic measures and vice versa. The third criterion is that the measures should be related to each other over time in unambigious fashion. In this design, it is possible to calculate (over trials) the acquisition rates for somatic control and perceptual discrimination. The fourth criterion is that the design should permit the monitoring of other physiological events simultaneously with the events under investigation. This requires that the hardware available be of sufficient complexity to permit more than one channel for monitoring. This is a relatively simple task with a computerized system.

The dependent variables for data analysis are the verbal and somatic responses per trial. Table 5 shows the possible outcomes per trial using these dependent variables. If the outcome is the cell labeled 'A', it means that the physiological change occurred in the desired direction and the subject also correctly discriminated the change. Cell 'B' means that the change occurred in the desired direction and that the subject did not correctly discriminate the direction of change. Cell 'C' means the change did not occur in the predicted direction but that the subject did make a correct discrimination. Cell 'D' means that the change did not occur in the predicted direction and the subject did not correctly discriminate the direction of change. Cells 'A' and 'D' provide evidence for the proposition that awareness is a necessary and sufficient condition for the development of somatic control. Taken together, they show that where the verbal response is correct, the subject is likely to be making the instructed somatic response and where it is incorrect, the subject is not. Cell 'B' provides evidence of response control without discriminative awareness but only if it can be demonstrated that there are no correlated stimuli from other body processes influencing the somatic response. Cell 'C' provides evidence that awareness and control are independent of each other since it implies that it is possible for someone to be correct in their ability to discriminate the physiological state but without making any somatic changes in the desired direction. For any trial block, there probably will be some entries in each cell.

What makes this interesting is the possibility of change in the proportion of entries in each cell as subjects undergo training. Consider the possible outcomes for a highly trained subject who is capable of a considerable degree of somatic control. Good arguments can be made for a predominance of entries in either 'A' or 'B'. It may very well be the case that with the development of visceral control, there is a concomitant development in greater awareness. Alternatively as happens with the development of motor skills, awareness may drop away with the development of visceral control. The design permits an analysis of this problem by examination of response rates across sessions of training.

Table 5. Outcome alternatives for proposed design.

VERBAL RESPONSE

		<u>Correct</u>		<u>Incorrect</u>
<u>Rein- forced</u>	<u>A</u>	Correct Reinforced	<u>B</u>	Incorrect Reinforced
<u>Not Rein- forced</u>	C	Correct Not Reinforced	D	Incorrect Not Reinforced

VISCERAL
RESPONSE

If, for example, the response rate for visceral perception increases to asymptote significantly in advance of the response
rate for somatic control, clear evidence is provided that the
development of somatic control is preceded by the development of
discriminative control. Conversely, if the rate for somatic control
increases to asymptote prior in time to the development of the rate
for the verbal response, one could make the argument that it is the
increase in somatic control which gives rise to the discrimination.
In addition, manipulations of the points in training at which knowledge of results and reinforcement are introduced, as well as the
schedules for each, should provide detailed information on the learning mechanisms involved.

Clearly there are limitation on designs such as this one. In
addition to the limitations imposed by A-B-A designs in general, the
design does not permit the experimenter to determine if there are
correlated somatic stimuli available to awareness which may be guiding
the discriminative judgment. A possible control for this would involve the use of monitoring other physiological channels while the
somatic response under training is being manipulated. Simple correlational techniques should demonstrate whether the verbal response is
being correlated with some somatic response other than the one under
investigation.

In summary, the argument has been made here that work in visceral
perception is of importance to both clinicians and researchers. Much
of the work done thus far has been done with between group designs.
These designs provide limited information because the variability
among subjects is so great as to render their measures of central
tendency useless. The alternative is to use single case designs of
the type shown here. Such designs have the capacity of illuminating,
in an unambiguous way, the mechanisms of visceral perception and
visceral control.

REFERENCES

Adam, G., 1980, "Perception, consciousness, memory: Reflections of a biologist," Plenum, New York.

Blanchard, E. B., and Young, L. D., 1974, Clinical applications of biofeedback training: A review of the evidence, Archives of General Psychiatry, 30:573-589.

Brener, J. M., 1977, Visceral perception, in: "Biofeedback and behavior," J. Beatty & J. Leguwie, eds., Plenum, New York.

Brener, J. M., 1982, Psychobiological mechanisms in biofeedback, in: "Clinical biofeedback: Efficacy and Mechanisms," L. White & B. Tursky, eds., Guilford, New York.

Chernigovskiy, V. N., 1967, Interoceptors, presented at: Annual Meeting of the American Psychological Association, Washington, D.C.

Gannon, L., 1977, The role of interoception in learned visceral control, Biofeedback and Self-Regulation, 2:337-347.

Herson, M., and Barlow, D. H., 1976, "Single case experimental designs," Pergamon, New York.

Lacroix, J. M., 1981, The acquisition of autonomic control through biofeedback: The case against an afferent process and a two-process alternative, Psychophysiology, 18:573-587.

Mandler, G., and Kahn, M., 1960, Discrimination of changes in heart rate: Two unsuccessful attempts, Journal of the Experimental Analysis of Behavior, 3:21-25.

Pennebaker, J. W., Gonder-Frederick, L. A., Stewart, H., Elfman, L., and Skilton, J. A., 1982, Physical symptoms associated with blood pressure, Psychophysiology, 19:201-210.

Pennebaker, J. W., 1982, "The psychology of physical symptoms," Springer-Verlag, New York.

Powers, W. T., 1973, "Behavior: The control of perception," Aldine, Chicago.

Sherrington, C. S., 1906, "The integrative action of the nervous system," University Press, Cambridge.

Vidargar, L. J., Lee, R. M., and Goldman, M. S., 1983, Discrimination of systolic blood pressure, Biofeedback and Self-Regulation, 8:45-62.

A MODEL OF PEOPLE AT HIGH RISK TO DEVELOP

CHRONIC STRESS RELATED SYMPTOMS

Ian Wickramasekera

Professor of Psychiatry and Behavioral Science
Eastern Virginia Medical School
Norfolk, Virginia

PROFILES OF STRESS SUSCEPTABILITY

Sir William Osler is reported to have said that "sometimes it is more important to know what kind of patient has a disease than what kind of disease the patient has". One implication of this statement is that certain types of personality features can potentiate or atten- uate either the symptoms or the etiology of a disease, or both. The first goal of this paper is to start to specify a promising set of personality features and also a set of situational events under which people who are either biologically prone to a disease or exposed to the relevant pathogens will become symptomatic. The second goal of this paper is to tentatively suggest some procedures to quantify these personality dimensions and these situational conditions. The third goal is to present evidence from my clinical practice and the research literature to support this model of the patient at high risk to devel- op chronic stress related illness. The present model (Wickramasekera, 1979, 1980a;1980b) is based on clinical observations made, and case study data collected, over the last 15 years in an increasingly spec- ialized clinical practice.

In modern industrialized society psycho-social stressors are probably the primary class of stressors that activate the fight or flight response (Cannon, 1932) and/or the general adaptation syndrome (Selye, 1956). As Mason (1971) has suggested, both physical and psychological stressors operate through a common psychological mecha- nism, the perception of "threat" to the well being of the animal. However, psycho-social stressors like a problem child, an unhappy marriage, the death of a spouse, an unpleasant job, an aging parent who resides with you, etc. have certain unique and different features from physical stressors. First, psycho-social stressors commonly elicit both avoidance and approach tendencies either sequentially or

simultaneously. For example, a divorce after many years of marriage
can be both a relief and a regret. Second, the sources of psycho-
social stress are often nebulous and difficult to recognize, and even
harder to define, unlike the threat from a sabre-toothed tiger.
Third, psycho-social stressors tend to be chronic and resistant to
rapid resolution by primitive defenses like either "fight or flight".
For example the problems posed by an adolescent or an aging parent
cannot be resolved by either physical attack or flight. In summary,
then, <u>ambivalence, ambiguity and chronicity</u> are special problematic
features of psycho-social stressors that interact with the following
special features of people at high risk potentiating the probability
of somatic disorders and disease. The following personality features
(Wickramasekera, 1979a;1979b) are particularly vulnerable to the above
unique features of psycho-social stressors. These personality features are 1) either very high or very low hypnotic ability, 2) neuroticism (Eysenck, 1960) and autonomic response specificity (Lacey,
1967), 3) the cognitive tendency to "catastrophize" (Ellis, 1962)
often based on pessimistic belief systems and 4) a deficit in adaptive
support systems and coping skills. Finally, if there is either 5) a
high frequency of <u>major life changes</u> (e.g., divorce, loss of employment, physical injury and/or a high frequency of <u>minor hassles</u> over a
short period of time, the coping resources of the person may be over-
taxed and the person may become symptomatic. The massed and chronic
elicitation of ambivalent feelings by ambigious and complex psycho-
social stimuli in people who are either very high or very low on
hypnotic ability, autonomically labile, (neurotic) and prone to cognitively catastrophize, places them at high risk for developing somatic
disorders or disease.

Hypnotic Ability

Hypnotic ability is a normally distributed, stable individual
difference variable (Hilgard, 1965; Barber, 1976) that is partly
genetically based (Hilgard, 1977). Current research suggests that
hypnotic ability is best considered a mode of information processing
that can occur in a variety of situations (e.g., hypnotic induction,
transferance) but particularly for this context under conditions of
<u>high</u> or <u>low</u> arousal (Wickramasekera, 1971; 1972; 1973; 1976; 1977).
It is most important to stop thinking of hypnosis as an event that
occurs only during a hypnotic induction in the same way that we do not
think of intelligence as an event that occurs only during an intelligence test. About 10% of the population are able to very readily and
profoundly access the hypnotic mode of information processing and an
equal percentage are almost never able to do so. There are 3 features
of the high use of this mode of information processing and 3 features
of the low use of this mode of information processing that place
people at high risk for developing somatic symptoms. High use of this
mode results in a relatively "unfiltered" perception of the world, a
tendency to amplify even minimal sensory stimuli, and to react to
those potentiated sensory events with inadvertent but, at least in

part,voluntarily induced exaggerated autonomic reactions. For highly hypnotizable people, "beliefs irrespective of their validity are more likely to have biological consequences" (Wickramasekera, 1979). The capacity for relatively "unfiltered" or non-critical analytic menta- tion has several demonstrated consequences. First, these people are much more responsive to social-psychological influence procedures (Wickramasekera, 1976) in the form of operant verbal conditioning (Webb, 1962; Wickramasekera, 1970; King and McDonald; 1976, Weiss et al, 1960) respondent conditioning (Das, 1958 a; 1958 b), and various types of short term psychotherapy (Laren, 1966; Nace et al, 1982). They are also more responsive to psychological pollution and are very much more likely to report "parapsychological events" (Wickramasekera, 1979; Wilson and Barber, 1982). Wilson and Barber (1982) report that 92% of their high-hypnotic responders (N=27) and only 16% of their low or moderate hypnotic responders (N=25) report psychic experiences like telepathy, pre-cognition, out-of-the-body experiences, etc. At least some of these people are prone to psychological pollution because they can voluntarily reset their perceptual filters outside the constraints of logical-critical analytic brain functions. There is also some evidence that they can acquire and retain information, without waking up, in stage I alpha free sleep and stage REM sleep (Evans, 1977). They can also fall asleep easily in the sleep laboratory, wake up at a preselected time before their alarm goes off (Evans, 1977) and influ- ence the content of their REM dreams (Stoyva, 1965). Essentially high hypnotic ability people seem to have superior voluntary control of altered states of consciousness (waking, sleeping, napping, dreaming, etc.) and this ability may be used to alter their own perceptual and cognitive filters in ways that could be physiologically adaptive or maladaptive.

People of high hypnotic ability can amplify or attenuate sensory signals. Factor analytic studies of hypnotic behavior indicate that the capacity to provide the mind with rich images and fantasies is a major factor in hypnotic ability accounting for close to 50% of the variance on standardized tests of hypnosis. In fact it has been found that people of high hypnotic ability are less tolerant of pain if their hypnotic ability is not used than people of low hypnotic ability (Shor, 1964). People of high hypnotic ability have an unusual capac- .ity for attention to and absorption in fantasy, and sometimes this ability can be used to amplify their response to even minimal physical or visceral sensations. Since hypnotic ability is positively correl- ated with standardized tests of creativity it is possible that this creative ability is at times used to elaborate "meanings" or amplify even minimal sensations. Since these people can hallucinate volun- tarily they probably would need little or no sensory basis on which to develop delusional pain or a false pregnancy. Studies of information processing in high and low hypnotic ability subjects have found that high ability subjects have superior sensory memory and a superior ability to transfer information from sensory to short term memory (Ingram et al, 1979; Saccuzzo et al, 1982). This ability may be used to learn and retain both respondent and operant pain and anxiety more quickly than other people. Several studies (Frankel and Orne, 1976; Foenander et al, 1980; Perry et al, 1982; Gerschman et al, 1979) have found that an unexpectedly large percentage (48%-58%) of clinical phobics are highly hypnotizable. It seems that for these subjects, images and/or fantasies become so vivid and real as to be confused with the world outside. Another study (Pettinatti et al, 1982) found that 57% of all bulimics were high on hypnotic ability and 0% were low. A subject of anorexics who use purging as opposed to abstention from food, are also high on hypnotic ability. It is likely that these purging behaviors are enacted in a dissociative state.

Factor analytic studies have found that a second major factor in hypnotic ability is the ability to make the mind blank (amnesic) and that this factor is orthogonal to the fantasy factor. This ability may be used to deny or delay the recognition or organically based sensory stimuli in the acute phase of a disease, resulting in development of a chronic disease. It may also be used to alter states of consciousness, and may be used or abused to decouple the verbal-subjective response system (contents of verbally mediated consciousness) from the motor or physiological response system resulting in somatization or conversion symptoms.

Much less is known about people of low hypnotic ability beyond the fact that they are limited nearly always to a rational critical analytic mode of processing information and that they are unwilling or unable to use fantasy or imagination. They also appear to condition more poorly in both the operant and respondent modes (Weiss, Ullman and Krasner, 1960; Webb, 1962; Wickramasekera, 1970; King and McDonald, 1976; Das, 1958a;1958b) and probably are deficient in forming anticipatory response. Within limits, fantasy and imagination may be useful to prepare the body for stress and to reset dysfunctional physiological systems. It is likely that their thinking is quite concrete and that they either do not have or do not use a rich vocabulary to label and discriminate their feelings and moods and consequently tend to attribute psychological changes to physical or biological events or to mislabel psychological changes as physical changes. One study (Frank el et al, 1977) found that 73% of low hypnotic ability people were rated as "alexithymic". Alexithymia is defined as "lacking words for moods" (Sifneos, 1972). Sifneos found that only 8% of superior hypnotic subjects were rated as alexithymic. There is a large literature linking alexithymia to psychophysiological disorders (Lesse, 1981).

Catastrophizing cognitions and pessimistic belief systems

Several large scale prospective longitudinal studies (Hinkel, 1961; Valliant, 1978;Stewart, 1962) have shown that pessimism, self doubt, passivity and dependence are good predictions of subsequent complaints of psychosomatic illness. Ellis (1962) has elucidated the role of catastrophizing cognitions in the acquisition and maintenance of psychopathology. It is likely that the cognitive tendency to catastrophize is at least partly based on pessimistic and nihilistic belief systems. It is likely that catastrophizing also plays a major role in <u>attending</u> to symptoms, <u>altering sensory thresholds</u> and <u>escalating</u> the levels of sympathetic arousal in stress related disorders. Catastrophizing has at least two response components. First, that of keeping the attentional focus on the sensory or visceral events that are antecedents or consequences of symptoms; and second, remembering or anticipating a wide range of negative physical and psycho-social consequences and antecedents of the symptomatic event. Chaves and Brown (1978) found that dental patients could be divided into catastrophizers and copers during an injection or extraction. Catastrophizing ideation was reliably associated with higher levels of distress and pain in the dental situation. Brown (1979) replicated the above clinical finding with experimentally produced pain. Brown and Chaves (1980) found that the bulk of chronic pain patients (low back and headache) are catastrophizers. Catastrophizers have significantly higher pain ratings than copers. Eighty-six percent of catastrophizers were prescribed anti-anxiety or anti-depressant medication whereas only 12% of copers were on this type of medication. Spanos et al (1979) also found that catastrophizers had higher pain ratings than copers. Copers can be defined as people who use pleasant or positive cognitive distractions to attenuate their response to unpleasant sensory events. Spanos et al (1981) found that catastrophizing (exaggerating) self-statements increased pain reports in experimental pain situations.

Neuroticism

Neuroticism is a dimension of personality that is based on the degree of reactivity of the sympathetic division of the ANS (Eysenck, 1960). Clinically the most promising aspect of sympathetic reactivity is <u>autonomic response specificity</u> (ARS). ARS (Lacey, 1967; Sternbach, 1966) refers to the frequent observation of a stable profile of sympathetic response regardless of variations in the character of the stressor (e.g., mental arithmetic or cold pressor). There appears for many people to be a constant prioritized order of physiological response magnitudes across a variety of stressors. The analysis of autonomic response specificity appears to provide a useful approach to the predictions of somatic symptoms. The psychophysiological stress profile procedure (Wickramasekera, 1976) is a method of testing individual subjects sympathetic reactivity and I have supplemented this procedure with a paper and pencil test like the Eysenck Personality Inventory.

The Psychophysiological Stress Profile is a standardized procedure we have developed to directly measure the magnitude and duration of a patient's physiological response to a standardized cognitive stressor (mental arithmetic). An on-line computer collects, reduces and prints data (high-low, numbers of data points, mean and standard deviation) on heart rate, blood pressure, frontalis EMG, skin

conductance, respiration and peripheral skin temperature under three
conditions. The first condition is a 15 minute habituation period,
the second is the stress (mental arithmetic) period and finally, a
recovery or return to baseline period. We also request the patient to
give us a subjective rating of their level of anxiety, on a subjective
unit of disturbance scale (SUD) ranging from 0 — 50 SUDs. Like
hypnotizability, neuroticism also appears to have a clear genetic
component (Shields, 1962) and is a stable individual difference
variable.

Major Life Changes and/or Daily Hassles

A massing of major life event changes appears to be associated
with a high probability of illness. A method of assessing the impact
of situational stress on health is the measurement of major life
changes (Holmes and Rahe, 1967). The major weakness of this method is
the empirical finding that the relationship between life event change
scores and health outcomes is too weak for individual prediction.
Major life changes are also infrequent. I have supplemented the major
life change procedure with the Hassle Scale. The Hassle Scale asses-
ses the ongoing daily stresses and strains of everyday life. For
example, getting caught in rush hour traffic, running out of gas,
noise, work overload, unexpected company, etc. The research of Kanner
et al (1981) and DeLongis et al (1982) demonstrates that massing of
daily hassles is strongly related to somatic health outcomes and that
this effect remained even after the effects of major life events was
statistically removed (DeLongis et al, 1982).

Deficient Support Systems and Coping Skills

The impact of a massing of major life event changes or minor
hassles or both, will depend not only on the degree of psychological
"threat" (Mason, 1971) provoked by a wide variety of physical and other
changes but also on the effective use of support systems and coping
skills by the patient. Support systems are essentially psychological
resources (wife, siblings, psychotherapist, church, friends) on which
the patient can lean and with whom he can abreact to cushion the
impact of stressors. Coping skills (religion, escape through fantasy
or reading, physical distractions, recreation, relaxation, meditation,
etc.) can also be used to distract the patient, change the meaning of
events, and reduce preoccupation during both the acute and chronic
phases of stressor impact.

From a review of hypnotic ability and neuroticism scores of a
consecutive series of all patients I have seen in the last 3 years
(N=103) the following preliminary observations may be made. It is
important to repeat that this report will be limited to the incidence
of only two of the 5 risk factors (hypnotic ability and neuroticism):
(1) 8% of the patients show neither sign (7 of 84); (2) 48% (40 of 84)
showed both signs; (3) 83% showed the hypnosis sign; (4) 65% showed
the neuroticism sign.

CONCLUSION

The impact of multiple life changes or multiple hassles will depend not only on personality features (high or low hypnotic ability, autonomic response specificity and neuroticism, catastrophizing ideational tendencies), but also on the patient access to and effective use of social support systems and personal coping skills. The patient at greatest risk is the one who is positive for all the personality features, and who is deficient in support systems and coping skills. The person at lowest risk is the patient who has none of these personality features and who has effective use of multiple support systems and coping skills. In the absence of positive physical findings but in the presence of 2 or more of the above psychophysiological findings one may reasonably make a diagnosis of psychophysiological disorder. To the extent that most or all of the above risk factors can be found in a given case, the probability that psychophiological factors are causing the disorder is high, or alternatively that psychophysiological factors are aggravating is a minimal physical cause. Clinical observations over the last 15 years have directed our attention to the above high risk factors. We have continued to focus on these 5 constructs over the years and attempted to quantify them with procedures of increasing validity and realiability. These five risk factors have been clinically important in that they have clarified difficult diagnoses, enhanced the prediction of clinical outcome and most importantly have provided broad targets for clinical intervention. These five risk factors after further validation may be the focus of primary prevention efforts starting in childhood or adolescence.

REFERENCES

Barber, T. X., 1976, "Hypnosis: a scientific approach," Van Nostrand Reinhold, New York.

Brown, J. M., 1979, Cognitive Activity, pain perception and hypnotic susceptibility, presented at: the Annual Meeting of the American Psychological Association, New York.

Brown, J. M., and Chaves, J. F., 1980, Cognitive Activity Perception and hypnotic susceptibility in chronic pain patients, presented at: Annual Meeting of the American Psychological Association, Montreal, Canada.

Cannon, W. B., 1932, "The Wisdom of the Body," Appleton-Century Crofts, New York.

Chaves, J. F., and Brown, J. M., 1978, Self-generated strategies for the control of pain and stress, presented at: Annual Meeting of the American Psychological Association, Toronto, Canada.

Acknowledgement: I wish to express my sincere gratitude to Ms. Anne Page for her assistance in the typing of this manuscript.

Das, J. P., 1958a, The Pavlovian Theory of Hypnosis: An evaluation, Journal of Mental Sciences, 104:82-90.

Das, J. P., 1958b, Conditioning and Hypnosis, Journal of Experimental Psychology, 56:110-113.

DeLongis, A., Coyne, J. C., Dakof, G., Folkman, S., and Lazarus, R. S., 1982, Relationship of Daily Hassles, Uplifts and Major Life Events to Health Status, Health Psychology, 1(2):119-136.

Ellis, A., 1962, "Reason and Emotion in Psychotherapy," Lyle Stuart Inc., New York.

Evans, F. J., 1977, Hypnosis and Sleep: The control of Altered States of Consciousness, in: "Conceptual and Investigative Approaches to Hypnosis and Hypnotic Phenomena," W. E. Edmonston, ed., Annals of the New York Academy of Sciences, 296:162-174.

Eysenck, H. J., 1960, "The structure of human personality (2nd Ed)," Methuen, London.

Foenander, G., Burrows, G. G., Gerschman, J., and Horne, D. J., 1980, Phobic behavior and hypnotic susceptibility, Australian Journal of Clinical and Experimental Hypnosis, 8:41-46.

Frankel, F. H., Apfel-Savitz, R., Nemiah, J. C., and Sifneos, P. E., 1977, the relationship between hypnotizability and Alexithymia, Proc. II, European Conference on Psychosomatic Research, Heidelberg, 1976, Psychotherapy and Psychosomatics, 28:172-178.

Frankel, F. H., and Orne, M. T., 1976, Hypnotizability and phobic behavior, Archives of General Psychiatry, 33:1259-1261.

Gerschman, J., Burrows, G. D., Reade, P., and Foenander, G., 1979, Hypnotizability and the treatment of dental phobic behavior, in: "Hypnosis," G. D. Burrow, D. R. Collison, and L. Dennerstein, eds., Elseiver, New York.

Hilgard, E. R., 1965, "Hypnotic Susceptibility," Harcourt, Brace and World, New York.

Hilgard, E. R., 1977, "Divided consciousness: multiple controls in human thought and action," Wiley Interscience, New York.

Hinkle, L. E., 1961, Ecological observations on the relation of physical illness, mental illness and the social environment, Psychosomatic Medicine, 23:289-296.

Holmes, T. H., and Rahe, R. H., 1967, The social readjustment scale, Journal of Psychosomatic Research, 11:213-218.

Ingram, R. E., Saccuzzo, D. P., McNeil, B. W., and McDonald, R., 1979, Speed of Information Processing in high and low susceptible subjects. A preliminary study, International Journal of Clinical and Experimental Hypnosis, 42-47.

Kanner, A. D., Coyne, J. C., Schaefer, C., and Lazarus, R. S., 1981, Comparison of two modes of stress measurement: Daily hassles and uplifts versus major life events, Journal of Behavioral Medicine, 4:1-39.

King, D. R., and McDonald, R. D., 1976, Hypnotic susceptibility and verbal conditioning, International Journal of Clinical and Experimental Hypnosis, 24:29-37.

Lacey, J. I., 1967, Somatic response patterning and stress: some revisions of activation theory, in: "Psychological Stress," M. H. Appley, and R. Trumball, eds., Appleton-Century Crofts, New York.

Larsen, S., 1966, Strategies for reducing phobic behavior, Dissertation Abstracts, 26:6850.

Lesser, I. M., 1981, A review of the Alexithymia Concept, Psychosomatic Medicine, 43,6:531-543.

Mason, J. W., 1971, A re-evaluation of the concept of "non-specificity" in stress theory, Journal of Psychiatric Research, 8:323-333.

Nace, E. P., Warwick, A. M., Kelley, R. L., and Evans, F. J., 1982, Hypnotizability and outcome in brief psychotherapy, Journal of Clinical Psychiatry, 43:129-133.

Perry, C., John, R., and Hollander, B., 1982, Hypnotizability and phobic behavior, 34th Annual S.E.C.H. Convention, Indianapolis.

Pettinati, H., Horne, R. L., and Staats, J. M., 1982, Assessment of hypnotic capacity in patients with Anorexia Nervosa, presented at: the 90th Annual Convention of the American Psychological Association, Washington, DC.

Selye, H., 1956, "Stress and Disease," McGraw-Hill, New York.

Shields, J., 1962, "Monozygotic twins brought up apart and brought up together," Oxford University Press, Oxford

Shor, R. E., 1964, A note on the shock tolerance of real and simulating hypnotic subjects, International Journal of Clinical and Experimental Hypnosis, 12:258-262.

Sifneos, P. M., 1972, "Short-term Psychotherapy and Emotional Crisis," Havard University Press, Cambridge.

Spanos, N. P., Radtke-Bodonik, L., Ferguson, J. D., and Jones, B., 1979, The effects of hypnotic susceptibility, suggestions for analgesia, and the utilization of cognitive strategies on the reduction of pain, Journal of Abnormal Psychology, 88,3:282-292.

Spanos, P. N., Brown, J. M., Jones, B., and Horner, D., 1981, Cognitive activity and Suggestions for Analgesia in the reduction of reported pain, Journal of Abnormal Psychology, 90,6:554-561.

Sternbach, R. A., 1966, "Principles of Psychophysiology," Academic Press, London.

Stewart, L. H., 1962, Social and emotional adjustment during adolescence as related to the development of psychosomatic illness and adulthood, Genetic Psychology Monographs, 175-215.

Stoyva, J. M., 1965, Posthypnotically suggested dreams at the Sleep Cycle, Archives of General Psychiatry, 12:287-294.

Valliant, G. E., 1978, Natural history of male psychological health IV: What kind of men do not get psychosomatic illness, Psychosomatic Medicine, 40:420-431.

Webb, R. A., 1962, Suggestibility and Verbal Conditioning, International Journal of Clinical and Experimental Hypnosis, 10:275-279.

Weiss, R. L., Ullman, L. P., and Krasner, L., 1960, On the relation-
        ship between hypnotizability and response to verbal operant
        conditioning, Psychological Reports, 59-60.
Wickramasekera, I., 1970, The effects of hypnosis and a control
        procedure on verbal conditioning, presented at: the
        meeting of the American Psychological Association, Miami.
Wickramasekera, I., 1971, Effects of EMG feedback training on
        susceptibility to hypnosis: preliminary observations,
        Proc. 79, Annual Convention of the American Psychological
        Association, 6:787-785.
Wickramasekera, I., 1972, A technique for controlling a certain type
        of sexual exhibitionism, Psychotherapy: Theory, Research
        and Practice, 9:207-210.
Wickramasekera, I., 1973, The effects of EMG feedback on hypnotic
        susceptibility: More preliminary data, Journal of Abnormal
        Psychology, 82:74-77.
Wickramasekera, I., 1976, "Biofeedback, Behavior Therapy, and
        Hypnosis," Nelson-Hall, Chicago.
Wickramasekera, I., 1976, Aversive behavior rehearsal for sexual
        exhibitionism, Behavior Therapy, 7:167-176.
Wickramasekera, I., 1977, On attempts to modify hypnotic susceptibil-
        ity: some psychophysiological procedures and promising direc-
        tions, Annals of the New York Academy of Sciences, 296:143-153.
Wickramasekera, I., 1979, A model of the patient at high risk for
        chronic stress related disorders, presented at: Annual
        Convention of the Biofeedback Society of America, San Diego,
        California.
Wickramasekera, I., 1980, Aversive Behavior Rehearsal: A cognitive-
        behavioral procedure, in: "Exhibitionism: Description,
        Assessment and Treatment," D. J. Cox, and R. J. Daitzman, eds.,
        Harland, New York.
Wickramasekera, I., 1980a, Patient Variables, In Behavioral Medicine
        and the Psychological Aspects of Health Care, invited address:
        Veterans Administration, North Central Regional Medical
        Education Center.
Wickramasekera, I., 1980b, Principles of Psychophysiology and the High
        Risk patient, invited address: University of Illinois College
        of Medicine, Peoria.
Wickramasekera, I., 1980a, Patient Variables, In Behavioral Medicine
        and the Psychological Aspects of Health Care, invited address:
        Veterans Administration, North Central Regional Medical
        Education Center.
Wilson, S. C., and Barber, T. X., 1982, The fantasy-prone personality:
        implications for understanding imagery, hypnosis and parapsy-
        chological phenomena, in: "Imagery: Current Theory Research and
        Application," A. A. Sheikh, ed, John Wiley, New York.

SOCIAL PSYCHOPHYSIOLOGY OF RELAXATION TRAINING

Margaret Cork and Tom Cox

Stress Research
Nottingham University
Nottingham, England

ABSTRACT

It has been demonstrated that similar psychophysiological effects can be produced both by relaxation training and by simply sitting quietly (Cork and Cox, 1982). In using subject samples that included individuals with neuroticism scores in the abnormal range, the data showed that control sessions and relaxation training were psychophysiologically equivalent regardless of subjects' level of neuroticism and prior experience with relaxation or meditation techniques. This pattern of effects may be partly determined by psychosocial factors together with other non-specific effects of the experimental environment (Orne, 1962).

INTRODUCTION

There is a growing body of literature which reflects the increasing use of relaxation training in behavioral medicine (see, for example, Silver and Blanchard, 1978), but sadly a large number of the published studies have proved scientifically inadequate. The research reviewed in this paper represents a number of investigations of the immediate (short term) psychophysiological effects of relaxation training. A series of studies are described which were initially designed to improve on aspects of the methodology used in previous studies; such as inappropriate control conditions, obtrusive measurements, brief (single session) exposure to relaxation training, and lack of pre-experimental measures.

A number of methodological improvements over previous work have also been recommended by Borkovec and Sides (1979), in their critical review article. These are outlined in a later section of this paper.

Inappropriate Controls

A number of studies have employed control conditions sufficiently disparate from the experimental condition that comparison is rendered invalid. For example, subjects received non-directive counselling in the control condition of the Brauer et al (1979) study. This bore little resemblance to the experimental conditions which involved live or tape-recorded relaxation instructions. Differences did occur between the two groups but only at six month follow-up and not immediately subsequent to therapy. Warrenburg et al (1980) compared long term transcendental meditation practitioners (LTM), and long term progressive relaxation practitioners (LPR) with novices. During the experimental condition experienced subjects practised their respective relaxation techniques whilst the novices practised progressive relaxation. During the control condition LTM and LPR subjects were instructed to sit quietly without practising their techniques, whilst novices sat quietly and listened to a piece of preferred relaxing music. The two groups (novice/long term) thus underwent different control conditions, which were not consistent in the way in which they differed from the relaxation condition. There was no significant difference between groups in the degree of somatic relaxation which occurred during the practise of their relaxation technique, and the changes which did occur were of small magnitude.

To conclude, when designing control conditions, it is good practise to systematically remove from the treatment condition those elements of relaxation under study, maintaining all else constant.

Obtrusive Measurements

When training subjects to relax it is essential that any measures taken should be relatively unobtrusive and not interfere with the relaxation process. Wallace et al (1971) conducted a study in which subjects wore tight oxygen masks and were fitted with arterial catheters and rectal thermometers, whilst they practised transcendental meditation. Although respiration rate, oxygen consumption, and carbon dioxide elimination all decreased (possibly as subjects became acclimatised to the apparatus) blood pressure, rectal temperature, arterial oxygen and carbon dioxide did not change. Investigations subsequent to this have not, for the most part, employed such obtrusive techniques.

Brief Exposure

For an acceptable validation of the effects of relaxation training it is sensible that subjects be exposed to the technique for an adequate length of time, that is, for at least more than one, brief, training session. Nonetheless numerous investigators have employed very brief training procedures. For example, in Beary and Benson's (1974) study, subjects participated in a single experimental session, incorporating five, twelve minute periods, only one of which comprised relaxation training. Oxygen consumption, carbon dioxide elimination, and respiration rate all decreased, compared to sitting quietly, possibly because subjects were deliverately concentrating on and regulating their breathing, rather than relaxing.

In Pollack and Zeiner's (1979) study subjects were exposed to one session comprising ten minutes baseline followed by thirty minutes relaxation. There were no psychophysiological effects of the relaxation training. Woolfolk and Rooney (1981) investigated the effect of an expectancy manipulation on the psychophysiological correlates of a non-cultic meditation technique. Subjects took part in a single session comprising fifteen minutes baseline and fifteen minutes meditation. Skin conductance, heart rate, and systolic blood pressure all decreased over time.

Any psychophysiological effects resulting from such brief exposure to meditation or relaxation training procedures may be attributed to:

(a) the effects of habituation and adaptation to the strange experimental environment rather than to the procedure itself, or

(b) deliberate regulation of breathing.

It is highly unlikely that one-trial learning would result from such brief exposure. It is likely that multi-session training is necessary before such effects could be attributed to the relaxation or meditation technique.

Pre-Experimental Measures

A number of investigators (Cauthen and Prymak, 1977, Lehrer, 1980, and Puente, 1981) have failed to assess subject's psychophysiological state prior to the experiment, simply measuring it following training. Such studies have commonly employed a between subjects design; comparing the psychophysiological effects of differing relaxation techniques, taught to separate subject groups. Differences between subjects at the end of the study may have resulted from inherent individual differences rather than from differential effects of the training procedures used.

RESEARCH AT NOTTINGHAM

Three studies, conducted at Nottingham, are reviewed below. The initial experiment examined the immediate psychophysiological effects of relaxation training, using a design which sought to meet the var-picked up on the differences between relaxation training and the control maneuvre, while the third study (a secondary analysis) considered the effects of individual differences in neuroticism on the response to relaxation training.

STUDY ONE:
THE IMMEDIATE PSYCHOPHYSIOLOGICAL RESPONSE TO RELAXATION TRAINING

The aim of this first study was to investigate the immediate (short term) effects of relaxation training. The relaxation training technique employed was designed by Benson and was first reported by Beary and Benson (1974). Benson (1975, 1981) suggests that four main factors are necessary for production of the "relaxation response"; a mental device (constant and repetitive attentional focus), a passive attitude, decreased muscle tonus (facilitated by a comfortable position), and a quiet environment. Validation of relaxation training was attempted using self-reported mood, and cardiovascular function as criteria.

Method

A within subjects design was employed with all subjects participating in five daily sessions incorporating five minutes baseline, twenty minutes relaxation training (days two to four) or the control maneuvre (days one and five), and five minutes recovery. The baseline and recovery periods were identical: subjects sat quietly, in a comfortable position, with their eyes closed. During the control sessions subjects also sat quietly and comfortably with closed eyes, whilst listening to a taped story. In the relaxation training sessions subjects listened to taped relaxation instructions (taken from Beary and Benson, 1974) and then practised the technique.

A hierarchical structure of independent variables was therefore employed. Baseline and recovery periods incorporated two factors; a quiet environment and decreased muscle tonus, as subjects rested quietly and comfortably. In addition to these factors, the control condition incorporated an auditory stimulus as an attentional focus. This was not identical to the mental device incorporated in the relaxation technique, as it was not repetitive, but both nonetheless presented a constant focus of attention. The relaxation condition employed all four factors deemed crucial by Benson in the production of a relaxation response.

The dependent variables selected for use were relatively nondisturbing. Systolic and diastolic blood pressure (SBP and DBP) were measured using a Visomat (R) electronic sphygmomanometer, Type 3003 (Boehringer-Mannheim). Heart rate was measured continuously, using a Medilog Cardiac Monitor, series 4-24 (Oxford Electronic Instruments). Self-reported mood was assessed using the Stress-Arousal Checklist (S.A.C.L.), devised by Mackay et al (1978). Adjective order was randomised for each administration to minimise response set effects.

Results and Discussion

The first question to be addressed was whether the relaxation technique employed had any immediate psychophysiological effects. A series of three-way split plot analyses of variance were computed on data from days one to four of the study, with one between subjects factor, sex of subject, and two within subjects factors, day and condition. Day one (control) represented no relaxation training, through to day four, which represented three days of training.

Systolic and diastolic blood pressure (SBP:DBP). The subject population were normotensive with initial blood pressure levels at the lower end of the range (mean: 109/73 mm Hg - normal range: 100/60 - 150/90 mm Hg). There was a significant sex difference in SBP, with males manifesting higher levels throughout the study (male subjects: 110/73 mm Hg; female subjects: 107/74 mm Hg). This is to be expected for individuals in the twenty to thirty-two year age range within which the present subject population fell (Bell et al, 1965). There was a significant decrease in SBP and DBP from days one to four with blood pressure levels decreasing significantly in the baseline, relaxation and control conditions and increasing during the recovery period. (This is illustrated in Figure 1). A Page's L non-parametric test of trend, carried out on systolic blood pressure difference data (pre minus post relaxation and control session scores) was significant (p<0.01). This indicated that the change (decrease) in SBP scores became significantly larger as the experiment progressed.

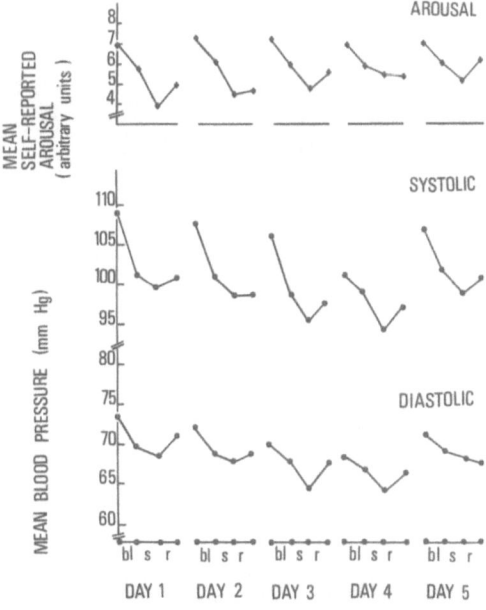

Figure 1. Levels of systolic blood pressure, diastolic blood pressure, and self-reported arousal, across the five treatment sessions.

Heart Rate. (The heart rate data are illustrated in Figure 2, below).
Mean and minimum heart rate varied significantly across conditions,
reaching their lowest levels during the relaxation session. There was
a significant interaction between day and condition, with a different
pattern emerging for control and relaxation days, during the recovery
period. On control day one heart rate decreased during the control
session and continued to decrease during the recovery period. On the
relaxation training days two through four heart rate decreased during
the relaxation session, reaching lower levels than occurred in the
control condition but then increased during the recovery period. This
may have resulted from subjects mental preparation for the end of the
daily session.

Maximum heart rate and heart rate variability both showed a
significant difference during each experimental session with the

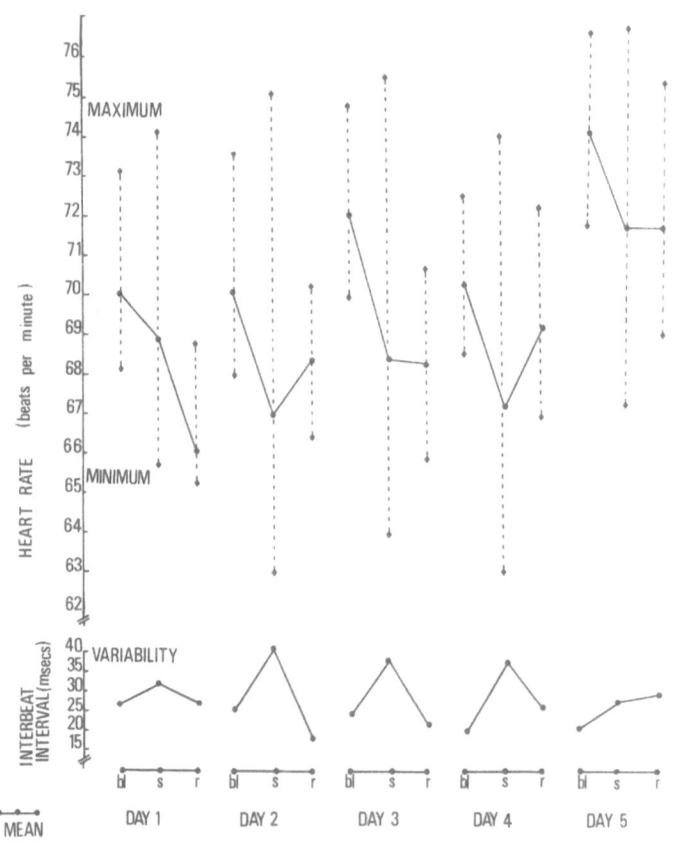

Figure 2. Mean, minimum and maximum heart rate, and heart rate
          variability across the five treatment sessions.

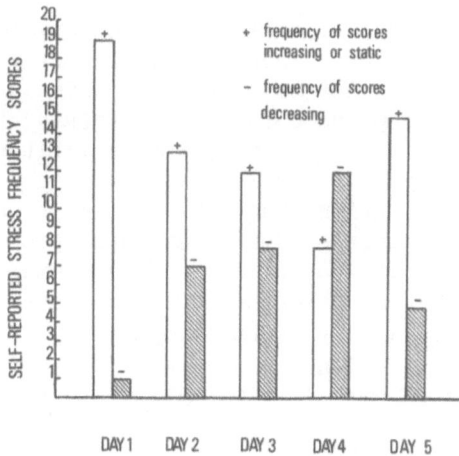

Figure 3. Self-reported stress frequency scores, during the control and relaxation conditions, across the five sessions.

highest values occurring in the relaxation condition. Deep regulated breathing can produce fluctuations in both vagal tone and heart rate. The heart speeds up during inspiration and then slows down during expiration, thereby producing sinus arrythmia. This is mainly a result of irradiation from the respiratory centre to the cardiac centre. During relaxation training the deeper breathing pattern leads to accelerated cardiac rate on inspiration, hence producing the highest values for maximum heart rate. Heart rate variability increases during relaxation as cardiac rate alternately accelerates and decelerates (Green, 1976).

Self-Reported Mood. Individuals subjective perception of arousal levels mirrored the cardiovascular indicants of their physiological state thus providing some validation for the use of self-report measures in psychophysiological studies. This is illustrated in Figure 1 above. Throughout the study levels of self-reported stress were very low, and a large number of zero scores were recorded. The data did not, therefore, fulfill the requirements of analysis of variance, and so a non-parametric chi-squared analysis was computed. This indicated an increase in self-reported levels of stress across the baseline and a decrease across the relaxation and control conditions. This pattern is shown in Figure 3.

Individual Differences. The subject group was comprised of people with previous experience of relaxation/meditation and novices. The practitioners showed a psychophysiological pattern of responding identical to that of the novices. Perhaps the effects of participating in an experiment were, therefore, stronger than practise in such techniques.

The results reported above suggest that relaxation training has immediate psychophysiological effects, characterised by a wakeful state in which parasympathetic nervous system activity predominates, and is accompanied by low levels of self-reported arousal. An interesting point emerged from the data. This pattern of effects also occurred in the baseline condition, and to a limited extent in the control condition. Levels of SBP at first showed a greater decrease during the baseline condition than during relaxation training. It is possible that the experimental design employed produced a floor effect. During the baseline condition SBP dropped by 6 mmHg on average, and homeostatic mechanisms may have then come into operation to inhibit any further decrease. This pattern changed as the experiment progressed. The degree of change across the baseline decreased, possibly as subjects became increasingly acclimatised to the experimental environment. By contrast, the extent of the change produced by relaxation training gradually increased, and, by the third day of training, SBP reductions were greatest across this period.

The finding that both baseline and relaxation conditions produced the "relaxation response" indicates that the four factors deemed crucial in the production of such a response may not all be essential. Perhaps only those factors common to both the relaxation training and baseline sessions are necessary in eliciting the "relaxation response". That is: subjects are sitting quietly, in a comfortable position, and in a quiet environment. Perhaps in addition the subjects perceived sitting quietly as "legitimate". That is, as an acceptable thing to do, by virtue of the fact that it was a requirement of the experiment.

According to definitions in the social psychology literature (e.g. Secord and Backman, 1974) an activity is perceived to be legitimate when individuals have positive prior expectations about it, which they feel are shared by the appropriate reference group. The subjects perception that their behaviour during the experiment was a legitimate or valuable use of their time may have therefore been an important factor in the production of the "relaxation response". The same pattern of effects occurred in the control condition but to a significantly lesser extent. Perhaps the additional factors incorporated in the relaxation technique (passive attitude and repetitive mental device) were of some effect in determining the degree of response, if not its pattern. Whether this occurred by means of a direct psychophysiological effect of the additional factors, or via an indirect effect mediated by the subjects perceptions of the legitimacy of the activity is not clear.

There were no significant differences between days one and five in any of the measures taken. The relaxation training therefore had no carry-over effects. This suggests that a short period of relaxation training may only produce immediate and transient psychophysiological effects. It is possible that resting comfortably and quietly, whilst an important factor in the production of a relaxed state in the short term, may be insufficient to produce longer term benefits. The additional factors deemed crucial by Benson may then come into operation.

STUDY TWO: A FOLLOW-UP INVESTIGATION OF THE DIFFERENCES BETWEEN CONTROL AND RELAXATION PROCEDURES

The second study was designed to follow up the differences between the relaxation training and the control maneuvres and to consider the role of "perceived legitimacy".

Method

A within subjects design was employed with subjects initially attending for two consecutive control days. Their psychophysiological state was measured on five separate occasions throughout each day, and, to ensure that resting levels of each measure were taken, subjects sat quietly for five to ten minutes prior to measurement. The aim of this procedure was two-fold. First, to acclimatise the subjects to the experimental environment, experimenter, and measurement procedures, and second, to provide a set of control data for diurnal variation. During the following week subjects attended four daily twenty minute experimental sessions. Pre and post session measures were taken. Subjects practised Bensons (1975, 1981) technique during two of the sessions, and were simply instructed to sit comfortably and quietly on the other two occasions. Order and timing (morning versus afternoon) of sessions was randomised across the subject sample.

The dependent measures were identical to study one with the exception of heart rate for which a continuous recording was not taken. Pulse readings were obtained together with blood pressure levels using an electronic sphygmomanometer; Digital-3 (R) Cossor (Medical) ltd.

Results and Discussion

A series of three-way within subjects analyses of variance were computed, with three factors; treatment condition (relaxation training/sitting quietly), day, and time (pre/post). There were significant differences between pre and post measures of SBP, pulse, and self-reported mood. However these occurred both when subjects were practising Bensons technique and when they were simply resting quietly. There were, therefore, no differential effects of the two procedures on psychophysiological state. This pattern of effects is illustrated in Figures 5 and 6.

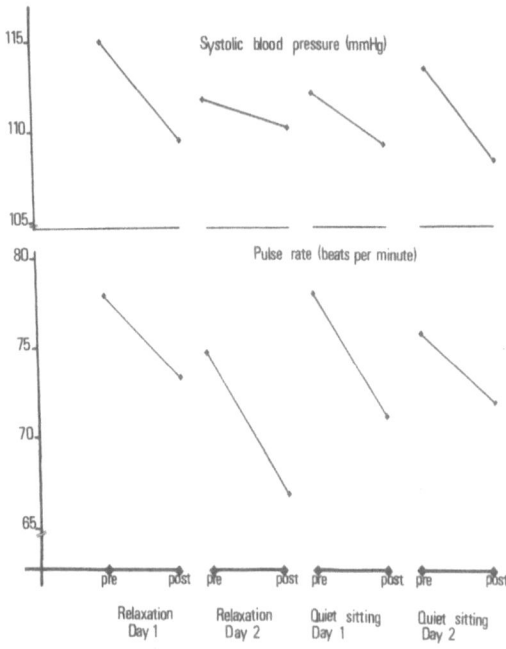

Figure 4. Levels of systolic blood pressure, and pulse rate, across
the relaxation training sessions, and the periods of quiet rest.

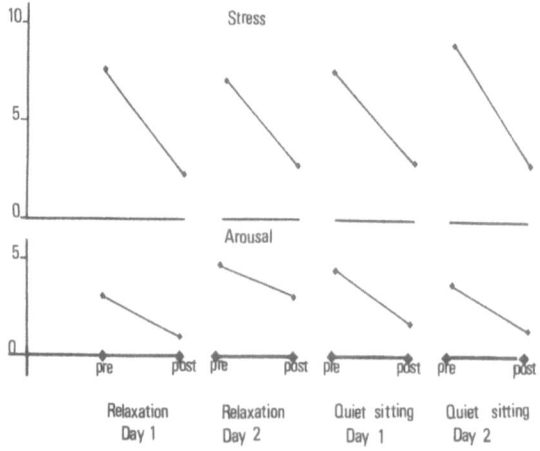

Figure 5. Self-reported stress, and arousal, across the relaxation
training sessions, and the periods of quiet rest.

Pre-session SBP levels and self-reported mood were not significantly different from the pre-study control data. However, the post-session measures of the same variables were, but again this occurred whether subjects were resting quietly or practising the relaxation technique. Both procedures thus produced a pattern of psychophysiological response significantly different from the subjects normal diurnal pattern.

Pulse rate showed a slightly different pattern of effects as it was significantly elevated (over and above control levels) prior to the first session of each procedure. The subjects did not know the nature of the techniques they were to be taught prior to the experimental session, and this may have produced something of an anticipatory effect.

Subjects perceptions of the legitimacy of the two procedures were evaluated post-experimentally using a visual analogue scale (0: non-legitimate; a complete waste of time, to 100: highly legitimate; an extremely valuable use of time).

The average legitimacy rating for meditation was 65.5, whilst for resting quietly it was 54.7. This difference, although it approached significance, (F=3.9, p<0.06) was small. Furthermore, the subjects psychophysiological state did not reflect any difference in the perceived legitimacy of the two techniques. Sitting quietly, when set in the legitimate context of the experimental environment, produces, in the short term, an identical pattern of effects to relaxation. It may only be in the long term that any differences between the procedures become manifested. The (small) differences in perceived legitimacy of the two procedures may have been reflected in psychophysiological differences if the experiment had continued for a longer period of time.

The results from the experiments reviewed therefore indicate that a period of sitting quietly is equivalent in its short term psychophysiological effects, to a period of meditative relaxation. These studies are consistent with other research which has shown control and relaxation procedures to have equivalent psychophysiological effects. Edelman (1970) reported two studies which indicated that the significant psychophysiological effects of progressive relaxation were no different from three control conditions; suggestions of relaxation, instructions for skeletal movement (both taken from the relaxation instructions) and continuous semi-classical music. Edelman attributed this to the normal, non-anxious nature of the subject population. The subjects with high anxiety levels were, however, affected no differently, by progressive relaxation to the subjects with low levels of anxiety. The brevity of training in these studies may have been responsible for the lack of difference between conditions (both employed single session designs).

Smith (1976) found transcendental meditation and a control procedure involving quiet sitting with closed eyes, produced identical psychological effects; a decrease in self-reported anxiety. Pollack and Zeiner (1979) found no significant differences in the psychophysiological effect of Bensons technique, uninstructed relaxation, and sitting quietly. The only procedure causing any significant change was sitting quietly which led to the greatest reductions in heart rate. This was also a single session experiment.

Bradley and McCanne (1981) found relaxation training (either progressive relaxation or Bensons technique) to have little effect, relative to a control condition, which involved sitting quietly with eyes closed, on subjects self-reported anxiety following a laboratory stressor (film). The relaxation procedures did, however, lower heart rate prior to and during the film to a significantly greater degrees than the control. Puente (1981) reported two studies evaluating the psychophysiological changes occurring during transcendental meditation, Benson's technique, and a no treatment control. They found that although the three conditions showed different physiological patterns, none of them was clearly superior in reducing tonic physiological arousal.

Although the investigations reported above are consistent with the authors' research in indicating no difference between control and relaxation procedures the question of why there was no difference has not been answered.

Borkovec and Sides (1979) have reviewed twenty five investigations of the physiological effects of progressive relaxation training and have addressed this question. They classified the literature according the whether relaxation was shown to be equivalent or superior to control conditions. They found that the two sets of studies differed significantly in: the number of training sessions used, in the use of taped versus live administration of training, and in the use of normal versus patient populations.

Among studies indicating equivalence between conditions the average number of sessions was 2.3 (s.d= 1.64), seventy percent used standardised taped instructions, with no opportunity for the subjects to exercise control over their progress during training, and eighty percent involved normal subject populations. Among studies showing progressive relaxation to be superior to control conditions they found the average number of training sessions employed was 4.57 (s.d.=3.02), seventy three percent were therapist conducted, and forty seven percent involved patient samples. They concluded that the likelihood of producing significantly larger physiological reductions with progressive relaxation than control procedures was much greater when multi-session subject controlled training was conducted with subjects for whom physiological activity contributed to a clinical problem.

The studies described above (Edelman, 1970; Smith, 1975; Pollack and Zeiner, 1979; Bradley and McCanne, 1981; Puente, 1981) which demonstrated equivalent psychophysiological effects of relaxation and control procedures, together with the authors'research, had one main factor in common: the use of a normal subject population.

Several investigators have indicated the superiority of relaxation training over control procedures when the subject sample has comprised highly anxious individuals (Lehrer, 1978; Borkovec et al, 1978; Schwartz, 1978). The subject samples used in the authors research, although not clinically defined, included significant numbers of individuals with neuroticism scores in the abnormal range (as defined by Eysenck and Eysenck, 1964). The third study therefore aimed to investigate whether individual differences in neuroticism levels mediated the response to relaxation training.

STUDY THREE: SECONDARY ANALYSIS

The aim of this experiment was to ascertain whether subjects with high neuroticism levels demonstrated a different pattern of response to non-neurotic (stable) individuals. It was hypothesised that neurotic individuals would respond differentially to control and relaxation conditions whilst those who were more stable would show no such difference.

Method

Data from the two studies were subjected to secondary analysis. The mean neuroticism scores for both subject samples were computed. They were slightly higher in the second subject sample (mean = 11.8) than the first (mean = 8.5). Experimental group was therefore retained as a factor in the analysis. Subjects with neuroticsm scores in the top third of the range (study one: 10-17, study two: 13-19) were defined as neurotic whilst those in the bottom third (study one:0-7, study two: 6-10) were defined as stable. Data were analysed using a four-way split plot analysis of variance, with two between subject factors, level of neuroticism (high/low) and experimental group (one/two), and two within subject factors, condition (relaxation/control) and time (pre/post). The condition factor was further differentiated so that data from the first control and relaxation days was analysed together, and independent of data from the second control and relaxation days. Any initial effects of the experimental environment could therefore be identified. The dependent variables selected for analysis were levels of self-reported stress and arousal, and SBP.

Results and Discussion

Self-reported mood during the relaxation condition was not significantly different from the control period in either the high neuroticism or stable (low neuroticism) group. There was a significant
interaction effect, however, between level of neuroticism and condition on SBP levels at the start of the experiment. The significant
effect was identified, by means of post hoc comparisons, to have
occurred during the control condition, as neurotic subjects had significantly higher SBP levels than stable subjects. As might have been
expected neurotic individuals reacted to the onset of the experiment
with increased sympathetic nervous system activity, indicated by elevated SBP levels. Nonetheless neither neurotic nor stable individuals manifested a difference in SBP levels between control and relaxation conditions.

The hypothesis that neurotic individuals would manifest different
psychophysiological responses to control and relaxation conditions,
whilst stable subjects would respond in the same way was not supported. The data showed that control sessions and relaxation training
produced equivalent effects, regardless of the subjects' level of
neuroticism. As a number of individuals had neuroticism scores in the
abnormal range this result is somewhat unexpected. Returning to
Borkovec and Sides (1979) review, other factors proposed as reasons
for the equivalence between control and relaxation procedures were,
the use of taped relaxation instructions and brief relaxation courses.
However the studies reported as indicating equivalence did not all
employ taped instructions, and although some were of brief duration,
the authors research involved multiple sessions. These additional
factors do not, therefore, totally explain the lack of difference in
the present study between control and relaxation procedures. Borkovec
and Sides (1979) indeed suggest that the use of such methods does not
preclude conclusions regarding other, nonphysiological mediators of
relaxation, such as perceived legitimacy.

CONCLUSION

The authors' research, in combination with other studies discussed in this paper, supports a social psychophysiological model of
relaxation training. The pattern of effects observed with training
may be partly determined by psychosocial factors such as the subject's
perceptions of the legitimacy of the exercise (relaxation and control
procedures). An individual may believe that sitting quietly, when it
occurs as part of an experiment, is a legitimate use of their time,
and thereby achieve some degree of relaxation.

The recognition that non-specific effects of the experimental environment play an important role in determining the outcome of the experiment is not a recent occurrence. Orne (1962) suggested that a subject's behaviour in any experimental situation will be determined by two sets of variables:

1) those which are traditionally defined as experimental variables, and,

2) the perceived demand characteristics of the experimental situation.

This latter set includes the nature of the experimenter, information conveyed during the initial contact between the subject and experimenter, the setting of the laboratory, the experimental procedure per se, together with explicit and implicit communications during the experiment proper. "Perceived legitimacy" would also be classed with the second group.

The quality of the subject-experimenter relationship was demonstrated by Cuthbert et al (1981) to be a highly significant variable in determining experimental outcome. In a series of experiments they found Benson's technique to be superior to heart rate biofeedback in affecting reductions in cardiac rate and lowering activation, under conditions of high experimenter involvement, when the nature of the subject-experimenter relationship was active and supportive. The two techniques were shown to be equivalent under conditions of low experimenter involvement, when the experimenter was formal and distant. This effect was dependent on the subject receiving specific knowledge of the results of their performance. Absence of such information led the relaxation subjects who had a formal relationship with the experimenter to achieve the best results. Cuthbert et al (1981) concluded this important paper by saying that the results demonstrate the great power in relaxation experiments of psychosocial and other moderator variables, and signal the practical difficulty of their control when the variables seem to be as potent in changing physiology as the primary training methods.

The subject's expectations of success also play an important role in determining the outcome of the experiment. Zuroff and Schwartz (1980) reported on a two year follow up of an investigation into the relative psychophysiological effects of transcendental meditation versus muscle relaxation. Subject's expectancy of benefit, following a nine week training course, was the only factor shown to be predictive of positive outcome, indicated by satisfaction with and frequency of practise. Agras et al (1982) demonstrated subjects expectations to play an important role in determining the extent of the blood pressure lowering effects of relaxation. Hypertensive subjects, told to expect immediate lowering of blood pressure, demonstrated a 17 mmHg

decrease during the training period, compared with 2.4 mmHg in subjects told to expect delayed lowering. They conclude by suggesting
that procedural differences between researchers in engendering expectations concerning the outcome of therapy, may indeed lead to
different results. This may be one reason for the large range of
results produced by investigations into the blood pressure lowering
effects of relaxation training.

The pattern of effects observed in the Nottingham experiments
may, therefore, have been determined in part by psychosocial factors.
Subjects expectations of success and their perceptions that both
control and relaxation procedures were a valuable use of their time,
in combination with a high degree of experimenter involvement, may
have contributed significantly to the production of a relaxation
response. Such non-specific effects of the experimental environment
may have overridden any potential differences between neurotic and
stable individuals, and between long-term practitioners of relaxation
and novices. The interpretation of short term laboratory investigations of relaxation training is therefore limited because of the
confounding effects of such demand characteristics. Due to the complexity of the experimental environment any psychophysiological effects observed cannot be attributed in total to the relaxation and/or
control procedures. As Cuthbert et al (1981) pointed out, control of
such non specific variables is extremely difficult as their potency in
changing physiology (or psychological state) appears to be equivalent
to that of the training procedures.

ACKNOWLEDGEMENTS

This work was supported by the Social Science Research Council
through a postgraduate training grant for the first author. The views
expressed do not reflect those of the sponsoring body. The authors
acknowledge the help and advice given by other members of the Stress
Research Group, and the assistance provided by the technical staff.
The results of the first study have been presented at the British
Psychological Society annual (London) conference, and accounts of the
three studies are being prepared for separate publication.

REFERENCES

Agras, W., Horne, M., and Taylor, C., 1982, Expectation and the blood-
     pressure lowering effects of relaxation, Psychosomatic
     Medicine, 44:389.
Beary, J., and Benson, H., with Klemchuck, H., 1974, A simple
     psychophysiological technique which elicits the hypometabolic
     changes of the relaxation response, Psychosomatic Medicine,
     36:115.

Bell, G., Davidson, J., and Scarborough, H., 1965, "Textbook of
    Physiology and Biochemistry," Livingstone Ltd., Edinburgh.
Benson, H., 1975, "The Relaxation Response," Morrow, New York.
Benson, H., and Goodale, I., 1981, The relaxation response: your
    inborn capacity to counteract the harmful effects of stress,
    Journal of Florida Medical Association, 68:265.
Brauer, A., Horlick, L., Nelson, E., Farquhar, J., and Agras, W.,
    1979, Relaxation therapy for essential hypertension: a veterans
    admission outpatients study, Journal of Behavioral Medicine,
    2:21.
Borkovec, T., Grayson, J., and Cooper, K., 1978, Treatment of general
    tension: subjective and physiological effects of progressive
    relaxation, Journal of Consulting and Clinical Psychology,
    46:518.
Borkovec, T., and Sides, J., 1979, Critical procedural variables
    related to the physiological effects of progressive relaxation:
    a review, Behavior Research and Therapy, 17:119.
Bradley, B., and McCanne, T., 1981, The effects of progressive
    relaxation, the relaxation response, and expectancy of relief,
    Biofeedback and Self-Regulation, 6:235.
Cauthen, N., and Prymack, C., 1977, Meditation versus relaxation: an
    examination of the physiological effects of relaxation training
    and of different levels of experience with transcendental
    meditation, Journal of Consulting and Clinical Psychology,
    45:496.
Cork, M., and Cox, T., 1982, The immediate psychophysiological
    response to relaxation training, presented at: British
    Psychological Society Annual Conference, London.
Cuthbert, B., Kristeller, J., Simons, R., Hodes, R., and Lang, P.,
    1981, Strategies of arousal control: biofeedback, meditation,
    and motivation, Journal of Experimental Psychology: General,
    110-518.
Edelman, R., 1970, Effects of progressive relaxation on autonomic
    processes, Journal of Clinical Psychology, 26:421.
Eysenck, H., and Eysenck, S., 1964, "Manual of the Eysenck Personality
    Inventory," Hodder and Stoughton, Great Britain.
Green, J., 1976, "An Introduction to Human Physiology," Oxford
    University Press, London.
Leavitt, F., 1974, "Drugs and Behaviour," Saunders Company, London.
Lehrer, P., 1978, Psychological effects of progressive relaxation in
    anxiety neurotic patients and of progressive relaxation and
    alpha feedback in non-patients, Journal of Consulting and
    Clinical Psychology, 46:389.
Mackay, C., Cox, T., Burrows, G., and Lazzerini, T., 1978, An
    inventory for the measurement of stress and arousal, British
    Journal of Social and Clinical Psychology, 17:283.
Orne, M., 1962, On the social psychology of the psychological
    experiment with particular reference to demand characteristics
    and their implications, American Psychology, 17:776.

Patel, C., 1981, Yoga and biofeedback in the management of
    hypertension, in: "Stress and the Heart," D. Wheatly, ed.,
    Raven Press, New York.
Pollack, M., and Zeiner, A., 1979, Physiological correlates of an
    experimental relaxation procedure with comparison to
    uninstructed relaxation and quiet sitting, Biology Psychology
    Bulletin, 5:161.
Puente, A., 1981, Psychophysiological investigations on transcendental
    meditation, Biofeedback and Self-Regulation, 6:327.
Schwartz, G., Davidson, R., and Goleman, D., 1978, Patterning of
    cognitive and somatic processes in the self-regulation of
    anxiety: effects of meditation versus exercise,
    Psychosomatic Medicine, 40:321.
Secord, C., and Backman, C., 1974, "Social Psychology," McGraw-Hill,
    Tokyo.
Shapiro, S., and Lehrer, P., 1980, Psychophysiological effects of
    autogenic training and progressive relaxation, Biofeedback
    and Self-Regulation, 5:249.
Silver, B., and Blanchard, E., 1978, Biofeedback and relaxation
    training in the treatment of psychophysiological disorders: or;
    Are the machines really necessary? Journal of Behavioral
    Medicine, 1:217.
Smith, J., 1976, Psychotherapeutic effects of transcendental
    meditation with controls for expectation of relief and daily
    sitting, Journal of Consulting and Clinical Psychology, 44:630.
Wallace, R., Benson, H., and Wilson, A., 1971, A wakeful hypometabolic
    physiological state, American Journal of Physiology, 221:795.
Warrenburg, S., Pagano, R., Woods, M., and Hlastala, M., 1980, A
    comparison of somatic relaxation and EEG activity in classical
    progressive relaxation and transcendental meditation, Journal
    of Behavioral Medicine, 1:73.
Woolfolk, R., and Rooney, A., 1981, The effect of explicit
    expectations on initial meditation experiences, Biofeedback and
    Self-Regulation, 6:483.
Zuroff, D., and Schwarz, J., 1980, Transcendental meditation versus
    muscle relaxation: a two year follow-up of a controlled
    experiment, American Journal of Psychiatry, 137:1229.

THE ROLE OF SUBSTANCE P IN NORMALIZING STRESS RESPONSE

Ann Karnitschnig

Eastern Virginia Medical School
Norfolk, Virginia

Throughout history and from the first stirrings of consciousness in the womb we are exposed to a bewildering array of polypeptides. Often we like the feelings we experience, i.e., they are pleasurable. The child is no longer considered a blank slate on which we program a personality, but a supersensitive being. We can only speculate how the unborn child feels with the adrenaline surge or the endorphins when his or her mother is exposed to danger or adventure.

All of us like the pleasurable feelings adrenaline and cortisone gives us, though physicians have traditionally denied the euphoria produced by cortisone. The adrenaline surge that a skiing addict may receive may be as expensive, as time consuming and as self-destructive as the heroin addict or alcoholic trying to get their own "high". Certain men even get "high" from killing, becoming heros in wars, mercenaries, etc. Everyone likes to get "high". Adrenaline and cortisone make us feel good, so all of us expose ourselves to stress.

Basically we have two competing systems. Noradrenaline from the sympathetic system and the locus coeruleus (the pleasure center) with the adrenal cortex producing the fight, flight or fright reaction, and the parasympathetic system mediated by acetylcholine carried by the vagus and inhibited by atropine. We have Hans Selye's non-specific response to any demand and get into first gear. This was needed by the caveman to survive an animal attack. If modern man is unable to switch off this first gear he may suffer organic disease ranging from hypertension to ulcers, or may be labeled as "psychosomatic", a word first used in 1818 by Heinroch. All stress reduction techniques basically teach people to shift out of first gear by controlling the autonomic nervous system (for instance in biofeedback) or by finding better ways to feel good (endorphins producing by jogging being preferable to heroin addiction).

The precursor for ACTH, which triggers all cortisone from the adrenal cortex, is Beta-lipotropin with its 91 amino acids. Beta-lipotropin is also the precursor for endorphins (enkephalin, the opiate within us). This is another pleasurable chemical produced by stress. The endorphins can prevent one from feeling pain during a particularly stressful situation. The effects of endorphins are calmness, a sort of serenity, accompanied by lowering blood pressure, respiration and heart rate, also a reduced intestinal motility, i.e., counteracting the sympathetic effect. Endorphins therefore normalize the stress response of adrenaline. Naloxone is the endorphin and opiate blocker.

There is some evidence that occasionally our bodies produce too many endorphins. Two cases are presented: cardiagenic shock and septicemic shock, where Naloxone dramatically raises and stabilizes blood pressure, and normalizes heart rate. On the whole, however, health care providers try to increase peoples' endorphins. It has been observed experimentally that acupuncture, biofeedback for pain relief and the placebo effect of the physician are blocked by Naloxone, so these can be explained by endorphin release. Exercise, certain foods and pleasurable interaction with friends are also endorphin producers.

Substance P (discovered in 1931 by Gaddum) is another chemical found in our bodies which is related to pain/pleasure. Substance P is found throughout the body -- in the heart, in the intestinal mucosa, in the central nervous system. A critical balance between endorphins and Substance P is thought to use a factor in the gate control theory of pain. A noxious stimulus, be it thermal, chemical or traumatic, releases Substance P in the dorsal root ganglion and the lamine II (the substantia gelatinosa) of the dorsal horn of the spinal cord. If enough messages come through, the person will feel pain. So humans need enough Substance P and not too many endorphins to experience pain (without Substance P experimental animals do not feel pain, which may have clinical significance).

A second important action of Substance P is the antidromic response. When cells release histamine, the body overcomes tissue damage by a sort of "rinsing out system". This is the histamine flare response first described by Sir Thomas Lewis. Thus a mouse that has been rendered deficient in Substance P does not feel pain -- nor does it have the ability to heal a cut or injury. How does all this correlate clinically? In human amputees, the spinal cord shows a striking depletion of Substance P on the side of the missing limb. There is a very rare familial hereditary disease called Familial Dysautonomia. It is characterized by autonomic instability, (biofeedback would not work) impaired pain, and temperature perception, also esophagal dysfunction, vomiting, ataxic and loss of the histamine flare response. In addition, Substance P is virtually undetectable in the substantia gelatinosa of victims of Familial Dysautonomia. So

here we have a disease where people cannot feel pain and do not have
the histamine flare response and lack Substance P. We can theorize
certinly that massage, acupuncture and manipulation may work because
of Substance P production. As with endorphin we have to consider what
happens if we have too much Substance P. Some chronic pain, for
instance some migraine headaches, could similarly be explained. It is
a slow, persistent response. Intracranial or intrathecal Substance P
in animals produces a scratching syndrome -- could itchiness in humans
be related to Substance P production?

When I was a medical student I saw several burns from counter
irritants. Mustard poltices were applied to relieve intractable pain.
So how do they work when applied to intact skin? Capsaisin - a mustard oil - in man produces intense burning followed by blister formation. In animals the same response occurs, and initially Substance P
is released in the dorsal horn, followed by a depletion. The histamine flare response is initiated -- so plasma pours into the tissues.
If the skin is denervated there is obviously no pain but also no
histamine flare response. If this counter irritant, capsaisin is
given to newborn animals systemically there is selective destruction
of Substance P sensitive neurons. So we have artifically produced a
Substance P deficient animal. There are trophic changes in the chronically denervated skin. The animals do not feel pain nor do they have
the histamine flare response -- so they do not heal. They are just
like victims of Familial Dysautonomia. These pain-free animals who
cannot heal themselves confirm the theory that Substance P is needed
for the healing process. Again, massage, acupuncture, manipulation
may be a little painful, but the trade off may be producing Substance
P. Could the cure of "rheumatism" by bee stings have anything to do
with Substance P?

The formula of Substance P is presented along with two Substance
P analogues which contain the same amount of amino acids with the same
last three amino acids. One analogue found in toad's skin is seven
and one-half times more potent, and one in the salivary gland of th
octopus is ten times more potent. Substance P is considered a neurotransmitter and is compared with other neurotransmitters -- acetylcholine, GABA (gamma amino butyric acid) dopamine, noradrenaline, gly-
cine, 5-hydroxytryptamine (serotonin) encephalin and glutamic acid.
Substance P is considered excitatory and a sensory signaller (pain
perception and axon reflex). Many neurotransmitters affect mood and
behavior (dopamine, noradrenaline, serotonin, GABA (endogenous tranquilizers) and encephalins (endogenous opiate). The termination of
action of Substance P is considered as re-uptake.

Rats stressed by mild tail pinch apparently deal with stress by
overeating (Moorely and Levine, 1980). A parenteral injection of
Substance P is compared to a control, where saline water is injected.
Fifteen minutes later food intakes are compared. The rats with Substance P injections ate only one-seventh of what the controls ate. In

another experiment, rats that had been starved for twenty-four hours ate the same with or without Substance P. Thus, Substance P select- ively supresses stress-induced eating in rats. Opiate like analgesics do not prevent these stressed rats from overeating. Someone even thought of trying to calm these rats down with tranquilizers and what happened? They ate even more when tranquilized. This correlates with my clinical impression that anxious people who overeat, probably eat even more if on tranquilizers. Thus, Substance P has a specific effect reducing stress induced eating. It is a regulatory peptide (regulide) with a slow prolonged action normalizing the stress response.

In another animal experiment, Hecht and Oehme (1980) stressed rats by sleep deprivation and immobilization. There were three groups — each with ten animals. One group lived normally without stress. The other twenty suffered twenty-four hours fixation followed by twenty-four hours free movement for four weeks. Sleep levels were then assessed by observing behavior and EEG readings. Ten animals given a Substance P analogue intrathecally and observed for two hours showed sleep patterns much more like the control, than the other stressed rats. Thus, Substance P seems to normalize sleep patterns in stressed rats. So if Substance P were to make human beings sleep better under stress and help us avoid overeating under stress, it might be a good chemical to have around. We obviously need some Substance P to heal our wounds even if we suffer a little pain in the process. Counter irritants — bee stings, massage, manipulation may help us produce the Substance P that we need to heal.

Children with their sensitive brains and neurons have to adjust to the adrenaline and cortisone surge of television, while regulating their anxiety with gamma amino butyric acid. They are modulating their moods with dopamine and serotonin, while killing pain with a critical adjustment of endorphins and Substance P. They will avoid overeating with a little Substance P, and fall asleep and sleep sound- ly with a little more Substance P.

There is now a new paradigm in medicine. Physicians do not cure diseases with drugs and surgery, but they teach their patients to heal themselves. Alcoholics and drug addicts are taught to make their own endorphins by getting high on life, because they have forgotten how to make their own endorphins. The doctor's bag of the future will be stuffed with tools to find out what neurotransmitters the patient needs and how he can make his own endogenously.

There are exciting new frontiers in medicine. Lack of dopamine from the substantia nigra produces Parkinsonism, which can now be treated with dopamine. It now seems possible that Alzheimer's Disease (presenile dementia) occurs because our bodies run out of acetylcho- line from the Nucleus basilis of Nigra. One cannot give people an acetylcholine pill, but a significant improvement occurs with prostig-

mine, which prevents the destruction of acetylcholine. A solution
might be in sight once we understand the problem. We think we can
retrain people to produce endorphins from the locus coeruleus. How
about Substance P? Should we inject it into obese patients? Or use
it for insomnia? Maybe capsaisin could be used for cancer victims
with intractable pain, but perhaps it might interfere with healing or
the immune system. Further study is needed to refine technology and
clinical efficacy in this area.

REFERENCES

Hecht, K., Oehme, P., et al, 1980, Effect of Substance P Analogue on
        Deprivation of Sleep of Wistar Rats Under Stress, in:
        "Neuropeptides and Neural transmission," Raven Press, New York.

Morley, John E., and Levine, Allen S., 1980, Substance P Suppresses
        Stress-induced Eating, European Journal of Pharmacology,
        67:309-311.

CLINICAL APPLICATIONS IN MEDICINE, PSYCHOLOGY-PSYCHIATRY,
SPEECH PATHOLOGY AND DENTISTRY

STRESS, TENSION AND RHEUMATIC DISEASE

Robert E. Rinehart

Director, Rinehart Clinic
Portland, Oregon

INTRODUCTION

An understanding of rheumatic diseases cannot be attained without considering the role of the neuro-motor system. It would seem, on the basis of available evidence, that such an understanding is essential if further progress is to be made in prevention and control of these conditions. We will consider here the behavioral mechanisms involved and how these relate to rheumatism (fibrositis, fibromyalgia etc.), degenerative joint disease (osteoarthritis, wear and tear arthritis), rheumatoid arthritis and a variety of other auto-immune disorders. Thorough understanding will require basic knowledge in kinesiology, neurology, immunology, endocrinology and physiology. Some knowledge of biochemistry is helpful. References are kept to a minimum, a few general texts and recent articles.

BACKGROUND

The history of rheumatic disease is replete with arguments and counter arguments regarding their origin and nature. These arguments are still extant, although a few conditions are well delineated (Rodman and Schumaker, 1983). The first to be understood was gout in 1847. This is not really a rheumatic disease in the usual sense, it is a disorder of metabolism. It was not until 1931 that it was demonstrated that a beta hemolytic streptococcus initiated rheumatic fever, as well as chorea, nephritis and heart disease. With development of the science of bacteriolgy, gonococcal arthritis was identified in 1879, soon followed by indentification of other types of specific infectious arthritis. Satisfactory treatment of these conditions was not attained until antibiotics became available in 1935.

These accounted for only a tiny portion of the disorders labeled
arthritis and rheumatism. There remained three classes of disease
about which little was understood. These were degenerative joint
disease (DJD), rheumatoid arthritis (RA) and a melange of "connective
tissue disease". Under this latter designation were fibrositis
(rheumatism), nonarticular rheumatism, scleroderma, lupus erythematosus, polymyositis, dermatomyositis, mixed connective tissue disease,
polymyalgia rheumatica and many others. It is our purpose here to
describe a common denominator which will explain these disorders.

BEHAVIOR

Behavior is the reaction of an organism to its environment (Rinehart, 1975). This is done by muscular activity controlled by the
nervous system. It can be instinctive or learned. Only a tiny portion of our behavior is instinctive. Most of it, such as the highly
skilled art of walking, is learned helter skelter without formal
training. If we are lucky we are taught some activities such as
writing, playing a musical instrument or driving a car. Initially the
development of behavior depends largely on the use of exteroceptors to
program the neuromuscular system. Once the behavior (habit) is learned, it is carried out largely under the guidance of proprioceptors
(muscle spindle nerve endings) without conscious thought.

It is obvious to all of us that there are "good" (adaptive) and
"bad" (maladaptive) behaviors. We are going to consider two forms of
maladaptive behavior which are the initiators of these common maladies. In most instances the relationship is simple, verifiable and
reproducible. There are a few gaps in our knowledge which should be
easily bridged by further investigation.

We will first consider the maladaptive behavior of habitual
muscular tension, responsible for the majority of rheumatic disease
(Dixon, Hangen and Dickel, 1958). This behavior develops as a response to our instinctive reaction to uncertainty, the fight-or-flight
reaction. This reaction consists initially of muscular bracing or
tension in preparation for action. Tensed muscles stimulate nerve
endings sending danger signals to the central nervous system and
brain. The signals stimulate the reticular formation, hypothalamus
and limbic system leading to manifestations of alertness, corticotro-
phin release, adrenal stimulation and manifestation or urgency, irritability and anxiety. These neural pathways are outlined by Campbell
(1966). Chronic overstimulation of these structures leads to progressive bodily failure through development of what Selye calls the
General Adaptation Syndrome (1952).

Another form of maladaptive behavior is more regional and not so
apt to lead to systemic symptoms. It develops from overuse of muscles
as a result of bodily deformity, repetitive occupational movements,
immobilization of a part after injury and many other reasons (Jesse,

1977). When one or more pairs of muscle becomes fatigued and cramped postural disturbances develop leading to further impairment. The two behavioral abnormalities, generalized and localized, often occur - together and can become extremely disabling.

RHEUMATISM

Long before these visceral and emotional disturbances become manifest the chronically stimulated muscles begin to tire. This may be a gradual and generalized process as in the "tired housewife syndrome". More often localized muscle spasm develops in muscles of scalp (headache), neck (pain in the neck) and low back (pain in the buttocks). The latter two sites are favorites because the initial bracing reaction starts in the neck and shoulders (to fight) and low back and hips (to run). Chronically cramped muscles develop nodular painful areas so frequently found in these locations and called "fibrositic nodules". Pain of course leads to further muscular bracing and the game is in full play (Dixon, O'Hara and Peterson, 1967). This manifestation of behavior leads to disabilities ranging from an occassional headache or stiff neck to a chronic generalized muscular rigidity known as the "stiff man syndrome". Lumped together these conditions are commonly called rheumatism.

DEGENERATIVE JOINT DISEASE

Joints occur in the body to allow one bone to slide on another. The ends of bones are covered with a semi-soft gristly material called cartilage. The surface is somewhat harder than the base, much like a finger nail. It contains tiny tubes. As pressure is applied and released nourishing liquid is pumped in and out from the surface (the under part is nourished by blood vessels in the bone). It can easily be visualized that constant pressure on cartilage will interfere with this pumping action, leading to "degeneration". As a rule this is a slow process and at times is relatively painless until the process is far advanced.

Initially the process is relatively painless, manifested only by minor lameness and soreness. Careful examination will demonstrate limitation of joint motion and some hard, cramped muscles. Eventually the synovial tissue of the joint, containing A cells which are the "clean-up crew" and B cells which manufacture lubricants, becomes in- flamed. Inflammation simply means an accelerated reaction of repair. In this case the A cells are presented with an excessive amount of destroyed tissue to dispose of causing synovial inflammation. Even- tually cartilage wears through, allowing the joint to communicate with underlying bone. Pressure forces fluid into the bone resulting in cystic changes. Bone destruction leads to a new source of inflam- matory tissue resulting in further pathologic changes. Chunks of cartilage can become loosened and form "joint mice".

RHEUMATOID ARTHRITIS

The process here is initiated in an identical manner. Muscle
spasm results in excessive joint wear resulting in excessive repair
(inflammation). Three additional factors are involved in varying
degrees. These include: 1) stress or the general adaptation syndrome
(Selye, 1952) 2) an inherited predisposition and 3) one or more
autoimmune reactions.

The stress reaction is obvious, manifested by elevated plasma
cortisol (early), a variety of visceral disturbances and some degree
of anxiety. My own experience shows persistent motor unit activity of
two varieties. There is generalized persistent activity seen in
habitually tense (stressed) individuals described by Jacobson (1938).
Also there is synchronous motor unit firing described by Dixon et al
(1967) in fatigue contracture.

An inherited predisposition to RA has not been identified. However most studies show a significantly higher incidence in relatives
of patients (Hollander, 1960). However when I have examined the
incidence of other diseases demonstrating an autoimmune component in
relatives, it appears to be exceedingly high.

The precise antigen initiating an autoimmune reaction in RA has
not been identified. It is probable that one or more components of
abnormal metabolism (inflammation) is responsible. When one considers
the effects of even minor stress on immunocompetence the conclusion is
inevitable that abnormal metabolites in this setting play a primary
role (Piley, 1981).

It is further probable that varying combinations of muscle tension, fatigue, spasm and immunoincompetence are responsible for
"rheumatoid variants" such as lupus erythematosus, scleroderma, myosi-
tis, polymyalgia rheumatic, mixed connective disease etc.

TREATMENT OF RHEUMATIC DISEASE

Habitual muscular tension, the initiator of stress and the general adaptation syndrome, produces three abnormal states of muscle.
These are fatigue, weakness and spasm (cramp). These three conditions
result in symptoms associated with rheumatism. Spasm leads to abnormal breakdown of muscle, cartilage and synovial tissue resulting in
accelerated repair, called inflammation. In some individuals immuno-
genic abnormalities arise causing the peculiar inflammatory reactions
seen in RA and rheumatoid variants.

Pharmacologic treatment of these conditions has never been satisfactory. Various anti-inflammatory drugs, of which aspirin is the

prototype, have been tried with temporary benefit. Drugs which in some way seem to affect the immune response, i.e. gold, chloroquine and a variety of immunosuppresive agents are often toxic. Cortisol and its derivatives, due to their immunosuppresive and stimulatory powers, produce immediate and spectacular improvement followed by exacerbation and total collapse. They simply prolong the stage of resistance in the general adaptation syndrome and lead to further debilitation.

Tried and true measures used for centuries have been rest, exercise and physical measures for relief of muscle spasm. Unfortunately these measures have been applied empirically, greatly limiting their effectiveness. Rest is the only known treatment for fatigue; exercise is essential to rehabilitate weak muscles; and physical therapy is the only physiological way to relieve fatigue-spasm. However even the most rigorous application of these essential measures will not be sufficient as long as we remain ignorant of habitual muscular tension and its relation to the general adaptation syndrome with its psycho-neuroendocrine manifestations.

Modern treatment should consist of the limited use of drugs only to ameliorate pain and/or potentially fatal immunologic complications. The basic program must involve rest (prolonged and complete), gradually increasing exercise (mild, rhythmic and useless muscular activity), physical measures for relief of muscle spasm and training to correct habitual tension.

The first three of these measures can be dealt with by any reasonably intelligent physical therapist with experience in dealing with rheumatic disease. A moderate amount of study will provide sufficient understanding to permit application of many currently used modalities.

Correction of habitual muscular tension will require a revolutionary change in thinking by the medical profession. It is highly unlikely, unless the public and third party insurance carriers become involved, that this will occur for another generation or two. At present the best we can hope for is an occassional alliance between physicians (internists, family practitioners and psychiatrists) and physical therapists. As skill develops and changes are noted public demand will increase.

We are fortunate in having three basic references of understanding and applying training methods to accomplish this end. Dixon et al (1958) describes the mechanism of development of habitual muscular tension. Jacobson (1938) has developed a training program for correction of this maladaptive habit pattern. Feldenkrais (1975) presents a training program broader is scope but fails to recognize the specific role of tension. Both of these programs need to be studied in real life application before an attempt is made to apply them.

REFERENCES

Campbell, H. J., 1966, "Correlative Physiology of the Nervous System,"
        Academic Press, London.
Dixon, H., Hangen, G. B., and Dickel, H., 1958, "A Therapy for Anxiety
        Tension Reactions," Macmillan, New York.
Dixon, H., O'Hara, M., and Peterson, R. D., 1967, Fatigue Contracture
        of Skeletal Muscle, Northwest Medicine, 66:813-816.
Feldenkrais, M., 1975, "Awareness Through Movement," Springer-Verlag.
Hollander, J. L., 1960, "Arthritis and Allied Conditions," Lea and
        Fabiger.
Jacobson, E., 1938 (Revised Edition), "Progressive Relaxation,"
        University of Chicago Press.
Jesse, J., 1977, "Hidden Causes of Injury: Prevention and Correction
        for Running Athletes and Joggers," The Athletic Press.
Riley, T., 1981, Psychoneuroendocrine Influences on Immunocompetence
        and Neoplasia, Science, 212:1100.
Rinehart, R., 1975, "Evolution of Behavior," Physical Biological
        Sciences Ltd.
Rodman, G. P., and Schumaker, H. R., 1983, "Primer on the Rheumatic
        Diseases, Eighth Edition," The Arthritis Foundation.

NEUROMUSCULAR PSYCHOPHYSIOLOGY OF DEPRESSION

Arnold H. Gessel

Private Practice
Philadelphia, Pennsylvania

That the motor system played a significant role in mental activity was the first premise that inspired Jacobson (1911) and the last to which he turned his attention (1982). Much of his work was devoted to investigating the relationship between the motor system and mental activity on the one hand, and physiology on the other. For our convenience, we arbitrarily divide the functioning organism into systems, which consist of structures of similar form, then try to figure out how the systems interact. Disease processes, the interest of physicians, rarely affect an isolated system. Therapeutic tradition generally proceeds on the assumption that one system can be addressed as an "entry key" that will positively influence the others. The psychotherapist addresses the cognitive system, the internist the chemical, the surgeon the structural. Medical progressive relaxation sometimes treats purely muscular symptoms, but usually uses the relationships between the motor system and other elements of the human being. The emotional experience has many components, salient among which are affective, neurochemical, visceral, cognitive, behavioral, and micro-behavioral. These elements vary widely in their interaction and independence. For example, one of my patients complained that each time he drove on the expressway he experienced tachycardia, sinking in the stomach, sweating, dizziness and faintness. When I spoke to him of his "anxiety", he protested that he felt no such thing -- he was aware only of the tachycardia, sweating, etc. As Lesse (1968) put it regarding depression, "The term depression as used by laymen refers only to a mood which in psychiatric circles is more specifically labelled as sadness, dejection, despair, gloominess or despondency. If this mood pattern is not noted overtly, the patient is not depressed in the layman's view. Unfortunately, this same type of scotoma occurs not infrequently among physicians in general and even among psychiatrists."

Schachter and Singer's (1962) experiment provides an illustration of the way that emotional responses, which we tend to consider as discrete entities, may be analyzed in terms of their components. The subject group was divided into a 3 x 2 set. They were given the instruction that they were participating in a test of the effects of a certain substance on vision. Groups A and B received an epinephrine injection with the above instruction, except that group B was also told that they might experience some side effects, namely those of an epinephrine injection. Group C received a placebo injection. Each group was then divided into two subgroups. Each subgroup was then subjected to the machinations of a stooge, ostensibly a fellow subject, during the waiting period for their participation in the "real" experiment. For one subgroup, the stooge clowned, joking, jumping on the table, making paper airplanes. For the other, the stooge complained with increasing hostility about the form they were filling out and the procedure in general. The analysis of the results showed not surprisingly that the epinephrine uninformed groups responded to the situation with the greatest emotion, but the quality of the emotion depended on the situation: the joking group became euphoric, the hostile group became hostile. The increase in arousal brought by an injection of epinephrine determined the level of an induced affect, but its content depended on the cognitive substrate.

Surprisingly, the least responsive group was the epinephrine-informed. The understanding that "my reaction is the result of an injection" nullified not only the arousal that would have been caused by the injected epinephrine, but also that generated by the subject himself in the situation.

While at the University of Pennsylvania School of Dental Medicine, I had the opportunity to work with a variety of patients suffering from complaints that appeared to be "psychological" or "psychosomatic". As time went on, I concluded that these various diagnostic categories afforded me a glimpse into the natural history of the neuromuscular psychophysiology of depression.

One group of patients was afflicted with the temporomandibular joint syndrome, or myofascial pain dysfunction syndrome. While there continues to be much controversy about the nature and origins of this condition, it is generally agreed that the primary symptoms arise from painful spasm of the muscles of mastication. It has also been observed that many of these patients also suffer other symptoms generally classified as psychosomatic (Alderman, 1971). I observed that certain of them were able to relieve their symptoms and adopt a more constructive life style when taught the techniques and principles of neuromuscular training, but others failed to benefit (Gessel and Alderman, 1971). This sort of result is customary, and it is usually triumphantly proclaimed that "X is helpful in N% of cases of . . ." The vital question is the difference between the groups. Two people who have the same "condition" are obviously not IN the same condition if they respond differently to the same conditions.

Two factors which have been found to be statistically related to treatment outcome in general are age and social assets. Indeed, the failure group has a higher median age, although the overlap was far wider than the difference. Work history however was critical, and all patients who had stopped working because of their problem, with one notable exception, failed to respond to any treatment. This is not the case with all illness, and it was not clear from the state of their muscles either why they felt they were prevented from working or why they might not respond. I had been impressed with the signs of depression in the failure group of my first series, so in repeating the work with a second, larger group, (Gessel, 1975) those patients who had failed to benefit from relaxations training were given a trial of tricyclic antidepressants in low doses. At this point categories began to emerge. Nondepressed to mildly depressed patients improved with relaxation. A moderately or greater degree of depression but still working responded to tricyclics and nonworkers to nothing.

The notable exception referred to was age 12 at the time of her disability, and age 17 when first seen. After the experimental proto-col had clearly failed to provide any relief, she engaged in a psycho-therapeutic relationship which terminated with a vigorous confronta-tion regarding the manipulative aspects of the sick role. She left angry, but later told me she had taken a job and married, and although she still had pain, had decided it would not rule her life.

Here is a series of troubled people presenting with the same basic symptoms, musculospastic pain, who had to be approached in four distinct ways: as a tension problem, as a depression, as a cognitive problem, and as a life style that was at best extremely resistant to change. The sequential ordering of the approaches confirms the sepa-ration of the groups; those further down the line had distinguished themselves by not reacting to the previous treatment. Furthermore, it is apparent that as one goes down the line, depression becomes in-creasingly severe. We may then take a look at some of the systems involved in the emotional syndrome which we call depression.

Medications work on the biochemical system. The utility of antidepressants in pain syndromes is presently recognized by every authority on pain control. The concept of pain control generally denotes a diagnostic and frequently a therapeutic nihilism. The awareness that some "pain" syndromes are musculospastic syndromes adds a conceptual dimension. How much of the effectiveness of antidepress-ants in pain control comes from muscle relaxation? That these drugs are useful in doses far smaller than is effective in the treatment of depression raises questions, but some similarity in biochemical mech-anism is suggested. The action of tricyclics is believed to relate to catecholamine metabolism in the nervous system. A theory of the relationship of depression to catecholamines has been developed, but is too extensive for further discussion here (Schildkraut, 1965). Another available bridge between muscular physiology and depression

lies in the sodium metabolism. Cardiologists treating hypertension
with diuretics observe a certain incidence of depressions which resolve upon loosening of the sodium restrictions, implying a relationship between sodium metabolism and the depressive picture. The pre-
menstrual syndrome, in which there is retension of water and sodium,
and which is frequently relieved with diuretics, has elements of
tension and depression. The advent of the use of lithium for manic-
depressive disease strongly underscores the importance of sodium in
affective states, since its effect is essentially to antagonize
sodium. Interestingly, Tyber (1974) has found lithium to have a
salutary effect on muscle spasm in the painful shoulder syndrome,
particularly when used in conjunction with amitriptylene. He found
that 76% of patients with this condition, in which there may be radio-
logic abnormalities, were significantly depressed. He reasons that
the available knowledge of the biochemistry indicates that abnormal
sustained contraction of muscle fibers are the result of unavailabili-
ty of ATP and disturbances of calcium transport. Changes in intra-
cellular sodium, such as occur in depression and are reversed with
lithium treatment, can lead to such disturbances. Thus another link
between depression and muscle spasm is forged in salt.

Another diagnostic group frequently seen in a dental clinic is
atypical facial neuralgia, or atypical facial pain. As Lesse (1967)
described this condition, the patient appears with a vague clinical
picture not resembling specific organic problems. Repeated consulta-
tions and treatment attempts are usual. The suffering described
customarily far exceeds that expected with a similar organic illness.
The history reveals many of the symptoms associated with severe depressions, but the depressive affect is strongly denied verbally and
frequently behaviorally, although the patient interviewer may hear
about suicidal intentions. The miserable, wretched quality of these
patients, the deterioration of their life quality and the lack of
anatomical credibility of their complaints sets them off from the TMJ
syndrome patients, although there is frequently diagnostic overlap.
They also tend to be older. In an attempt to relieve the affects of
these patients, Webb and Lascelles (1969) administered monoamine
oxidase inhibitors with a surprising benefit as far as the pain was
concerned. Subsequently, the condition was considered to be in the
category of "atypical" or "masked" depression (DaLessio, 1968). An
examination of Travell's description (1960) of patients with referral
of pain to the face from trigger points in the head and neck reveals
sufficient similarity that they appear to be the same people who in
another context would be diagnosed as atypical neuralgia, implying
that many patients with this condition are suffering a musculospastic
disorder, not a "pain problem".

An awareness of the potential involvement of the motor system and
its interaction with depressive mechanisms provides a practical conceptual framework in working with some difficult painful conditions.
Malmo's elegantly direct observations (1975) provide additional in-

sight into the mechanisms involved. Beginning with measurement of muscle activity in an unused arm during a task requiring attention, he demonstrated that muscle tension rises continually in a gradient as the task continues, and drops to baseline when the task is done. The more highly motivated the subject, either from internal interest or external reward, the steeper the gradients. They are selective, appearing in certain muscles, with individual differences among subjects, and particularly steep at the site of a musculospastic problem such as a tension headache if one exists. A finding of particular interest regards the resolution of the accumulated gradient to baseline levels. When the subject perceives the task as completed, there is a prompt drop to pretask levels. When, however, the instructional set is manipulated to the effect that the task must be stopped, but has not been successfully completed, the gradients do not resolve, providing a neat and discreet scientific demonstration of the relationship between frustration and chronic muscular tension.

For an explanation of these events, Malmo looked at the arousal mechanism. The reticular activating system is that part of the brain stem that keeps us more or less awake by its outflow to the cortex. The cortical neurons require facilitative activation from the reticular core for good performance, an optimal level of arousal. Both too low or too high activation lead to deterioration. The cortical circuits adapt to a steady level of stimulation, a process similar to addiction. The everyday experience of this is boredom; when the environment offers an unvarying pattern and level of stimuli, we become sleepy. Therefore, an increasing activation from the reticular core is necessary for steady performance. This increasing reticular output goes not only to the cortex, but flows directly to the motor system, giving rise to gradients of activity.

There may be another function of these gradients. Sensory inputs from muscle fibers enter the CNS at various levels. Some may feed the reticular activating system, contributing to its ability to provide an increasing output of the cortex. An increasing bracing of the motor system may be initiated from volitional centers (that is, purposefully) as a response to a felt need for awakeness, as when a highly performance-oriented subject finds his attention flagging at a repetitive and not very interesting task. This bracing would be carried out purposefully but at the same time outside of conscious awareness. There is no contradiction in these terms.

The considerations dealt with so far have leaned heavily on chemical and biological matters. There has been much growth of knowledge in recent years in these areas, particularly in regard to their role in depression. One also picks up psychological insights along the way. The youngest and most ambitious of the patients, those showing classic Type A behavior, are prone to experience depressive affects and to express lack of worth as training in relaxation proceeds. These are patients one feels comfortable working with. Their competence and commitment extends to the treatment situation, their

cooperation is unambivalent, and they readily comprehend the principles I present both at the intellectual and at the "guts" level.
After several training sessions, they are likely to complain about
themselves, of becoming lazy, of not getting anything done, and of
becoming emotionally sensitive; one patient cried over the fate of the
animals while watching a film on environment, a feeling she considered
foreign to her, while another reported crying spells but stoutly
denied any accompanying affect. These experiences tended to occur
when symptoms first began to ease, and when I judged that the patients
were beginning to "let go." To understand these events, we must evoke
the concept of defenses. Freud conceived of the mental apparatus as
three compartments, id, ego, and superego, each operating at a different level, and for the most part in conflict with each other. We
tend to construct images of ourselves and our relationships with which
we are satisfied, and of these we are aware. There are other more
animal impulses which are at variance with our "civilized" selves,
which we tend to keep out of awareness. The mechanism providing this
oblivion is referred to as defense. It was Wilhelm Reich (1948) who
first brought strong attention to the role of the motor system in
defense formation. These extremely capable and achieving young people
appear to have an underlying negative, depressed viewpoint about
themselves, against which they have the defenses of activity and
muscular tension.

After a number of years in clinical practice one begins to see
beyond the individual symptom picture the continuum of a natural
history. Returning to our young incipient TMJ patient, we see her
faced with a repetitive task, to prove her worth, to herself and
ostensibly to others. Motivation is high, as the stakes are significant. Much of her vigor and striving are diverted to dealing with a
sense of lack of worth, loneliness and despair, but there is hope; if
you do enough of the right thing, you will be loved and happy. The
effort leads to unremitting gradient formation, as the task is too
ephemeral for closure and resolution to take place. Insidiously,
fatigue develops in the most used muscles, and the pain-spasm cycle
begins. Now a patient, she may be treated mechanically with local
maneuvers with disappearance of the immediate symptoms, giving the
appearance of a good result, but attitudes won't change and the overall efforts will continue. The chronic lack of satisfaction, along
with the growing realization with the progress of time that the old
goals will never be achieved, coupled with the wearing-down effects of
relatively constant pain which become quite severe at times, along
with the metabolic changes of the chronic stress response, contribute
to the transition which develops. The patient becomes more disillu-
sioned, frustrated, angry and tired. Efforts diminish and with them
stimulation and arousal. The illness becomes more prominent, a part
of life, and the patient becomes more dependent on and devoted to the
illness, until consultation and complaining become the major occupations. The clinical picture goes from TMJ syndrome to atypical facial
pain. Depression in situ becomes masked depression, becomes manifest
depression.

The inevitable conclusion is that many, although certainly not all, of the patients considered deteriorate over time. The pathologic process begins as a voluntary, purposeful, although unconscious effort. As these efforts are sustained, the effects of the chronic use of acute mechanisms begins to manifest, and effects spread beyond the voluntary motor system. Once higher set-points (Malmo, 1975) are established in various systems and begin to interact with each other the problem begins to become grossly complicated as self-sustaining neurological and chemical loops heat up and increasingly distort normal functions. It is obvious then that much more can be done for the patient early in the game; although the "disease" is the same, the therapeutic problem is much simpler. This point must be kept in mind in research-outcome studies. The next question is how early, and for the answer to that one need only refer to the old adage about the ounce of prevention (Anon.).

REFERENCES

Alderman, M. M., 1971, "Oral Medicine," Lippincott, Philadelphia.

Anon., Date ?, Old Adage.

DaLessio, D., 1968, Some Reflections on the Etiologic Role of Depression in Head Pain, Headache, 8:28.

Gessel, A., and Alderman, M. M., 1971, Management of the Myofascial Pain Dysfunction Syndrome of the Temporomandibular Joint by Tension Control Training, Psychosom., 12:302.

Gessel, A., 1975, Electromyographic Biofeedback and Tricyclic Antidepressants In Myofascial Pain Dysfunction Syndrome, JADA, 91:1048.

Jacobson, E., 1911, On Meaning and Understanding, Am. J. Psychol., 22:553.

Jacobson, E., 1983, "The Human Mind: A Physiological Clarification," Thomas, Springfield.

Lesse, S., 1967, Hypochondriasis and Psychosomatic Disorders, Am. J. Psychother., 21:607.

Lesse, S., 1968, The Multivariant Masks of Depression, Am. J. Psychiat., 124:35.

Malmo, R., 1965, "On Emotions, Needs, and Our Archaic Brain," Holt Rinehart, New York.

Reich, W., 1948, "Character Analysis," Vision Press, London.

Schachter, S., and Singer, J.E., 1962, Cognitive, Social and Physiological Determinants of Emotional State, Psych. Rev., 69:379.

Schildkraut, J., 1965, The Catecholamine Hypothesis of Affective Disorder, Am. J. Psychiatr., 122:509.

Travell, J., 1960, Temporomandibular Joint Pain Referred from the Muscles of the Head and Neck, J. Prosth. Dent., 10:745.

Tyber, M. A., 1974, Treatment of Painful Shoulder Syndrome with Amitriptyline and Lithium Carbonate, CMA Journal, 11:137.

Webb, H. E., and Lascelles, R. G., Treatment of Facial and Head Pain Associated with Depression, Lancet, Feb 17, 1965:355.

"STRESS HEADACHES" FROM CAUSES IN THE MUSCULOSKELETAL SYSTEM

AND THEIR TREATMENT BY PHYSICAL MEANS

John Mennell

Professor of Physical Medicine (Retired)
University of California at Davis
Vero Beach, Florida

SYNOPSIS

It is generally accepted that musculoskeletal dysfunctions may manifest themselves clinically as referred pain mimicking organ disease. Headaches and head pain may have their source in clinically active "trigger points" in muscles of the head and neck. Most often these symptoms are called "tension headaches" and, in a manner of speaking, there may be some truth to this description.

But the myofascial trigger point, when it is clinically active, is a diagnosable entity and therefore it is a cause of headache which responds satisfactorily to treatment.

For some reason, the irritable trigger point as a cause of pain has been overlooked or ignored by clinicians. To Dr. Janet G. Travell, Professor Emeritus of Medicine, George Washington University School of Medicine, Washington, D. C., should go the credit for this work.

BACKGROUND

Forty one years ago a paper entitled "Pain and disability of the shoulder and arm; Treatment by intramuscular infiltration with procaine hydrochloride" was published. Its senior author Janet G. Travell, (1942) reported a major break-through in the causation of referred pain from myofacial sources to which attention was not drawn by the patient because usually there was no subjective pain in that location. These sources of pain, particularly in muscle, have come to be known as trigger points.

Thirty five years ago at the 7th International Conference on Rheumatism in New York I gave a paper drawing attention to referred pain in patients whose pain source was in the cervical spine. For this knowledge I claim no originality (Mennell, 1980).

But both these papers stressed the clinical observation that symptoms of pain in the forequarter of the body and in the head might have specific sources in the musculoskeletal system which are not localized within the pattern of the patient's pain.

In 1967 Dr. Travell gave a paper at the 8th annual meeting of the American Association for the Study of Headache entitled "Mechanical Headache" and in 1980 I reiterated both of our hypotheses at the 22nd annual meeting of the American Association for the Study of Headache.

At the first and second International Congresses for the Study of Pain, Travell and I contributed to both programs in the same vein (Travell, 1976a; Mennell, 1976; Travell, 1976b; Mennell 1978). In 1974 we made an educational film on trigger points and their treatment that has been widely shown at international meetings. We have also contributed extensively to the medical literature on trigger points with acceptance by medical, osteopathic, dental and podiatric journals.

I have taken time to give this review of trigger points covering nearly half a century because I am astounded at how casual has been the attention to this work in spite of its wide dissemination. Can it be that those who espouse the so-called "scientific method" in medi-cine tend to ignore common sense, logic and sound observations of the clinician unless the clinical work is supported by at least statisti-cal research? Or can it be that physical treatment is denigrated in an economic society, the written prescription being so much easier to use and so much less time-consuming?

NATURE OF TRIGGER POINTS

Trigger points in muscle are clinical entities. Research has yet to establish a scientific basis for the clinical observations which have been made regarding them. Currently investigations are being undertaken by Travell (1981) and Simons (1981a; 1981b) using electro-myography and electron microscopy as research tools. Though repeated-ly finding the same changes in areas of muscle in which trigger points are clinically found, the changes are non-specific in nature. So we are still left with our clinical observations which I commend to you. These observations do not fly in the face of what is known in the fields of muscle physiology, neuroanatomy and neurophysiology (Reynolds, 1981). Our reliance on the modified gate theory of Melzack (1981), on the classical work on the refactory period in the reception of afferent sensory nerve impulses (Sherrington, 1894), and the observed facts on the function of the spindle apparatus in muscle (Granit, 1955) are all within the bounds of Travell's hypotheses and observations.

I wish again to add to this work, my own observations on the subject of referred pain from mechanical dysfunction of synovial joints (Mennell, 1960; Mennell, 1981; Zohn and Mennell, 1976; Mennell, 1952) in the cervical spine as an added cause of mechanical headache which is readily relieved by joint manipulation. Trigger points in muscle and mechanical dysfunction in synovial joints have to be considered with one another; they may occur separately or together and the clinical differentiation is sometimes difficult.

DIAGNOSIS

The diagnostic criteria of an active trigger point causing pain are:

(1) The pattern of pain from a trigger point in any given muscle is the same from one person to another.

(2) Given a pattern of pain which is characteristic of a trigger point, the source of the pain can predictably be found. The source - the trigger point - is scarcely ever near the place where the patient claims to be hurting. This is very clear in the illustrations on which comment is made later.

(3) On examination for the trigger point by palpation the area of muscle in which it is situated is locally tender, even though the patient is not aware of its presence or location and does not complain of pain in the area.

(4) There appears to be a ropey band in the muscle at this point. The examining fingers stimulate the tender "point" and reproduce the distribution of the patient's pain. Under the palpating fingers the muscle twitches: this, Travell (1976a) calls the "twitch response". The contraction of the ropey band in the muscle and of the muscle itself are not to be confused with the patient's jumping because of pain.

Thus we have four predictable facts in arriving at a diagnosis of the trigger point as the cause of pain: the pattern of pain is predictable when it arises from any given trigger point; given a pattern of pain, the location of the trigger point is predictable; irritating the trigger point by palpation predictably reproduces the patient's pain; and a "jump response" is elicited in the involved muscle on physical examination of it (Travell, 1981). All these things give rise to another predictability when the diagnosis is established, and that is that the relief of pain is predictable with proper treatment using either the "fluori-methane" spray and muscle stretch (Travell, 1952) or by accurate injection therapy using 0.5 percent procaine in physiological saline without epinephrine (Weeks and Travell, 1957). Passive stretching of the involved muscle insures that it eventually resumes at rest its normal (maximal) length, which is a pain-free state, and this is essential to the success of either method of treatment.

THERAPY

There are other prerequisites for the therapeutic success which we expect. These are:

(1) Correcting any occupational causes or bad habits, for instance, of posture or other physical causes of muscle imbalance, muscle strain or muscle spasm. In addition to such relatively obvious precipatating causes of trigger points, deficient nutrition, subclinical endocrine dysfunction such as marginal hypothyroidism, subclinical gout, or occult collagen vascular disease may maintain irritability in a trigger point and be the cause of therapeutic failure. Also, any musculoskeletal pain problem may be perpetuated if the patient has an unrecognized focus of infection: Most patients in these categories have a chronic deficiency of their daily fluid intake as well.

(2) Strengthening the muscle in which the trigger point was located, by gentle stretching (lengthening) exercises following or during the application of gentle moist heat is essential to the success of any therapeutic program. Most therapeutic exercise programs are designed around a muscle-shortening and accelerating type of exercise. This is useless in strengthening muscle which has been weakened by shortening because of spasm with or without atrophy. Starling's Law is the basis of this observation which says, in brief, that a long muscle is a strong muscle, and a shortened muscle is a weak muscle. In the circumstances which we are discussing, then, lengthening, decelerating exercises should be used.

(3) There is a principle of treatment in all musculoskeletal conditions and, indeed, in all therapeutic endeavors, which is too often ignored these days: this is that the structure in which any primary pathological condition is situated has to be rested from function (but not totally immobilized) whilst healing is taking place (Hilton, 1892). At the same time all structures which are not primarily involved in the primary pathological condition must be maintained in as normal a physiological state as possible. Any treatment program in the healing phase must include passive movement of the

structure involved in the primary pathological condition within the limit of pain unless some special technical reason makes this impossible. The correct dosage of movement is essential for success of treatment and knowledge of this is an important acquired therapeutic skill for a professional therapist. After healing has taken place a restorative program of therapy is usually needed and the patient should be instructed in prophylactic measures to be taken against recurrence.

(4) Emotional stress and tension may initiate a musculoskeletal response of spasm which, unrelieved, may be the initiator of the trigger point syndrome. But, conversely, unrelieved musculoskeletal pain may initiate stress and tension thus starting a vicious circle of events which is very difficult to break. Suffice it to say that psychological treatment fails if there is a physical basis for pain unless the physical cause is successfully eliminated (Mennell, 1980).

Even when the diagnosis of an irritable trigger point as the cause of pain is correct, treatment using the "spray-stretch" method may still fail if any of the following circumstances prevail: (Mennell, 1972; Mennell, 1975).

(1) If too much spray is used and the muscle in which the trigger point is situated is chilled, the muscle goes into spasm and the patient's pain becomes worse;

(2) If the muscle being treated is overstretched (the stretching must always be done, during the application of the spray, passively and gently) the pain of over-stretching, or just the over-stretching itself, further irritates the trigger point and makes the patient worse.

(3) If muscle spasm is substituting for the ligament integrity which normally supports a joint and the spasm is overcome by treatment, joint instability returns and the patient's symptoms become worse. In this instance the vapocoolant spray-stretch procedure becomes a diagnostic tool, as well as a therapeutic modality.

(4) If part of the pain complex is arising from radiculitis or neuritis, the stretch of the muscle may also stretch the involved neurogenic element and then the patient's pain is worsened. Again the vapocoolant spray-passive stretch becomes a diagnostic tool.

(5) If the muscle can only be stretched to an adapted resting length instead of to its normal maximal length, in a limb, for instance, shortened by healing of a fracture of a long

bone with shortening of it, or by amputation, or by epiphys-
eal trauma or disease, the result of treatment is
disappointing.

(6) If a satellite trigger point is mistaken for the parent
trigger point and is inactivated instead of the primary
source, treatment is likely to fail. In looking at the
patterns of pain referred to the head in the illustrations
which follow it is clear that there are overlapping patterns
of pain: i.e., two or more trigger points may partly have the
same pain pattern. Of course, treating the wrong muscle
results in failure to relieve the primary pain;

(7) If the musculoskeletal symptoms and, indeed, the signs are
somatic manifestations of visceral pathology then, not only
does the treatment of the muscle fail to accomplish more than
transient relief of that component of the symptom complex,
but it also may mask the diagnostic signs of a serious visceral pathological condition with catastrophic results for
the patient (Mennell, 1975).

SOURCES AND LOCATIONS OF MUSCLE PAIN

Amongst the common physical causes of pain from muscle which are
mistaken for headaches, migraine, postconcussion syndrome or
psychoneurosis are:

    (1) whiplash injuries (Mennell, 1966);
    (2) surgical procedures on the head and neck;
    (3) dental procedures;
    (4) general anesthetic procedures;
    (5) visual problems;
    (6) chilling of neck muscle by air-conditioning or draughts;
    (7) viral infections;
    (8) mechanical joint problems in the upper spine;
    (9) wearing a sling;
    (10) habitual postural problems or bad work habits and
    (11) many other such simple or complicated etiological factors.

As listed in Table 1, the muscles most commonly involved in the
trigger point syndrome causing pains in the head are:

    (1) the sternocleidomastoids;
    (2) the trapezii;
    (3) the masseters;
    (4) the temporalis muscles;
    (5) the splenius capitis muscles;
    (6) the splenius cervicis muscles and
    (7) the cervical strap muscles.

Table 1. Muscles producing overlapping pain patterns

(This table draws attention to potentially overlapping areas of referred pain from different muscles causing symptoms mistakenly diagnosed as headaches. Please refer also to the illustrations that follow in this paper.)

TYPE OF PAIN	PAIN SOURCE MUSCLES	NUMBER
EARACHE	Sternomastoid, Masseter, External Pterygoid	3
TEMPORAL HEADACHE	Sternomastoid, Temporalis, Trapezius	3
SUBOCCIPITAL HEADACHE	Trapezius, Posterior Cervical Straps	2
OCCIPITAL HEADACHE	Trapezius, Sternomastoid, Occipitalis, Posterior part of Temporalis	4
JAW AND TOOTHACHE	Masseter, Temporalis, Trapezius	3
HEADACHE BEHIND EYE	Temporalis, Trapezius, Sternomastoid, Splenius Cervicus	4
HEADACHE ABOVE EYE	Sternomastoid, Frontalis, Temporalis, Masseter	4

Somewhat surprisingly the scalene muscles do not refer pain into the head. Because the external pterygoid muscle's trigger point refers pain in an overlapping pattern with that of the masseter, and may also give rise to pain over the eye, it must be remembered here.

In the illustrations the trigger points in the various muscles are indicated by an "X" in Figure 1-7. The pain patterns of which the patient complains are indicated by the heavy black areas. The stippled areas may be the location of pain, but this distribution of pain is less frequent and the pain in them may be less intense (Travell and Wenzler, 1952; Travell, 1960).

The composite pain reference patterns from trigger points in the clavicular and sternal divisions of the sternocleidomastoid muscle are illustrated in Figure 1. The sternal division refers pain mainly to the face. The clavicular division refers pain mainly to the forehead bilaterally, deep into the ear, and to the teeth.

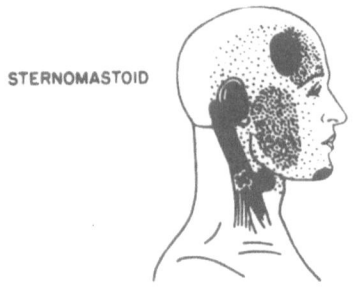

Figure 1:
The patterns of referred pain from
trigger points in the sternocleidomastoid muscle. The common frontal head pain (headache) may be
referred to the opposite side of
the head from the involved muscle.

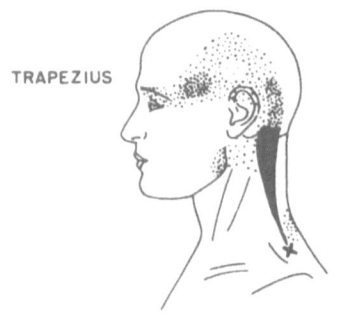

Figure 2 (a):
The pattern of referred pain from
the shawl area trigger point in
the trapezius muscle. Note it is
chiefly suboccipital but it is
frequently referred to behind the
eye.

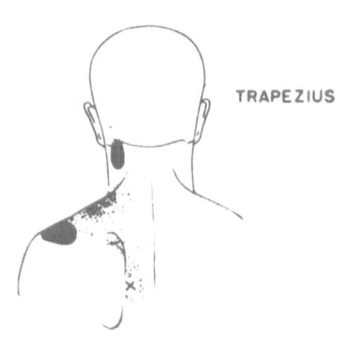

Figure 2 (b):
The pattern of referred pain from
the vertebral border of the scapula trigger point in the trapezius. Suboccipital head pain ·may
be quite intense.

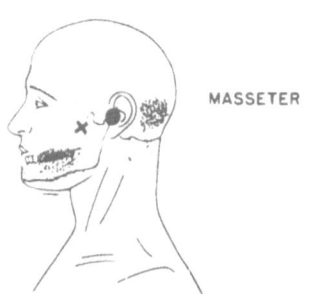

Figure 3:
The pattern of referred pain from
trigger point in the masseter
muscle. Loss of temporomandibular
joint function is quite common
when this trigger point is
irritable.

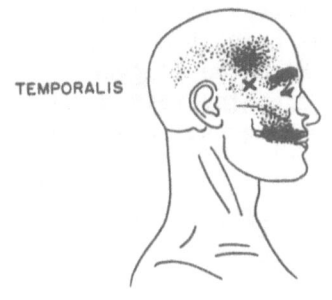

Figure 4:
The pattern of referred pain from
a trigger point in the temporalis
muscle. "Toothache" may be
intense in the upper jaw.

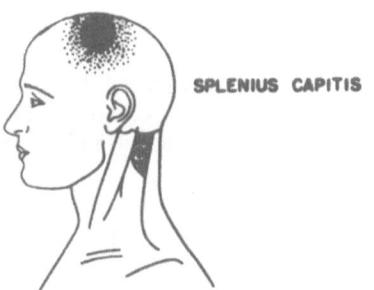

Figure 5:
The pattern of referred pain from
a trigger point in the splenius
capitis muscle.

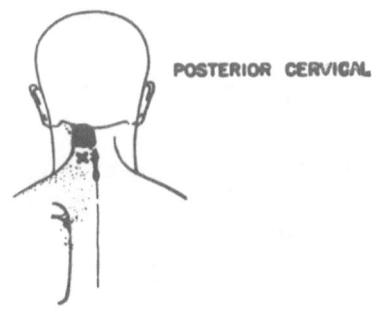

Figure 6:
The pattern of referred pain from
a trigger point in the posterior
cervical strap muscle.

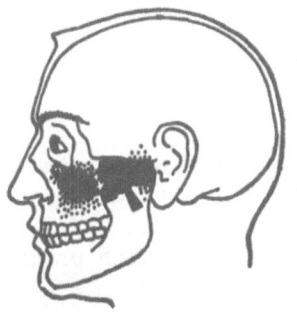

Figure 7:
The pattern of referred pain from
a trigger point in the external
pterygoid muscle; "sinus" pain may
be intense.

There are two main trigger points in the trapezius muscle. They are illustrated in Figures 2(a) & 2(b). From the trigger point in the shawl area, pain is chiefly referred to the back of the neck and to the suboccipital and occipital areas of the head, behind the eye and sometimes to the jaw and even the teeth. From the trigger point in the muscle at the vertebral border of the scapula, pain may be intense in the suboccipital area and may be referred to the shawl area and the tip of the shoulder.

The pain reference patterns from the trigger point in the masseter muscles mimic occipital head pain; there may also be intense ear and tooth pain in the lower jaws (Figure 3).

The pain reference patterns from the trigger point in the temporalis muscle include temporal head pain, supraorbital pain, pain behind the eye; tooth and upper jaw pain may be intense (Figure 4).

The pain reference patterns from the trigger point in the splenius capitis muscle are at the top of the head. Vertical headache is usually ascribed to psychoneurosis. This pain pattern is illustrated in Figure 5.

The main pain reference pattern from the common trigger point in the posterior cervical strap muscles is to the suboccipital area, as in Figure 6.

The pain reference patterns of the external pterygoid muscle is included because of its overlap of pain from both the sternomastoid and masseter muscles in the auricular area. "Maxillary antrum" pain (sinus pain) is characteristic (Figure 7).

I should especially draw attention to the black areas which involve the teeth and jaws, and deep in the ears. It is interesting that the pain in these locations is indistinguishable from tooth pain and earache due to local pathology. Not only are we dealing with referred pain but with referred deep tenderness as well; the teeth react to changes of heat and cold (Travell, 1960; Travell, 1976a) and chewing, but edentulous patients may also suffer from these pains.

It must be remembered that the headache from a sternomastoid muscle may be felt by the patient to be on the opposite side of the head from the source.

Limitation of movement of the neck is often associated with these headaches. This may occur without coincidental joint dysfunction being present. However, both conditions may coexist and joint dysfunction may be the primary cause of muscle spasm with or without trigger points in the involved muscles.

Pain in the head from joint dysfunction of the occipitoatlantal facet joints or of the atlantoaxial facet joints takes on the characteristics of radiating pain in the distribution of the occipital and auricular nerves which arise from the second and third cervical nerve roots, rather than the characteristics of referred pain which, when associated with trigger points, is in a pattern which is not segmental. Only by clinical examination can these causes of the loss of movement and pain be differentiated (Mennell, 1966).

TREATMENT TECHNIQUE

Time does not permit me to go into the details of properly administering treatment for clinically active trigger points (Travell, 1952). However, I have trouched on therapy earlier in this paper, but insufficiently to allow therapeutic trials on patients with any real expectation of success. The proper use of vapocoolant spray-passive stretch looks easy but the detailed technique of its use must be learned.[*]

Four words should be said about the use of ethyl chloride in treating trigger points, expecially those around the head and neck: "do not use it". Ethyl chloride is flammable; it is explosive in the right proportion with air; it is a general anesthetic and highly toxic; it is too cold; in short, it is dangerous under the conditions we are discussing. Travell has not used it in over 30 years. I have never used it.

Nor is injection therapy as easy as it sounds in the treatment of these myofascial trigger points (Weeks and Travell, 1957). I have mentioned earlier that Travell favors the use of 0.5 percent procaine without epinephrine. One of her reasons for this is that procaine has a little recognized curare-like action which surely is of additional benefit in the over-all success of this form of treatment. Another point which she stresses is that there must be no bleeding at the injection site since blood irritates trigger points. Epinephrine has a systemic effect which may be hazardous. Hydrocortisone injection into trigger points is unnecessary and irritates them. Distention of the injected area by use of excess of injected fluid makes it difficult to locate the trigger points and irritates them. Failure to use passive muscle stretching and a well designed physiotherapeutic restorative program after injection is a cause of disappointing results.

---

[*] Dr. Mennell's film "Spray-Stretch" illustrating how to use the spray is available through its distributor:
Richard Lambert, P.O. Box 701, Stinson Beach, California 94970

CONCLUDING COMMENTS

Each modality of therapy is fundamentally designed to produce relaxation and I wish to remind you of a wonderfully sage saying of P. G. E. Nixon (1980) at The First International Interdisciplinary Conference on Stress and Tension Control; he said:

"Bringing down the blood pressure with a drug is not the same as removing exhaustion and the smell of defeat. Inhibiting the heart's response with a beta-blocker is not the same as dealing with frustration, exhaustion and despair. But the more health professionals give impersonal treatments and select and organize themselves to have neither taste nor time for personal commitment the greater will be the field for the counsellor".

For Dr. Nixon's "blood pressure" and "the heart's response" we could substitute "headache"; and the longer we follow the example of the typical cardiologist, the longer will our consumers (our patients) have to remain pill poppers.

REFERENCES

Granit, R., 1955, A Discussion of Aims, Means and Results of Electro-physiological Research into the Process of Reception, in: "Receptors and Sensory Perception," Yale University Press, New Haven.

Hilton, J., 1892, in: "Rest and Pain," W. H. A. Jacobson, ed., G. Bell and Sons, Ltd., London.

Melzack, R., 1981, Relation of Myofascial Trigger Points to Acupuncture and Mechanisms of Pain, Archives of Physical Medicine and Rehabilitation, 62:114-117.

Mennell, J. M., 1952, 'Rheumatic' Symptoms Arising from the Cervical Spine and their Treatment, in: "Rheumatic Diseases: Postgraduate Medicine and Surgery," American Rheumatism Association, N. B. Saunders Company, Philadelphia, p129.

Mennell, J. M., 1960, "Back Pain - Diagnosis and Treatment Using Manipulative Techniques," Little, Brown and Company, Boston.

Mennell, J. M., 1966, Assessment of Residual Symptoms from a "Whiplash Injury" in: "Proceedings of the Fourth International Congre of Physical Medicine," Excerpt, Medical Foundations Internationa Congress Series 107:528-529.

Mennell, J. M., 1972, Treatment of Myofascial Pain Secondary to Facet Joint Dysfunction by Cold, Manuelle Medizin, Germany, 10:76-81.

Mennell, J. M., 1975, The Therapeutic Use of Cold, Journal of the American Osteopathic Association, 74:81-93.

Mennell, J. M., 1976, Spray-Stretch for Pain and Muscle Spasm, (Abstract of Scientific Film), presented at: 1st International Congress for the Study of Pain, Florence, Italy.

Mennell, J. M., 1978, Manipulation Therapy for Low Back Pain, in:

"Advances in Pain Research and Therapy, Vol. 3," J. J. Bonica, J. C. Leibaskind, and D. G. Alba-Fessard eds., Raven Press, New York, pp 685-697.

Mennell, J. M., 1980, Diagnosis and Treatment of Myofascial Pain Arising from Trigger Points, in: "Stress and Tension Control," F.J. McGuigan, W.E. Sime, and J.M. Wallace eds., Plenum Press, New York, pp 83-89.

Mennell, J. M., 1981, Pain Associated with Cervical Spine Problems - An Analytical Study of 100 Consecutive Patients, Archives of Physical Medicine and Rehabilitation, (Abstract of Poster Presentation), Annual Meeting, American Academy of Physical Medicine.

Nixon, D. G .F., 1980, We All Need Homeostasis, in: "Stress and Tension Control," F.J. McGuigan, W.E. Sime, and J.M. Wallace eds., Plenum Press, New York, pp 67-79.

Reynolds, M. D., 1981, Myofascial Trigger Point Syndrome in Practice of Rheumatology, Archives of Physical Medicine and Rehabilitation, 62:111-114.

Sherrington, C. S., 1894, On the Anatomical Constitution of Nerves of Skeletal Muscles, with Remarks on Recurrent Fibres in the Ventral Spinal Nerve-root, J. Physiology (London) 17:211-258.

Simons, D. G., 1981, Myofascial Trigger Points: A Need for Understanding, Archives of Physical Medicine and Rehabilitation, 62:97-99.

Simons, D. G., 1981, Letter to the Editor, Pain, 10:106-109.

Travell, J. G., 1952, Ethyl Chloride Spray for Painful Muscle Spasm, Archives Physical Medicine and Rehabilitation, 33:291-298.

Travell, J. G., 1960, Temporamandibular Joint Pain Referred from Muscles of the Head and Neck, The Journal of Prosthetic Dentistry, 10:No. 4:745-763.

Travell, J. G., 1967, Mechanical Headache, Headache, 7:1.

Travell, J. G., 1976, Myofascial Trigger Points: Clinical View, in: "Advances in Pain Research and Therapy, Vol. 1," J. J. Bonica, D. Alba-Fessard eds., Raven Press, New York, pp 919-926.

Travell, J. G., 1976, The Quadratus Lumborum Muscle: An Overlooked Cause of Your Back Pain, Archives of Physical Medicine and Rehabilitation, 57:566-580.

Travell, J. G., 1981, Identification of Myofascial Trigger Point Syndrome: A Case of Atypical Facial Neuralgia, Archives of Physical Medicine and Rehabilitation, 62:100-106.

Travell, J. G., Rinzler, S. H., and Herman M., 1942, Pain and Disability of the Shoulder and Arm Treatment by Intramuscular Infiltration by Procaine Hydrochloride, Journal of the American Medical Association, 120:417-422.

Travell, J. G., and Rinzler, S. H., 1952, Myofascial Gensis of Pain, Postgraduate Medicine, 11:425-434.

Weeks, V. D., and Travell, J. G., 1957, How to Give Painless Injections, in: "American Medical Association Scientific Exhibits," Ginne and Stratton, New York, pp 318-322.

Zohn, D. A., and Mennell, J. M., 1976, "Musculoskeletal Pain," Little, Brown and Company, Boston.

BIOFEEDBACK TREATMENT OF PRIMARY RAYNAUD'S[*]

Keith Sedlacek

Director, Stress Regulation Institute
St. Luke's-Roosevelt Hospital Center
New York, New York

SYMPTOMS AND CAUSES

There have been three basic treatments suggested for primary Raynaud's: surgical, pharmacological, and more recently, behavioral techniques, particularly biofeedback. There has also been one non-treatment which is basically the suggestion of environmental manipulation, that is, moving to a warmer climate, or taking long vacations in warm areas during the cold winter months. This would also include protecting the affected areas via scarves, gloves, caps, and other protective devices. In this paper, we will briefly review the diagnosis of Raynaud's and outline the types of treatment approaches which have been used for Raynaud's, of which biofeedback has been the most successful.

Raynaud's disease, which was first described in 1862, usually affects women in a ratio five times more frequently than men. Its symptoms most commonly consist of a blanching, mottled, acyanotic appearance of the fingers, usually bilateral. It is often progressive in nature and in some severe cases, approximately 0.5%, an amputation of the digit or affected area may be necessary, (Harrison, 1977). The most commonly affected areas are fingers, toes, nose, and earlobe. The triphasic response of Raynaud's is often described as: 1) a vascular spasm which causes a white blanching, 2) patchy areas of anoxia which give the blue response, and 3) the rubor or red phase. Often patients report only one or two of these three phases of the classic triphasic response.

---

[*] Parts of this paper appeared in "Biofeedback: Principles and Practice for Clinicians - 2nd Edition," 1983, J.V. Basmajian, editor, Williams & Wilkins Co. and in "Biofeedback for Raynaud's Disease," Psychosomatics, August 1979, vol 20. No. 8. They are included here by permission of the editors and publishers.

153

Primary Raynaud's disease is distinguished from secondary Raynaud's by absence of gangrene, consistent pain and the presence of an organic change in the arteries or arterioles. The original work by Raynaud described the condition as: 1) intermittent attacks of change in color; 2) symmetrical or bilateral involvement; 3) absence of clinical evidence of occlusive lesions of the peripheral arteries; and 4) trophic changes limited to the skin (Raynaud, 1888). Work by Allen and Brown (1932) indicated a difference between primary Raynaud's, and the secondary complications of such diseases as scleroderma, ulcers or arthritis and included two more criteria: 5) the disease or syndrome should be present for a minimum of two years and 6) there is no evidence of any other disease which could cause the condition.

Lewis (1949) has suggested that a local fault of the blood vessels was responsible: his cases were usually quite advanced and he ignored the effects of emotional stimuli upon this syndrome. Allen and Brown suggest that emotional or cold factors may trigger the vasomotor activity and concluded that the uncomplicated Raynaud's syndrome is due to the hyperactivity of the vasomotor system. The original work by Raynaud also suggested that the symptoms were not organic in etiology. It is important to recognize that secondary Raynaud's, also called Raynaud's pheonomenon or syndrome, can be caused by severe injuries to the arteries by trauma or toxic metals. Thus, arterial occlusions, such as arterial sclerosis or embolism, or thromboangitis obliterans or injury may cause secondary Raynaud's. Nerve lesions, such as thoracic outlet and causalgia may also cause symptoms of secondary Raynaud's as well as intoxication due to ergot and heavy metals. A third category of causes of secondary Raynaud's is post-traumatic Raynaud's phenomena which often occurs among people who use pneumatic and vibratory tools. The fourth group is often described as miscellaneous, which includes scleroderma, disseminated lupus, rhematoid arthritis, dermatomyositis, leukemia, myoloma and allergic arteritis.

The arterial supply and the anastemosis play a critical role in the symptoms of primary Raynaud's. Thus, shunting is an important factor in the vasospastic attack and to intervene we must be able to alleviate drastic blood flow changes. Work by Mittleman and Woff (1939) noted that variations in skin temperature do seem to be correlated with changes in affective states. This literature as well as work by Weber (1920) and Eng (1925), suggests that there is a decrease in skin temperature as a reaction to stress and fear. In work using different stressors there is shown to be a wide variation between each individual in change of skin temperature. Thus, as much of a change as 13° C. can be developed using different stresses. Even more important, in terms of the clinical treatment of Raynaud's, Mittleman and Woff found major drops in temperature in subjects who were placed in stressful situations regardless of whether the subjects were aware of their emotional state. Thus, the subjects who were under emotional stress and who could not relax showed a lower fingertip temperature.

While Raynaud's patients did not show a major difference from the controls with regard to the fall in the fingertip temperature with stress, they did show an association under stress of pain and skin color changes (pallor). Lowering the environmental temperature was not sufficient to precipitate the cyanosis and pain. However, emotional stress was sufficient to produce this cluster of symptoms. Thus, it is important to note that this syndrome, or primary Raynaud's (disease) is often a combination of both cold stress and emotional stress. This is demonstrated by the fact that symptoms were more severe when the cold was accompanied by emotional stress.

This important work by Mittleman and Wolff suggests that there is an important interaction between environmental temperature and emotional state. Further support for an affective component is derived from the fact that major drops in the peripheral skin temperature did not occur if the sympathetic nerve supply to the extremities was interrupted. Other work by Craig (1944), Graham (1955), and Peacock (1959) suggests the importance of psychogenic factors and the involvement of the sympathetic nervous system.

Primary Raynaud's appears to be a local reaction in the extremities or exposed skin which is most likely explained by a vasomotor change-vasoconstriction. Cold and emotional stress are the most commonly recognized stimuli. Temperatures below 58° F. will often cause the vasospasm. However, the disorder or disregulation processes can be brought on simply by an air conditioned room, a cool breeze or by reaching into an ordinary household freezer. While there have been great differences in opinion in the past as to the cause of primary Raynaud's, it appears to be clear now that there is no organic involvement.

SURGERY

The first major treatment attempts were to surgically cut the sympathetic fibers in order to produce chronic vasodilation. Recent reviews of sympathectomy offer inconsistent results and it is a very expensive and painful procedure with permanent side effects. This has been described by Ruch, Patton, Woodbury and Towe (1965). Allan and Brown have also suggested that the effect may last only for a period of months. Dale's data (1978) states that "sypathectomy is not effective for long".

Dale (1978) developed a diagnostic test using angiography to visualize the arteries of the wrist, hands, and fingers, allowing differentiation between ischemia due to organic occlusion and those due to vasospasm. His angiography technique suggests using a rapid injection of 20 cc of Conray 60 and then taking films at one second intervals. Saline is then flushed through the system while the films are developed and inspected. Then 20 mg of Priscoline (Tolazoline) are injected interarterially to produce maximum vasodilatation, in a period of approximately one to two minutes. Then a second injection of 20 cc of Conray 60 is made and films taken at one second intervals. These films are then studied to learn whether organic occlusion is present.

Using this diagnostic procedure, Dale (1978) reports that of 26 patients with secondary Raynaud's of the wrist, hands and fingers who had thoracic sympathectomy, 25 had good long term results, with one deriving no particular benefit. During a 10 year period, 24 patients with primary Raynaud's were seen and examined and all of the brachial arteriograms showed patent vessels. Of four sympathectomies performed on these patients only one had significant benefit. It is important to note that no deaths nor serious complications were found in these 30 patients.

Dale comments that thoracic sympathectomy produces good results when organic occlusions are demonstrated by angiography, but the operation should be avoided in their absence (1978). He also suggests the misconception that primary Raynaud's disease responded well to sympathectomy was due to the confusion of diagnoses between primary Raynaud's and secondary phenomenon. Other authors such as DeTakats (1959) also stated that lack of organic obstruction is a poor sign for surgery. Johnston, Summerly and Bernstingle (1965) reported "no cures" among 75 patients operated on 3 to 24 years earlier. They state that "although immediate postoperative results may be encouraging, the late results are not" and "it is doubtful that sympathectomy should ever be advised in Raynaud's disease (Johnson, 1965). Thus Dale (1978), in his article, suggests that "transaxillary thoracic sympathectomy is therefore now urged for Raynaud's syndrome with multiple organic occlusions of the hand or finger but not for primary Raynaud's disease".

This author would agree with this statement since long term studies do not support sympathectomies for primary Raynaud's and do appear to be useful in treatment of secondary Raynaud's.

PHARMACOLOGICAL AGENTS

Chemical compounds have often been used as an attempt at treating the vasospasm of primary Raynaud's and the associated conditions and diseases which cause secondary Raynaud's. The point of using vasodilators is to relax the smooth muscles in the wall of the blood vessel causing their vasodilation and therefore a decrease in symptoms. Thus, medications such as reserpine, guanethidine and phenoxybeziamine have been used. They may be given in doses of 0.25 for reserpine, and 10 mg daily of guanethidine or phenoxybeziamine. Pentoxifylline as well as Calcitonin (0.25–0.5 mg subcutaneously or intramuscularly) two to three times per week has also been tried with

some suggestive results (Molti, 1978). Other compounds such as nico-
tinic acid derivatives (e.g. tetranicotinoy fructose 250 mg 1-20
tablets per day) have also been used (Stachelin, 1977). While providing some brief relief, as measured by angiography, this has not
proven to be very useful clinically since the effect lessens dramatically within 1 to 5 days.

One study by Porter (1975) suggested that vasodilators taken just
before exposure to cold may decrease vasospasms and this may be of
practical use for the patient. Also other aids such as aspirin and
phenobarbitol have been used with varying degrees of success.

In many cases, the vasodilators or catecholamine depleting agents
produce side effects such as fatigue and low blood pressure that are
annoying to the patients. These chemical agents are often used in
medical practice; yet there is little evidence that they produce a
good result in primary Raynaud's. The main problem is that the medications have a transient effect on the patient's symptoms and disturbing side effects. Many clinicians have stopped using the sympathectomies and thus have tried vasodilators since they felt they had
no other treatment to offer the patient. Most patients with primary
Raynaud's get little lasting clinical effect.

BEHAVIORAL TREATMENT - BIOFEEDBACK

Over the last six years, there has been an accumulation of evidence in favor of the use of behavioral self-regulation techniques,
particularly what has been described as 'biofeedback' for the treatment of primary Raynaud's.

Work by Taub and Stroebel (1978), Surwit (1973; 1977), Lynch and
Miller (1976), Taub (1976; 1977) and this author (1976a; 1976b; 1979)
have demonstrated that subjects can learn to produce voluntary vasodilation. Experimental subjects and patients can specifically learn
to warm the areas involved in primary and secondary Raynaud's. Taub
has also demonstrated in work for the armed forces that even with
severe cold stresses produced by a body suit that students who are
trained with biofeedback and hand warming techniques are more able to
maintain peripheral warmth in the hands and fingers. Taub also has
demonstrated that normal subjects can self-regulate increases in hand
temperature, retain the ability for over one year, and perform other
tasks while maintaining hand temperature increases without temperature
feedback.

Stroebel and his group at the Institute of Living were probably
the first to demonstrate with large numbers of experimental patients
and clinical patients that this technique could be learned and maintained at followup. I would like to describe the general techniques
used in biofeedback treatment for primary Raynaud's at St. Luke's

Roosevelt Hospital and in my private practice. We also treat secondary Raynaud's symptoms, and have had similar success in reducing
symptoms of chronic arterial constriction.

Biofeedback is the use of modern instrumentation to give prac-
tically instantaneous information about specific physiologic processes
not clearly or accurately perceived but which are under the control of
the nervous system. People can develop control over specific physiologic processes with the aid of this specific information. Research
and clinical work indicate that research animals and human beings can
learn to self-regulate functions such as heart rate, blood pressure,
blood flow, skin temperature, sweat gland activity, and intestinal
processes.

To use biofeedback treatment, the autonomic system must be re-
trained in such a way that voluntary self-regulation is finally
achieved. We suggest to the patient that biofeedback is very similar
to bowel training; it is a "learned" voluntary self-regulation. If we
use the analogy of the skills necessary to learn to ride a bicycle,
then the biofeedback equipment is analogous to training wheels. With
sufficient practice and awareness of autonomic balance, the skill
becomes almost automatic (that is, one can talk, sing, or look around
while riding a bicycle). In an analogous way, the psychophysiological
skill can become "automatic" after a period of training and then be
transferred to everyday stressful situations.

Although there have been many reports and some overselling of
biofeedback in the popular press, there is good clinical evidence for
its usefulness in many vasoconstrictive and muscular disorders (classical migraine headache, primary Raynaud's disease, hypertension, tension headache, fecal incontinence, and torticollis).

The most widely used and most effective biofeedback methods have
used electromyography (EMG) and skin temperature feedback, which correlates closely with peripheral blood flow. Skin temperature feedback
is the attachment of a thermoprobe or thermistor to the finger or hand
so that knowledge of skin temperature can be fed back to the patient.

We find that a certain intensity of treatment (two sessions a
week of approximately 30-60 minutes duration, for 10 to 30 sessions)
is usually necessary. Thus, a two or three month treatment with
specific psychophysiologic retraining and awareness training is required. We cannot overstate the clinical importance of the "behavioral treatment package" consisting of: biofeedback session in the
office or hospital; biofeedback home practice twice a day for 15
minutes, progressive muscle relaxation, autogenic phrases and mental
imagery. I use a series of home practice tapes for my patients as
well as small temperature rings, small thermistors or liquid crystals
which 'feedback' finger temperature changes. This allows us to have
the patients record their finger temperature at home so we can follow

their progress in finger and hand warming in their home practice sessions. This gives the patients and therapists information as to the success or lack of success in transferring these learned skills of vasodilation into the home and work environment.

Since good protocols and equipment are now available, the biofeedback therapist is the other important variable in biofeedback treatment. A report by Taub (1977) suggests the importance of the therapist's role in skin temperature regulation. Taub had an experimenter who adopted an impersonal attitude toward the subject and who was not convinced of the feasibility of this learning (hand warming). This experimenter was able to train only two of 22 subjects to regulate skin temperature. Another experimenter employed the same techniques and in the same laboratory, but had no doubts as to the feasibility of the skin temperature training. She adopted a friendly approach toward the subject and succeeded in training 19 of 21 subjects. As this study suggests, an untrained person or a "rigid experimenter" may achieve strikingly different biofeedback treatment results. After over 15 years of biofeedback, it is clear that there is no "magic" in the machines.

TREATMENT - PSYCHOPHYSIOLOGICAL RETRAINING

Before starting training, we carefully explain the goals for the biofeedback treatment to the patients: they will be taught to self-regulate their vascular spasm and thus relieve the painful symptoms. We explain that many may experience relief from their symptoms in two to three months (10 to 30 sessions); however, their exercises and attention to their stress reactions must be continued for four to eight months in order to establish a healthy homeostatic balance between the parasympathetic and sympathetic branches of the autonomic nervous system. We explain that the activation of the emergency fight-or-flight response is a link-up to their particular vascular symptoms. Via their cool or cold hands, they can monitor their reactions. With the equipment, the training, and the home practice, they can learn to regulate their physiological disregulation. Thus, the goal is a rebalancing (homeostasis) of the autonomic nervous system which affects vasoconstriction.

It appears with clinical biofeedback that the more information that can be provided about the system, organ or troubled area, the better the results. Thus, the frontalis electromyography site is used for treatment of tension headaches and an anal placement is used for fecal incontinence. In Raynaud's, we have an excellent site, i.e., the affected area - fingers, toes, earlobes, and nose.

For the first four sessions, we use electromyography (EMG) biofeedback training (forehead placement) and then for the next four sessions we use thermal biofeedback training. This basic eight session sequence is used for all patients, for scientific relaxation. Most patients learn to relax the skeletal muscles and begin to warm their hands by the first 8 to 10 sessions. With Raynaud's patients, we then continue thermal training at the site of the dysfunction. We use the back of the finger since this is easily taped on and does not interfere very much with the patient's adjustments of hand position during the 30-60 minute session. Right and left finger temperatures may vary during individual sessions. During training in this office it is good to also train the non-dominant hand for at least two to three sessions, providing a check on the generalization of the treatment. Specific warmth phrases adapted from autogenic (Tuthe and Schultz, 1969) training are used such as "my right hand is warm and heavy". There are commercially available tapes that are useful to lend to the patient. We also provide them with a page of these autogenic phrases for their home practice. The physician, nurse, technician or trainer should be able to warm their own hands as well as repeat the phrases to the patient. This demonstrates the skill is learnable, effective and that the therapist is capable of producing vasodilatation involuntarily.

COMMON EXPERIENCES WITH BIOFEEDBACK TREATMENT

One of the purposes of EMG relaxation training is to teach people awareness of deep muscle relaxation. Thus with four to five sessions, people are able to notice specific sensations as their muscles relax. In the first stage of skeletal muscle relaxation, a patient's muscle may feel loose, limp and heavy. In the second state of muscle relaxation, people often report that a limb or their whole body is floating or drifting. They also report having calmer sensations inside their body. As a result of muscle relaxation, patients often report sleeping better, feeling calmer, and having more energy.

Since the skin temperature biofeedback tells the patient the direction and the speed of changes when a person does warm their hands, this 'warmth' sensation may be used to check their response during their home practice session and throughout the day. By teaching the patient these relaxation skills and having them properly identify them, the patient in essence develops their own biofeedback signal via these checks (rate and depth of breathing, decrease of muscle tension and 'warmth' feelings). We have our patients simply place their fingertips to their foreheads and they can learn to clearly discriminate between fingertips which are:

   1) ice cold, 2) cool, 3) lukewarm, 4) warm or 5) 'toasty' warm.
       - corresponding roughly to skin temperatures of -
   1) 65-73, 2) 73-80, 3) 80-86, 4) 86-92, and 5) 92-96 °F.

One of the important major differences between biofeedback and other verbal relaxation techniques is that the patient and doctor can con-

firm the physiological change. Thus, the patient can begin to discriminate and properly identify thoughts, feelings, and physical sensations that accompany muscle and nervous system activation and relaxation. Thus a verbal technique (relaxation) can, via the equipment, become a bio-behavioral treatment (biofeedback).

In general, biofeedback treatment is comfortable and relaxing for
the patient. However, there are certain disturbing effects of relaxation in biofeedback treatment. When they get a floating feeling,[*]
some patients, particularly those who have had training in the mental
health field, frighten themselves because they believe they are experiencing 'depersonalization'. This sensation of floating or drifting may be the readjustment or equalization of muscle tone level and
of the autonomic system. Other possible disphoric feelings are: itchy
skin, fasiculations of muscle groups, startle reactions as the muscles
relax dramatically and 'pictures' that appear to be similar to hypno-
gogic imagery. These pictures or images usually appear like a clear
daydream or dream sequence during the biofeedback session or may also
occur in the home practice session. Many of these images often provide insight into the patient's psychophysiological overreactions.

We also have observed differences in right and left hand temperature of up to 10° F. Patients report being aware of this right/left
difference. In fact, they often notice that the warmth is easier to
feel on one side of the body than the other. In some cases, they
notice one area relaxing first on one side and not on the other. The
"spotty" effect seems as if they are making contact in a new way with
certain parts of their body. When the new "feeling" and contact is
made with a bodily area, it is incorporated into a mind/body connection and rarely reappears in as striking a fashion. Some patients
experience difficulty relaxing and learning biofeedback skills because
they equate deep relaxation and peripheral vasodilatation with being
lazy or with images of death. Supportive psychotherapeutic skills of
the trainer and physician can help these patients recognize and move
through these reactions and gain physiological and mental self-
regulation.

We found with 20 patients, some of whom have now been followed
for more than five years, that 80% succeeded in learning self-regulation of the symptoms caused by vascular spasm (Sedlack, 1979). The
number of sessions required for treatment ranged from 12-36, with an
average of 20.8. The age range was from 20 to 73 and four of the
patients were men. All the men obtained relief. The patients practiced home exercises for 15 minutes twice every day with shortened
versions of the exercises that are taught to them in the office. We

* Editor's note: Such experiences are common in the very early stages
of learning Progressive Relaxation when the learner is beginning to
detect subtle tensions. They readily disappear as they continue to
relax.

shorten these exercises to five minutes, then two minutes, and suggest a brief 10 second relaxation at the three month follow-up visit. Twelve to 36 months later, all but two of the successful patients have maintained self-regulation.

If patients stop their home practice or are overloaded with stressful situations, symptoms often return. Of the 20 patients, one woman has stopped her home practice and her symptoms have recurred; she has not returned for further treatment. Another woman who had a relapse was treated with an additional six biofeedback sessions and regained self-regulation. The latter is very similar to our experience with migraine headaches, where the headache pain provides reinforcement for patients to return to regular home practice. We find that most patients who return to regular daily practice will regain self-regulation. Most patients are able to reduce vasospasm from 1-7 per week (4-28 per month) to 1-4 per month. Skin ulcerations show a reduction in frequency and in time it takes to heal.

PATIENTS WHO FAIL

In my practice, I find that 5-10% of the patients, even after relief of symptoms with the biofeedback technique, will stop their biofeedback and stress management techniques and their symptoms will return. Further therapy is recommended for them: assertiveness therapy, psychotherapy, group therapy, movement therapy, psycho-analysis, etc. Of our four failures, two have accepted the recommendation for further therapy and have begun these therapies. Three successful patients were also referred for other forms of therapy which they have followed up on.

Our results are very similar to those reported by Taub and Stroebel (1978), although we treat for more sessions. They treated more than 80 of his patients with a two year follow-up and reported that most (70-80%) of them were successful, using from 6-10 sessions with "booster sessions" available. After treatment, many patients can grasp a cold steering wheel, even make a snowball or pull frozen goods from the freezer without further attacks.

Let me describe one of our cases and the treatment result.

CASE REPORT

A 42 year old married Caucasion professional woman with two children was seen for treatment of primary Raynaud's disease. She had bilateral triphasic symptoms for more than seven years. Initially, she had noticed the characteristic triphasic response, but had no pain with these symptoms until the third year. The previous year she had to avoid touching the steering wheel of her car with her bare hands, because the cool or cold wheel could bring about a painful "attack". She also bought a heater to warm her room at work because of the increased frequency of the attacks.

She was seen for one hour sessions twice a week for eight weeks. Her initial finger temperature ranged from 72 to 80' F in both hands. By the 15th session, she reached our criterion of voluntarily raising the finger temperature to above 93' F for at least 15 minutes. Her highest recorded finger temperature (dominant hand, second finger) was 95.6' F. Her hand temperature now usually ranged from 84 to 90° F. We then tapered off the sessions to one per week for three more weeks. After 21 sessions she could achieve the same temperature regulation without using the temperature feedback equipment.

This woman learned to relax and to vasodilate her peripheral arteries and arterioles in order to stop her symptoms. She also became aware of aggravations concerning her spouse. It became clear the the repression of some of her feelings, especially anger and guilt, could bring about the conditions necessary for vascular spasm as well as cool or cold temperatures. She requested psychotherapy and we referred her for treatment twice a week. She was one of our first three patients and continues to do well at five-year follow up.

The above case is fairly typical of our patients, with the exception that she is among the 30 to 35% who, we believe, need further therapeutic work. The difference between these stress-related disorder (psychosomatic) patients and others is that in my experience they more readily request or accept suggestion for further exploration of their mental, emotional or physical reactions.

We have been pleased by this new understanding of body-mind relationships and the more mature attitude these patients take in dealing with their problems. They see and feel the importance of resolving some of their conflicts and emotional overreactions. They no longer are as defensive about symptoms, and gain a new self-confidence and willingness to try new situations. They also begin to adopt a less critical and compulsive way of dealing with themselves and others.

It appears that biofeedback has provided an excellent treatment for primary Raynaud's disease. Biofeedback retraining also suggests that we have all received a psychophysiologic training from our families, schools, society, and life experiences, and that our over-response to this training may predispose us to a series of disorders or diseases. How personality traits, patterns of cognition, and life stresses are acted out against particular organs or tissues - cranial arteries, coronary arteries, skin, stomach, intestines, and uterus - is still not understood.

While genetic predisposition may account for some symptoms, I believe we can continue to expand the research begun by Cannon (1914; 1935) and Selye (1946) by looking at many disorders as having a common underlying physical basis: autonomic dysfunction and exhaustion. When there is a specific social attitude or training about these physiological reactions, we can anticipate physical dysfunction to continue without a specific retraining.

Biofeedback behavioral treatment can be viewed as a method of physiological retraining to learn skills that were previously not developed or were developed through fear and social learning into an activated autonomic response (sympathetic firing). This pattern of increased sympathetic firing or tonus may bring about a vasoconstrictive reaction (response). Once the physiological overreactions or dysfunctional patterns are changed by biofeedback training, then the perceptual set of attitudes and fear responses can be re-evaluated and reorganized. The patient needs to become aware of particular stresses that trigger vasoconstriction. In most cases, this means regular daily practice of biofeedback skills and awareness training until the perception of fear and the autonomic overresponse have been altered. Thus, we suggest a range of 10 to 30 sessions for most of these vasoconstrictive disorders and as many as 30 to 60 sessions for the more difficult patients, or for patients who have more than one stress related disorder.

GENERAL PRECAUTIONS IN USING BIOFEEDBACK

With effective biofeedback training, we expect and we see changes in hormone requirements. Thus, relative caution needs to be exercised with people who are taking replacement medications such as thyroid, insulin and steroids. Because of lowered sympathetic firing, patients need to be cautioned that biofeedback training might alter the effect of anticoagulants and other medications. Biofeedback patients must be closely monitored by a physician so that medication can be properly titrated.

One other relative contraindication for biofeedback is in patients over 65 years of age, because of the possibility of hypotensive episodes. We have yet to observe a hypotensive episode in treating over 300 patients and more than 80 hypertensive patients. Other researchers, such as Stroebel, however, have reported observing hypotensive episodes.

General caution should be exercised in treating psychotic patients who are not well medicated and who do not have specific symptoms that can be treated by biofeedback. One of these successfully treated Raynaud's cases in our series had a history of psychosis. She continues to do well at 18 month follow up. Well compensated or adequately medicated psychotic patients with specific symptoms appear to do well with specific biofeedback treatment (Wentworth-Rohr,

1978). Caution should be exercised if the patient has a general symptom (such as sexual anxiety) which may involve or become involved in their delusional systems.

Some of the causes for failure in biobeedback patients are depression, secondary gain, and other psychologic resistance. If a patient is clinically depressed, we medicate them with anti-depressants. In up to 20% of our cases, we may uncover a masked depression as the symptom begins to be relieved. It is sometimes necessary to continue biofeedback and/or other therapies for as many as 30-60 sessions for successful results. In general, home biofeedback equipment is not necessary. With muscle rehabilitation cases, which may require 30 to 100 sessions, home equipment is sometimes used.

Biofeedback has demonstrated itself as a useful clinical treatment of primary and secondary Raynaud's. It also appears to be promising as a treatment for other vascular disorders such as angina, claudication, and Buerger's disease.

Some of the real difficulties in the emerging field of biofeedback treatment of Raynaud's is inadequate diagnosis, insufficient protocols, insufficient duration of treatment and poorly trained technicians, nurses, or physicians. These factors account for many of the mixed results that are cited in discussions about biofeedback treatment of Raynaud's.

The crucial factors of biofeedback treatment are the biofeedback package and the awareness training. This awareness training helps the patients transfer the biofeedback skills into everyday living. This transfer and "automatic" response is a necessary end point of successful biofeedback treatment.

SUMMARY

Primary Raynaud's seems to be best treated with a trial of approximately 20 biofeedback treatment sessions. Sympathectomies should be reserved for patients with secondary Raynaud's with demonstrated organic blocks. Vasoconstriction alone is not sufficient evidence for a sympathectomy. Vasodilators may be useful for brief periods in which cold exposure is necessary. Biofeedback treatment with an experienced clinician offers the patient the most effective treatment for primary Raynaud's at this time. There are few negative side effects of treatment and most patients experience a marked reduction of symptoms. Healing of skin ulceration occurs more rapidly and patients seem encouraged to develop new ways of maintaining healthy physical and mental attitudes in their daily life.

Behavioral treatment "packages" such as biofeedback should continue to provide many other useful educational and stress management techniques as well as a medical tool for diagnosis and treatment.

REFERENCES

Allen, E. V., and Brown, G. E., 1932, Raynaud's disease: A clinical
    study of one hundred and forty-seven cases, Journal of the
    American Medical Association, 99:1472-8.
Cannon, W. B., 1914, The interrelations of emotions as suggested by
    recent physiological researches, American Journal of
    Psychology, 25:256-82.
Cannon, W. B., 1935, Stresses and strains of homeostasis, American
    Journal of Medical Science, 189:1-14.
Craig, J. B., 1944, The psychogenesis of Raynaud's syndrome, Diseases
    of the Nervous System, 5:142-6.
Dale, W. A., 1978, Differential management of Raynaud's syndrome based
    upon angiograms, Connecticut Medicine, 42:447-51.
DeTakas, G., 1959, "Vascular Surgery," Philadelphia, W. B. Saunders.
Eng H., 1925, "Experimental Investigations of the Emotional Life of the
    Child as Compared to that of the Adult," (Translated by G. H.
    Morrison), Oxford University Press, London.
Graham, D. T., 1955, Cutaneous vascular reactions in Raynaud's disease
    and in states of hostility, anxiety and depression, Psychosomatic
    Medicine, 17:201-7.
Harrison, T. R., 1977, "Principles of Internal Medicine," 8th edition,
    McGraw-Hill, New York, p. 276.
Johnson, E. N. M., Summerly, R., and Birnstingl, M., 1965, Prognosis in
    Raynaud's phenomenon after sympathectomy, British Medical
    Journal, 1:962.
Lewis, T., 1949, "Vascular Disorders of the Limbs: Described for
    Practitioners and Students," Macmillan, London.
Luthe, W., and Shultz, J. H., 1969, "Autogenic Therapy, Volume 11,
    Medical Applications," Grune & Stratton, New York.
Lynch, W. C., Hama, H., Kohn, S., and Miller, N. E., 1976, Instrumental
    control of peripheral vasomotor responses in children,
    Psychophysiology, 13:219-21.
Mittelmann, B., and Wolff, H. G., 1939, Affective states and skin
    temperature: Experimental study of subjects with 'cold hands'
    and Raynaud's disease, Psychosomatic Medicine, 1:271-92.
Molti, A., 1978, Experimentally controlled evaluation of vasoactive
    drugs, Angiology, 22:89-94.
Peacock, J. H., 1959, A comparature study of the digital cutaneous
    temperatures and hand blood flows in the normal hand, primary
    Raynaud's disease and primary acrocyanosis, Clinical Science,
    18:25-33.
Porter, J. M., Snider, R. L., Bardana, E. J., Rosch, J., and
    Eidemiller, L. R., 1975, The diagnosis and treatment of Raynaud's
    phenomenon, Surgery, 77:11-23.
Raynaud, M., 1888, New Researches on the Nature and Treatment of
    Local Asphyxia of the Extremities, (Translated by T. Barlox),
    in: "Selected Monographs," The New Sydenham Society, London.
Ruch, T. L., Patton, H. D., Woodbury, J. W., and Towe, A. H., eds.,
    1965, "Neurophysiology," 2nd edition, W. B. Saunders,
    Philadelphia.

Sedlacek, K., 1976, EMG and Thermal Feedback for Treatment of
    Raynaud's Disease, presented at: Seventh Annual Meeting of the
    Biofeedback Research Society, Colorado Springs.
Sedlacek, K., 1976, EMG and thermal feedback as a treatment for
    Raynaud's disease, Biofeedback and Self-Regulation, 1:318.
Sedlacek, K., 1979, Biofeedback for Raynaud's disease, Psychosomatics,
    20:537-41.
Selye, H., 1946, The general adaptation syndrome and the disease of
    adaptation, Journal of Clinical and Endocrinological
    Metabolism, 6:117-230.
Stachelin, A., 1977, Treatment possibilities of arterial circulation
    disorders using calcitonin, Schweizerische Medizinische
    Wochenschrift, 107:1865-7.
Surwit, R. S., 1973, Biofeedback: A possible treatment of Raynaud's
    disease, Seminars in Psychiatry, 5:483-90.
Surwit, R. S., Pilon, R. N., and Fenton, C. H., 1977, Behavioral
    Treatment of Raynaud's Disease, presented at: Annual Convention
    of the Biofeedback Society of America, Albuquerque.
Taub, E., and Emurian, C. S., 1976, Feedback aided self-regulation of
    skin temperature with a single feedback locus, Biofeedback
    and Self-Regulation, 1 (2).
Taub, E., 1977, Self-regulation of human tissue temperature, in:
    "Biofeedback: Theory and Research," G. E. Schwartz, and
    J. Beatty eds., Academic Press, New York.
Taub, E., and Stroebel, C. F., 1978, "Use of Biofeedback in the
    Treatment of Vasoconstrictive Syndromes," Biofeedback Society
    of America, Denver.
Weber, E., 1920, "Einfluss psychischer vorginge auf den Korper,"
    Springer, Berlin.
Wentworth-Rohr, I., 1978, The reduction of anxiety related symptoms in
    schizophrenia through biofeedback-behavior therapy techniques,
    presented at: Ninth Annual Meeting of the Biofeedback Society
    of America, Albuquerque.

BIOFEEDBACK: TREATMENT OF ESSENTIAL HYPERTENSION

Keith Sedlacek

Stress Regulation Institute, St. Luke's-Roosevelt Hospital
New York, New York

Jonathan Cohen

Graduate Center, City University of New York
New York, New York

INTRODUCTION

In the last two decades there have been a number of reports
suggesting that non-pharmacological procedures can reduce blood pressure in hypertensive adults. It is well known that fifteen to twenty
percent of the adult population suffers from some form of hypertension. There is relatively clear evidence demonstrating an inverse
correlation between level of arterial blood pressure and the expected
length of life. Anti-hypertensive pharmacological interventions are
not always fully successful and are often accompanied by troubling
side effects.

It was initially believed that blood pressure biofeedback, meditation and other relaxation procedures would become effective alternative treatment modalities for essential hypertension. An initial wave
of clinical, Phase I, type studies in the 1960's seemed to support the
notion that biofeedback could be used as an effective adjunctive
treatment for essential hypertension. However, with the development

Acknowledgement: This work was part of Dr. Cohen's doctoral thesis at
the City University of New York. The authors would like to thank Paul
Wachtel, Ph.D. and Louis Gerstman, Ph.D. for their helpful suggestions
during the study. We are also grateful for grants from St. Luke's
Medical Center, New York, NY (grant #637-496-911) and an NIMH Pre-
Doctoral Research Fellowship to Dr. Cohen. Dr. Carl Boxhill and Ms.
Dylan Landis were also helpful in the execution of the project.

of more rigorous Phase II (or controlled follow-up studies) in the
1970's, blood pressure biofeedback treatment of essential hypertension
in the laboratory was shown to be generally neither clinically significant nor relatively cost efficient (Shapiro, Schwartz, Ferguson,
Redmond and Weiss, 1977).

Studies with other forms of biofeedback (e.g., EMG and thermal)
and relaxation procedures have often yielded clinical results that are
more impressive than blood pressure feedback for essential hypertensive adults. Patel (1973) reported that out of twenty hypertensive
patients who were treated with biofeedback (EMG and GSR) and Yoga
exercises, five patients ceased to need anti-hypertensive medication
altogether and seven others reduced their needs for medication by 33%
to 60%. The average systolic blood pressure reduction was 20 mm Hg.
In one of the first controlled follow-up studies in this area, Patel
and North (1975) reported that a Yoga relaxation/biofeedback (GSR and
EMG) group (N=17) reduced blood pressure (average reduction of 20/40)
and twelve of these patients lowered their need for anti-hypertensive
medication by 40%. The placebo therapy group (N=17) did not significantly reduce blood pressure levels or need for medication. These
changes persisted over a twelve month period. A more recent clinical
pilot study using EMG and thermal biofeedback in conjunction with a
variety of cognitive and somatic procedures (e.g., visualization,
autogenic training and progressive relaxation) also resulted in impressive clinical blood pressure reductions similar to Patel's findings (Sedlacek and Cohen, 1978). However, many non-blood pressure
biofeedback studies have lacked controlled procedures or other methodological procedures (e.g., base-line measures, follow-up evaluation)
that would permit firm conclusions to be drawn about the effectiveness
of this method of treatment.

This study was designed to provide information about the effectiveness of non-pharmacological treatment programs for essential hypertension, the carry-over effects from practice sessions, the comparison
of treatment programs with no treatment and the long term effects of
treatment. It is important to note that in contrast to recent studies
that have been experimentally rigorous and experienced by the patient
as "an experiment," the present study has sought to be relatively
rigorous (experimentally) and at the same time experienced by the
patients as a clinical situation. Thus patients were seen on an
active floor of the Department of Rehabilitation Medicine, not in a
research laboratory. A recent report that did not find biofeedback
and relaxation techniques useful as a method of reducing blood pressure, noted that the "experimental lab" setting may have detracted
from clinical efficacy (Frankel et al, 1978).

Although the motivation to learn autonomic self-regulation procedures and reduce blood pressure is a fundamental psychological
dimension that may effect autonomic self-regulation training programs,
it has not been systematically evaluated in hypertension studies

(Willamson and Blanchard, 1979). This study was designed to assess
the relationship between form of treatment program, amount of blood
pressure reduction (or increase) and motivation to achieve and learn
to reduce blood pressure. Recent research in motivation and achievement behavior shows that how individuals allocate (attribute) the
causes of success and failure in a given learning situation, affects
the motive to achieve and learn (Weiner et al, 1971). It is only
recently that investigators have begun to examine how this attribution
process may affect the motivation to learn autonomic self-regulation
skills (Peek, 1977; Plotkin, 1979). Individuals seem to attribute the
causes of success and failure to four elements: ability, effort, task
difficulty and luck (Heider, 1958; Weiner et al, 1971). Weiner and
his co-workers have shown that these four elements are comprised
within two causal dimensions: locus of control (internal versus external) and stability (fixed versus variable: see Table 1).

Locus of control refers to the self (internal) versus environmental (external) responsibility for an outcome (Rotter, 1966) while
the stability of attribution reflects its perceived fluctuations over
time (Weiner, 1972). Recent research examining locus of control
beliefs and self-regulation training programs suggests that internally
oriented individuals are more likely than an externally oriented
person to believe in and attempt to exercise self control over bodily
functions. Furthermore, internal subjects have displayed better feed-
back-assisted control in EMG reductions, (Carlson, 1977) GSR control,
(Wagner et al., 1974) alpha-wave activity, (Goesling, 1974; Johnson &
Meyer, 1974) and heart rate speeding (Gatchel, 1975; Ray, 1974).
However, these findings have not been replicated for heart rate
slowing (Ray, 1974).

This may be in part due to the fact that although locus of
control is a valid psychological determinant of success and failure,
while task difficulty and luck are relatively non-psychological
"external" (environmental, or in the above mentioned case, physiological) determinants of outcome. Weiner and his co-workers (1971)
have shown that perceived ability and task difficulty remain relatively constant over time (fixed elements), while effort and luck may
shift from moment to moment (variable elements). See Weiner (1972)
for a more extended discussion of these issues.

Table 1. Perceived Determinants of Success and Failure.

Locus of Control

Stability	Internal	External
Fixed	Ability	Task Difficulty
Variable	Effort	Luck

Thus, we planned to examine for I/E locus of control attitudes and amount of effort that individuals report they would "put into the program" before treatment began. This study was designed to assess these four factors: effort, ability, difficulty and luck retrospectively, and to begin to study some of the cognitive determinants of successful self-regulation training.

In addition to being able to assess motivation by how much effort a given individual "puts into" a learning task, task difficulty and the extent to which luck plays a role in successful learning, may all affect motivation. For instance, it has been shown that tasks of intermediate difficulty are most likely to elicit effort attributes which in turn increase motivation (Weiner, 1972).

METHODS

Subjects

Male and female volunteer subjects with essential hypertension were referred by their private or clinic physician to a "Hypertension Study." After a preliminary telephone screening, potential subjects were interviewed by a psychiatrist for possible inclusion in the study. On the basis of specific inclusion and exclusion criteria, to be described below, thirty-three patients were accepted into the study and baseline recordings of blood pressure began. During an initial medical and psychiatric interview, patients were informed of the nature of the study and medical history was gathered focusing on factors relevant to hypertension; all patients were taught to use a sphygmomanometer and instructed to record their blood pressure daily. These were used to evaluate any differences between home and office (taken by the therapist and bi-monthly by a 'blind' nurse). Subsequently, a description of the intent of the study was mailed to the volunteer's physician. Volunteers had to have their own physician agree to their participation in the study. The investigators did not assume the role of primary care physician.

For inclusion in the study, patients had to have been diagnosed as essential hypertensive for at least two years. Furthermore, patients were excluded if there was evidence of organic etiology for their hypertension, major complications related to the disease, or other serious medical or psychological illness (i.e. psychosis). Potential subjects whose histories met these criteria were seen for an extensive psychological assessment and bilateral blood pressure measurement while sitting and recumbent. Subjects were requested to continue their current dietary and medication practices and inform the investigators of any change in the regime during the study. In the course of pre-treatment evaluation, three subjects dropped out leaving ten subjects in each group. See Table 2 for a description of patient characteristics.

Table 2. Characteristics of Patients.
("BP" is baseline blood pressure.)

Biofeedback Treatment Group			Relaxation Response Treatment Group			Control Group		
Age	Sex	BP	Age	Sex	BP	Age	Sex	BP
53	F	154/110	48	F	145/90	33	F	144/98
65	F	130/90	68	F	130/84	28	F	130/92
46	F	122/82	63	F	146/96	39	F	220/120
49	F	130/104	40	F	120/100	31	F	140/92
53	F	136/92	44	F	140/98	37	F	120/90
33	F	160/94	45	M	134/94	53	F	132/84
41	M	150/92	44	M	140/106	34	M	140/100
72	M	150/96	43	M	160/100	37	M	140/100
36	M	141/86	56	M	158/100	54	M	140/98
26	M	160/100	31	N	140/90	32	M	140/96

Therapists

The two authors, an advanced doctoral student in Clinical Psychology and a psychiatrist, conducted the treatment. The doctoral student saw 90% of the biofeedback group patients, 50% of the relaxation response patients, and 50% of the waiting list control group patients. The psychiatrist saw the remaining patients. Statistical analysis (ANOVA) revealed no differential effectiveness for either therapist in terms of blood pressure changes or changes in medication.

Blood Pressure Monitoring

Blood pressure was derived by taking three measurements using the standard method. These readings taken at the beginning of each session, were averaged and used as the blood pressure for that session. The patients who were instructed to take their blood pressure at the same time each day, the "blind" nurse (who evaluated the patient's blood pressure every other week before a treatment session), and the therapist who evaluated blood pressure at the beginning and end of each treatment session used this method.

Pre-treatment evaluation of motivation and locus of control

In a pre-treatment interview, patients in the two active treatment groups completed a questionnaire that included assessment of how
much effort they planned to put into the treatment program (1 = 'none'
to 7 = 'very much') and Rotters I-E Scale (Rotter, 1966). Scores on
this locus of control scale range from 0 (internal) to 23 (external).

Post-treatment evaluation of motivation and other subjective factors

In the post-treatment interview, patients in the two active
treatment groups completed a questionnaire assessing a variety of
motivational and treatment experience dimensions. The interview was
conducted by a "blind" research assistant who had no clinical contact
with patients. Patients were asked to rate how much **effort** they had
put into the treatment program (1 = 'none' to 7 = 'very much'), how
much **ability** they feel they have to perform the relaxation and/or
self-regulation tasks (1 = 'have no ability' to 7 = 'very high amount
of ability'), how **difficult** the relaxation tasks were after two sessions and at the end of the treatment preogram (1 = 'not at all
difficult' to 7 = 'extremely difficult'). They were also asked to
evaluate the extent to which they believe luck contributed to their
treatment experience. In addition, patients were asked to describe
whether they had become more sensitive to anything and their overall
evaluation of the usefulness of (or lack of) the treatment program (0
= 'not at all'; 1 = 'slightly'; 2 = 'moderately'; 3 = 'very helpful').

Procedures

There are certain commonalities among the three groups. Patients
in all groups were seen for psychiatric and psychological evaluation
and taught to use the sphygmomanometer and to record their blood
pressures daily. In the two treatment groups, sessions lasted approximately twenty to thirty minutes. Electromyography recordings from
surface electrodes at the frontalis site and finger temperature readings were made during sessions. A Cyborg Bio-Lab Model BL533 or a
Biofeedback Instrument Company Model P775-1 were used for the EMG
recordings. A Cyborg Bio-Lab Temperature Trainer Model BL562 or a
Model P442 were used for the temperature recordings. Half of the
patients in the Biofeedback and Relaxation Response groups reclined
in a semi-recumbent lounge chair and half laid down during the sessions. There was no significant difference in physiological or
psychological change scores between the two conditions.

Baseline Evaluations

Before the ten week treatment (or Waiting List Control) began,
all patients were individually seen for baseline blood pressure evaluation. Bilateral sitting and recumbent blood pressure measurements
were taken on three different days over a two week period before the

treatment programs began. (The baseline blood pressures in Table 2 represent the average of two blood pressure recordings taken one and two weeks before the beginning of the 10 week treatment period).

The Treatment Groups

The Biofeedback Self-Regulation Treatment Group. Patients were seen individually twice a week for ten weeks. Following a "hook-up" period, the sessions began by discussing the patient's treatment experience and then listening to two sets of pre-recorded relaxation, visualization and self-regulation audio tapes (developed by K. Sedlacek, M. D. and T. Budzynski, Ph.D.). These tapes were used while the patients received EMG feedback for four weeks and then peripheral (finger) temperature feedback for the next four weeks. For the last two, either EMG, thermal or no feedback was used, depending on the patient's wish. At the end of each treatment session the patient's experience was discussed with the therapist. The nature of these supportive psychotherapeutic interactions ranged from the relatively superficial to the highly personal.

Patients were instructed to practice specific exercises at home twice a day for fifteen to twenty minutes. Blood pressure was recorded by the therapist before and after each session as well as by a "blind" nurse every other week. A one-way analysis of variance revealed that there was no significant difference between "blind" blood pressure assessment, home pressure levels and the therapist's recording of blood pressure (p>.10). It was decided that all pressures reported would be the pre-session blood pressure gathered by the therapist, since this would most closely correspond to blood pressure readings in a physician's office. At the initial treatment session the operation of feedback hardware was explained to the biofeedback group patients. They were given the expectancy of learning to relax via lower EMG signals, warming their hands (vasodilatation) and thus would obtain therapeutic improvement.

The Relaxation Response Treatment Group. Relaxation Response training was chosen as a control condition because 1) it was believed to be active enough to serve as an adequate attention placebo control (Paul, 1960) condition, and 2) there have been many reports of its usefulness in the lowering of blood pressure (Benson, 1975).

Patients in this group were given instructions outlined by Benson. They were seen once a week for five weeks and then five weeks later. Initially patients were taught to practice the relaxation response which involved sitting quietly, relaxing their muscles and then reciting the number "one" silently to themselves on each exhalation. They were instructed to concentrate on this recitation and passively bring their attention back to it if their minds wandered. After learning this simple meditative procedure, patients in this group practiced it for fifteen minutes and then discussed their experience with the therapist. Subsequent sessions began by the patient discussing the treatment experience with the therapist, practicing the relaxation response, and then again discussing their experience with their therapist. Thermal and EMG levels were recorded during all treatment sessions. As in the Biofeedback Group, the nature of this supportive psychotherapy varied. Patients were instructed to practice the relaxation response procedure at home twice a day for fifteen to twenty minutes. Similar expectations for improvement in hypertension as they mastered the relaxation response procedures were given.

The Waiting List Control Group. Since essential hypertension can run a somewhat fluctuating course, it was thought useful to include a group who received no treatment other than monitoring their blood pressure but who expected to receive treatment. If patients had sought hypertension treatment during an acute phase of their disorder and had improved spontaneously, this condition should detect the phenomena. The patients in this condition were told that due to the large number of volunteers in the project, not everyone could be treated. They were asked to continue keeping their daily blood pressure records and told that they may have an opportunity to receive whichever treatment seemed superior. Physiological and psychological evaluations were completed one week before the study began and after the ten week period.

Follow-up Evaluation

All patients were seen four months after the completion of the initial ten week program for a brief interview that consisted of blood pressure assessment as well as a brief overall psychiatric evaluation, and assessment of the patient's daily blood pressure 'logs'.

Home Practice

As noted above, patients in the two active treatment groups were asked to practice twice a day for fifteen to twenty minutes. Patients practiced the procedures they had learned in the clinic, and were not given "home practice" cassette tapes. Each day, patients were asked to fill out a brief questionnaire designed to enhance motivation, monitor home practice and record their experiences. The questionnaires were brought to the clinic once a week and discussed briefly. These questionnaires were used to evaluated percentage of home practice compliance.

Pre-Treatment Baseline Data

To determine if any of the three groups differed in average physiological (blood pressure and/or medication) levels, prior to the begining of training, a one-way analysis of variance was performed between groups assessing baseline blood pressures and amount of medication. Results of the one-way analysis of variance showed the base-

line blood pressure differences, amount of medication and pre-treatment physiological measures were not significantly different between
groups (p > .10).

Medication

Although the average amount of medication did not differ between
groups, random assignment of patients into the groups resulted in the
only patients not on medication happening to fall into the Biofeedback
Group. However, amount of blood pressure change between patients on
no medication versus patients on medication in the Biofeedback Group
and the Biofeedback and Relaxation Response Group combined, revealed
no significant differences (p > .10).

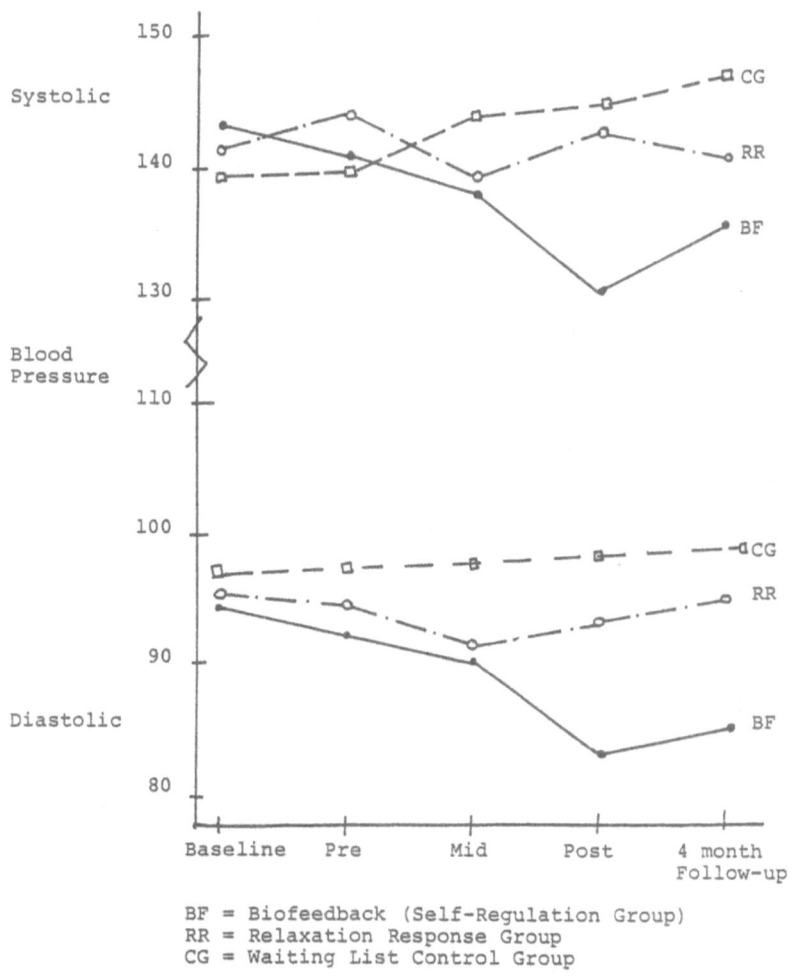

BF = Biofeedback (Self-Regulation Group)
RR = Relaxation Response Group
CG = Waiting List Control Group

Figure 1. Average blood pressure by treatment group across time.

In the Biofeedback Group, 43 percent of the patients (three out of seven) reduced their hypotensive medication by 50 percent or more (two patients reduced it by more than 66 percent, one patient by 50 percent). Only seven out of ten patients in this group were on medication at the outset of the study. One patient in the Relaxation Response Group increased medication by 50 percent. In the Waiting List Control Group, one patient increased medication by 50 percent and one patient decreased medication by 20 percent. These changes in antihypertensive medication underscore the clinical significance of the diastolic blood pressure change reflected in the Biofeedback Group.

Table 3. Mean Blood Pressure for each Group across Time
(N=10 in each group)

		Biofeedback/ Self Regulation	Relaxation Response	Control Group
Baseline				
Systolic	mean	143.6	141.3	139.1
	SD	13.2	12.0	10.2
Diastolic	mean	94.6	95.8	97.0
	SD	8.3	6.4	9.5
Pre				
Systolic	mean	141.1	144.2	139.8
	SD	13.8	8.1	9.1
Diastolic	mean	92.6	94.8	97.3
	SD	7.4	8.4	9.8
Mid				
Systolic	mean	138.1	139.2	144.0
	SD	13.4	10.1	11.3
Diastolic	mean	90.3	91.8	97.4
	SD	6.8	5.8	8.5
Post				
Systolic	mean	130.8	142.8	144.9
	SD	11.7	12.3	10.9
Diastolic	mean	83.2	93.8	98.1
	SD	6.3	8.7	9.7
Follow-up				
Systolic	mean	135.8	141.0	147.2
	SD	13.7	8.0	9.8
Diastolic	mean	85.4	95.1	98.4
	SD	5.5	7.6	7.2

Blood Pressure

Examination of the average baseline blood pressure to post-treatment for diastolic blood pressure shows that Biofeedback/Self-Regulation significantly reduced blood pressure at the end of the ten-week treatment program ($F = 8.23$; $p < .01$). Figure 1 and Table 3 show that the Biofeedback Treatment Group decreased blood pressure more than the other groups. A Scheffe test confirms that the Biofeedback Group has lower diastolic blood pressure recordings at the end of the the week treatment program than the other groups ($p < .05$). There was no statistically significant decrease in systolic blood pressure for any of the three groups, baseline to post-treatment. The Relaxation Response Group and the Waiting List Control Group did not show significant decreases in blood pressure, baseline to post treatment. The avearge Biofeedback Treatment Group reduction was 20/12 (baseline blood pressure = 150/95; post treatment blood pressure = 130/83).

At a four month follow-up evaluation, a one way analysis of variance showed that the Biofeedback Group continued to reflect a significant decrease in diastolic blood pressure (baseline blood pressure = 150/95 to follow-up blood pressure = 136/85; $F = 7.77$; $p < .01$).

ADDITIONAL FINDINGS

In a pre-treatment interview, patients rated how much effort they planned to put into the treatment program, and completed a Locus of Control Scale. There was no significant difference between the two active treatment groups in terms of how much effort they planned to put into the treatment program, or internal/external locus of control ($p > .10$). Patients in the two groups indicated that they were uniformly highly motivated (Biofeedback Group mean = 6.7; Relaxation Response Group mean = 6.8, where the upper end point of the scale, "7" reflected a "very high degree of motivation"), and somewhat more "internal" than "external" on the locus of control scale (Biofeedback Group mean = 9.8, Relaxation Response Group mean = 9.4).

As shown in Table 4, in a post-treatment interview (one week after the end of the ten week program) patients in the two active treatment conditions generally reported an increased ability to relax and cope more effectively with stressful situations. There was not an overall difference in terms of how helpful patients in the two groups rated the respective treatment programs to be. However, patients in the Biofeedback Group reported becoming significantly more sensitive to a variety of dimensions (i.e., sensations, emotions, environmental stimuli that is stressful, etc.) than patients in the Relaxation Response Group.

Table 4. Post-treatment Comparison of the Biofeedback and Relaxation
Response Groups' Assessment of Motivation and Other Attitude Factors

Factor	Biofeedback Group	Relaxation Response Group	t	p
Effort	5.7	4.8	1.19	NS
Ability	5.2	5.0	0.33	NS
Difficulty –				
after two sessions	3.7	4.3	0.64	NS
post-treatment	2.2	3.2	1.65	p < 0.1
four month follow-up	3.4	3.8	0.45	NS
Luck	0.0	0.0	0.0	NS
Percentage of Home Compliance –				
during the program	66%	58%	0.77	NS
during a one-month post-treatment period	48%	44%	0.41	NS
More Sensitive	1.9	0.6	3.67	p < 0.01
Overall Evaluation of Program Helpfulness	2.3	1.7	1.21	NS

Interestingly, four out of the ten patients in the Waiting List
Control Group reported feeling more relaxed and "aware of their
physiologic reactions" as a function of their daily blood pressure
evaluations. In fact, three of these ten patients described the
effect of taking their own blood pressure on a daily basis in a way
that was remarkably similar to how many of the patients in the
Biofeedback/Self-Regulation Treatment Group described their experience. They talked about how valuable it was to receive feedback as a
way of increasing their understanding of their responses to various
kinds of stimuli.

As part of our assessment of motivation in the post-treatment
interview, patients were asked how much effort they had put into the
treatment program. All 20 patients in the two active treatment groups
rated themselves as being "highly motivated to become involved with
the treatment programs and practice" before the treatments began. As
is shown in Table 2, there was not a significant difference reported
between the two treatment groups in terms of effort expended. Although patients in the two groups differed in terms of how successful
they were in reducing blood pressure, they did not rate themselves as
being significantly different in terms of ability to perform the

relaxation tasks. Although interestingly, there was not a significant difference between the two groups in terms of <u>how difficult</u> the relaxation exercises were after two sessions, there was a trend in the direction of the Biofeedback Group experiencing the relaxation exercises as slightly easier at the end of the treatment program ($p < .1$). Four months later, there was no significant difference in terms of how difficult they found the relaxation procedures at that time.

DISCUSSION

In this study, a combination of EMG and thermal biofeedback and various somatic (e.g., progressive relaxation) and cognitive regulation procedures resulted in significant and clinically meaningful reduction in diastolic blood pressure in a group of patients with essential hypertension. Although overall systolic blood pressure in this group became lower in a clinically meaningful fashion (pre-systolic blood pressure = 141.1; post-systolic pressure = 130.8), these changes were not statistically significant. Three of the seven patients in this group reduced their need for antihypertensive medication, underscoring the clinical significance of these findings. These clinical results are particularly significant in light of recent research that shows anti-hypertensive medication adversely affecting cognitive functioning (Solomon, 1979). In contrast to many recent non-pharmacological treatment programs that have found statistically but no clinically significant blood pressure reductions (Seer, 1979), the present results reflect clinically meaningful blood pressure reductions.

Patients in the Relaxation Response Group did not evidence significant reductions in blood pressure. However, it is interesting to note that after five weeks of treatment, or half way through the ten week treatment program, this group began to evidence blood pressure reduction. Patients in this group met with the therapist once a week up until this mid-point, and then worked on their own for five weeks. These results raise the important clinical question of how long and how many times a week essential hypertensive adults need to meet with a therapist. Five, once a week, meetings with the therapist may enable some patients to reduce blood pressure but may simply be an insufficient amount of time to fully learn relaxation and autogenic self-regulation procedures to reduce blood pressure in a sustained fashion. The Waiting List Control Group did not reduce blood pressure and, in fact, evidenced a slight trend toward increasing blood pressure.

These results resemble Patel's (1973) and Patel and North's (1975) findings, who noted reductions in the blood pressure of hypertensive adults who utilized a variety of relaxation/self-regulation procedures including biofeedback (e.g., EMG and/or thermal) and several somatic and cognitive relaxation/autogenic self-regulation

procedures. These results are in contrast to a variety of recent
studies that have relied on diastolic and/or systolic blood pressure
feedback and found little success in being able to aid patients in
reducing blood pressure (e.g., Schwartz and Shapiro, 1973; Surwit and
Shapiro, 1976; Miller, 1975; Frankel et al, 1978; Blanchard and Young,
1973). It is important to note that in addition to using blood pressure feedback, these studies differ from Patel and North's (1975) and
Green's (Green et al, 1980) and the present study in their attempt to
stricly adhere to protocol and maintain strict control procedures that
may have interfered with maximum clinical efficacy (Frankel et al,
1978). This may account for why some patients were unable to learn
how to relax and achieve clinically meaningful blood pressure reduction in recent studies (Surwit et al, 1978; Hager and Surwit, 1978).
In the present study, Biofeedback Group patients were encouraged to
utilize relaxation/self-regulation practices (EMG or thermal biofeedback as well as types of somatic/cognitive self-regulation techniques)
that seemed most helpful for them during the last two weeks of the
treatment program. In contrast to many other recent studies, patients
in the Biofeedback Group in this study were presented with a variety
of relaxation/self-regulation practices and instructed to utilize
whichever ones that seemed most helpful to them in their home practice. Perhaps, in addition to fostering a more positive doctor/patient
working alliance, this spectrum approach may increase the likelihood
that a given individual will discover a successful self-regulation
strategy for themselves. Although this approach introduces a variety
of uncontrolled experimental parameters, it may be an important ingredient in a successful clinical autonomic self-regulation treatment
program. On the other hand, this lack of experimental control and
consistency makes it impossible to know which ingredients were most
important in contributing to the clinically meaningful reductions in
diastolic blood pressure in the Biofeedback/Self-Regulation treatment
group.

The present study suggests that essential hypertensive adults can
reduce diastolic blood pressure in a non-pharmacological treatment
program. However, it is unclear whether this was due to the biofeedback and/or relaxation procedures and/or greater number of sessions
and hence, time spent with a therapist.

Although Benson has reported successful anti-hypertensive results
with the Relaxation Response procedure, the present study does not
support these claims. This is particularly striking in light of the
fact that Benson reports being able to teach the Relaxation Response
in one session and significant blood pressure reductions, often in
only four weeks (Benson, H., personal communication to Cohen J.,
8/19/80). In the present study, a therapist worked with the Relaxation Response group patients for 5 sessions over a 5 week period.
Although patients reported that they often felt relaxed and found the
continued meetings with the therapist supportive, there were no clinically significant blood pressure reductions. In light of the fact

that patients in this group evidenced a blood pressure reduction trend for as long as they were seeing a therapist, the present results may support the notion that extended training is useful and necessary for producing large magnitude blood pressure reductions (Williamson and Blanchard, 1979).

The present results do not support the clinical usefulness of the Relaxation Response. However, some patients apparently are able to use this procedure (Benson, 1975) and future research will need to address the question of what types of people will be able to profit from a general relaxation technique and whether the Relaxation Response generally produces clinically significant blood pressure reductions as opposed to statistically significant reductions.

This study did include several desirable elements of design that had been absent from many prior studies. The practice of assessing blood pressure in a time, place and with personnel away from the clinical setting (i.e., daily measurements by the patients themselves and biweekly "blind" nurse assessments) was intended to allow more valid evaluations of the carryover effects of blood pressure. Without the evaluation of carryover of effects, the patient and his physician might erroneously assume that the patient was protected because blood pressure lowering was conditioned to specific settings and occured when blood pressure lowering maneuvers were actually practiced. Patients in the present study were instructed not to practice their home practice exercises before blood pressure measurements. These control procedures have been absent from all but a few hypertensive self-regulation studies (Benson et al, 1974; Pollack et al, 1977; Frankel et al, 1978) and may have partially accounted for the clinically positive results in the Patel and North study (1975).

The assessment of patient compliance with home practice has generally been absent in previous studies with the exception of the work of Taylor, (1977), Kristt and Engel, (1975), and Frankel (1978). Clinicians typically describe patient compliance as an important and necessary factor that contributes to successful learning. Patients in the two active treatment groups in the present study did not significantly differ in terms of how often they practiced the home exercises during the ten week program and for four weeks after the treatment program. Interestingly, in a recent study that did not result in meaningful blood pressure reductions, systematic assessment of patient compliance revealed a 91 percent average level of compliance (i.e., they reported carrying out 91 percent of scheduled home practice exercises). In the present study, biofeedback group patients reported a 66 percent and in Relaxation Response Group patients, a 58 percent average level compliance. In the present study, and similar to Frankel and his coworkers (1978) findings, there were no correlations between amount of home practice and amount of blood pressure reduction. Thus, it may not be how many times a patient practices that is important in successful blood pressure reduction, but rather, how the

time is spent. This study utilized both an active attention control
group (the Relaxation Response Group) and a no-treatment control group
as aids in differentiating non-specific from treatment effects. In
studies with no control groups, spontaneous blood pressure decreases
might be interpreted as a result of the active intervention of the
treatment personnel. The active attention control group, however, can
only be considered a partial control for the impact of therapist
contact because (as noted above) the Relaxation Response Group
patients saw a therapist fewer times than the Biofeedback Group
patients.

This study also included a follow-up evaluation. Relatively few
studies have included long-term follow-up evaluations which provide an
opportunity to evaluate the true clinical significance of a treatment
program (Frankel, 1978; Williamson, 1979; Green, 1980). It is interesting to note in the present study that although diastolic blood
pressure is still significantly lower four months after the completion
of the treatment program, there is a slight drift upwards.

This study is one of the first to evaluate motivational factors.
All patients reported that they were highly motivated to learn and
practice relaxation procedures before the treatment program began.
There was no significant difference between the treatment groups in
terms of Locus of Control attitudes and motivation to become engaged
in the treatment programs. Hence, to the extent that these measures
rate motivation, the blood pressure reduction in the Biofeedback Group
cannot be attributed to this dimension alone. There was not a significant post-treatment difference in terms of how much effort (an
"internal-unstable" factor) they reported putting into the treatment
program. As noted above, percent of home practice compliance (another
indirect measure of motivation) was not significantly different between the groups. There were no significant correlations between
amount of blood pressure reductions and amount of effort put into the
program or percentage of home practice compliance. Thus, in the
present study, motivational factors did not appear to influence or
interact with blood pressure reductions. On another attitudinal dimension, there was no difference between the two treatment groups in
terms of how much ability (an "internal-stable" factor) they felt they
had to learn and practice the relaxation procedures. This is somewhat
surprising in light of the fact that the Biofeedback Group demonstrated greater ability to reduce blood pressure. An important difference between the Biofeedback and Relaxation Response treatment
groups was that patients in the former treatment group had the opportunity to perceive immediate physiological changes and thus become an
active agent in the process of change. Thus, it is interesting and
somewhat surprising that the Biofeedback Group did not perceive themselves as having greater ability to perform the relaxation exercises
than the Relaxation Response control group. One may speculate that

the Relaxation Response group rated themselves as relatively high on this dimension because of a "naive overestimation" of skills when there is no direct monitoring of relevant physiological information, whereas the Biofeedback Group would more correctly estimate the acquired skills.

Previous research (Weiner, 1972) suggests that when individuals feel more of an active agent, they report increased motivation to exert greater effort to produce the desired effect. This was apparently not the case in the present study. The two groups did not significantly differ in terms of how difficult (an "external-stable" factor) they felt the relaxation exercises to be at the beginning, end, and follow-up points of evaluation. However, there was a trend in the direction of the Biofeedback Group experiencing the exercises to be slightly less difficult at the end of the treatment program. One may speculate that this was due to the fact they received more practice and support from therapists and/or biofeedback and as a result acquiring relaxation skills more rapidly than patients in the Relaxation Response group. Patients in both active treatment groups believed that luck (an "external-unstable" factor) did not play a role in their treatment program experience. These findings are consistent with past research that shows that success makes people feel that the task is less difficult, and that luck did not play a significant role in their learning. To more fully understand the role of these attitudinal factors (ability, effort, difficulty and luck) future research will need to select individuals who are fairly high and low in regard to their pre-treatment attitudes and examine what effect they have on learning.

SUMMARY

In summary, our overall findings do support the practical clinical usefulness of a combination of EMG and thermal biofeedback used in conjunction with various relaxation/self-regulation procedures within a context of a short-term (20 sessions) treatment program for essential hypertensive adults. In contrast to other studies that showed substantial within session reduction of blood pressure with a nonpharmacological treatment program, this study demonstrated a carry-over of effects toward other situations. Furthermore, the significant diastolic blood pressure reductions were still evident four months after completion of the treatment program and were accompanied by significant reductions in anti-hypertensive medication in 43 percent of the biofeedback group patients. Motivational factors that were measured did not seem to influence or be related to blood pressure reduction.

REFERENCES

Benson, H., Beary J. H., and Carol, M. P., 1974, The relaxation
    response, Psychiatry, 37:37–46.
Benson, H., 1975, "The Relaxation Response," William Morrow, New York.
Blanchard, E. B., and Young, L. D., 1973, Self control of cardiac
    functioning: A promise as yet unfulfilled, Psychol. Bull.,
    79:145–163.
Carlson, J. G., 1977, Locus of control and frontal electromyographic
    response training, Biofeedback Self-Reg., 2(3):259–271.
Frankel, B. L., Patel, D. J., Horwitz, D., Friedewald, W. T., and
    Gaarder, K. R., 1978, Treatment of hypertension with
    biofeedback and relaxation techniques, Psychosom. Med.,
    40:276–293.
Gatchel, R., 1975, Change over training sessions of relationship
    between locus of control and voluntary heart rate control,
    Percept. Motor Skills, 40:424–426.
Goesling, W., May C., Lavont, D., Barnes, T., and Carrerira, C., 1974,
    Relation between I/E locus of control and operant conditioning
    of alpha through biofeedback training, Percept. Motor Skills,
    39:1339–1343.
Green, E., Green, A., and Norris, P., 1980, Self-regulation training
    for control of hypertension, Primary Cardiol., 6(3):126–137.
Hager, J. L., and Surwit, R. S., 1978, Hypertension self-control with
    a portable feedback unit or meditation-relaxation,
    Biofeedback Self-Reg., 3:269–276.
Heider, F., 1958, "The Psychology of Interpersonal Relations," John
    Wiley, New York.
Johnson, R. K., and Meyer, R. G., 1974, The locus of control construct
    in EEG alpha rhythm feedback, J. Consult. Clin. Psychol.,
    42:913.
Kristt, D. A., and Engel, B. T., 1975, Learned control of blood
    pressure in patients with high blood pressure, Circulation,
    51:370–378.
Miller, N. E., 1975, Clinical applications of biofeedback: Voluntary
    control of heart rate, rhythm and blood pressure, in: "New
    Horizons in Cardiovascular Practice," H. I. Russek, ed.,
    University Press, Baltimore, 239–249.
Patel, D., 1973, Yoga and biofeedback in the management of
    hypertension, Lancet, 2:1053–1055.
Patel, D., and North, W. R. S., 1975, Randomized controlled trials of
    yoga and biofeedback in management of hypertension, Lancet,
    2:93–95.
Paul, G. L., 1969, Behavior modification research: Design and tactics,
    in: "Behavior Therapy: Appraisal and Status," C. M. Franks
    ed., McGraw Hill, New York.
Peek, C. J. A., 1977, A critical look at the theory of placebo,
    Biofeedback Self-Reg., 2:327–335.
Plotkin, W. B., 1979, The alpha experience revisited: Biofeedback in
    the transformation of psychological state, Psychol. Bull.,
    86(5):1132–1148.

Pollack, A. G., Weber, M. A., Case, D. B., and Laragh, J. H., 1977,
    Limitations of transcendental meditation in the treatment of
    essential hypertension, Lancet, 1:71–73.
Ray, W. J., 1974, The relationship of locus of control, self-report
    measures and feedback to the voluntary control of heart rate,
    Psychophysiol., 11:527–534.
Ray, R. W. J., Raczunski, J. M., Rogers, T., and Kimball, W. H.,
    eds., 1979, "Evaluation of Clinical Biofeedback," Plenum
    Press, New York.
Rotter, J. B., 1966, Generalized expectancies for internal vs.
    external control of reinforcement, Psychol. Mono.,
    80(609):1–28.
Schwartz, G. E., and Shapiro, D., 1973, Biofeedback and essential
    hypertension: current findings and theoretical concerns,
    Sem. Psychiatr., 5:493–503.
Sedlacek, K., and Cohen, J., 1978, Biofeedback treatment of essential
    hypertension, (Abstract) Biofeedback Self-Reg., 3:2.
Seer, P., 1979, Psychological control of essential hypertension:
    Review of the literature and methodological critique, Psychol.
    Bull., 86(5):1015–1043.
Shapiro, A. P., Schwartz, G. E., Ferguson, D. C., Redmond, D. P., and
    Weiss, S. M., 1977, Behavioral methods in the treatment of
    hypertension: a review, Ann. Int. Med., 86:626–636.
Solomon, S., Hotchkiss, E., Saravay, S., and Bayer, C., 1979,
    Impairment of cognitive function by antihypertensive
    medication, (Abstract),Psychosom. Med., 41(7):582–583.
Surwit, R. S., and Shapiro, D., 1976, Biofeedback and meditation in
    the treatment of borderline hypertension, Psychosom. Med.,
    38:54.
Surwit, R. S., Shapiro, D., and Good, M. I., 1978, Comparison of
    cardiovascular biofeedback, neuromuscular biofeedback, and
    meditation in the treatment of borderline essential
    hypertension, J. Consult. Clin. Psychol., 46:252–263.
Taylor, C. B., Farquhar, J. W., Nelson, E., and Agras, S., 1977,
    Relaxation therapy and high blood pressure,
    Arch. Gen. Psychiatr., 34:339–342.
Wagner, C., Bourgeois, A., Levenson, H., and Denton, J., 1974,
    Multidimensions locus of control and voluntary control of GSR,
    Percept. Motor Skills, 39:1142.
Weiner, B., Frieze, I., Kukla, A., Reed, L., Rest, S., and Rosenbaum,
    R. M., 1971, "Perceiving the Causes of Success and Failure,"
    General Learning Press, New York.
Weiner, B., 1972, Causal ascriptions and achievement behavior: A
    conceptual analysis of effort and reanalysis of locus of
    control, Journal Per. Social Psychol., 21(2):239–248.
Weiner, B., 1972, "Theories of Motivation: From Mechanism to
    Cognition," Markham Pub., Chicago.
Williamson, D. A., and Blanchard, E. B., 1979, A review of the recent
    experimental literature, Biofeedback Self-Reg., 4:1.

BEHAVIOURAL STYLE, MYOCARDIAL INFARCTION,

AND CARDIOVASCULAR RESPONSIVENESS

David A. Hill

Reader, Health Sciences
Ulster Polytechnic
Belfast, Northern Ireland

Tom Cox

Director, Stress Research
University of Nottingham
Nottingham, England

INTRODUCTION

The study of the effects of rehabilitation by exercise of post coronary males at Craigavon Hospital, County Armagh, (Hill, 1979) investigated changes in various circulatory variables, and changes in anxiety as measured by the IPAT questionnaire. This study used a double crossover design, in which patients attended for rehabilitation exercises for eight weeks, and functioned as control subjects for a further eight weeks. Throughout the sixteen weeks of treatment, observation, and testing the group of twenty four patients showed significant reductions in anxiety scores ($p < 0.01$). The reductions, however, were not related significantly to the exercise rehabilitation periods of eight weeks, and it could not be assumed that the rehabilitation programme had been responsible for this reduction. Furthermore, the group norms for anxiety were below the norm for British males, and it therefore seemed unlikely that anxiety as measured by the IPAT questionnaire should be considered a major factor in the assessment and treatment of patients who had sustained a myocardial infarction (See Table 1). It was believed that an investigation into the relationships between behavioural styles and coronary heart disease (CHD) would be more informative.

Table 1. Anxiety Scores as a Function of Testing Period

		IPAT Anxiety Scores		
Group		Initial	Intermediate	Final
Early exercise Group	Mean	28.9	25.5	23.6
First 8 weeks	Dev	13.9	12.1	14.1
Late exercise Group	Mean	21.8	19.9	19.0
Second 8 weeks	Dev	8.7	9.5	8.3

The components of the type A coronary prone behaviour pattern have been frequently documented. The type A pattern was first implicated in the occurrence of coronary heart disease by Friedman and Rosenman (1959). This pattern is characterised by excessive amounts of competitiveness, aggressiveness, hostility, impatience, and time urgency evoked by a range of environmental circumstances (Rosenman, Friedman and Straus, 1964). Since then, both prospective and retrospective studies have established the association of the type A pattern with clinical manifestations of CHD, independently of the standard risk factors of smoking, diet, and lack of exercise. (Brand, Rosenman, Sholtz and Friedman, 1976; Friedman and Rosenman, 1959; Jenkins, 1976; Rosenman, Brand and Jenkins, 1975). Knowledge of the type A pattern was shown to be a slightly stronger predictor of recurring myocardial pathology than any other single predictor variable available (Jenkins, Jyzanski and Rosenman, 1976).

Attempts at intervention with the aim of changing the type A pattern have produced beneficial changes in physiological and psychological variables, but the durability and generalisability of these changes have not yet been fully assessed (Roskies, Spevak, Surkis, Cohen and Gilman, 1978). Psychoanalytic approaches and behaviouristic approaches were attempted, and the behaviouristic methods, which included Jacobson techniques of relaxation, were found to be more effective.

METHOD

The decision was taken to test for behavioural differences between a group of male patients who had sustained a myocardial infarction, and a control group not suffering from any known cardiovascular disorder.

Twenty two male subjects admitted to Coronary Care Units at three major Belfast hospitals from December 1980 to February 1982 formed the experimental group. The control consisted of twenty two indigenous males, the two groups being matched for age.

The instrument used for comparison was the Nottingham Activity Survey (NAS) questionnaire developed by Tom Cox and others at the Stress Research Centre in the Department of Psychology, University of Nottingham (Cox, Watts and Barnett, 1981). The NAS was originally designed to investigate behavioural styles and orientation to work. It was based on the studies carried out on coronary prone behaviour by Rosenman and Friedman and by Jenkins in the United States. The NAS was developed from the Jenkins Activity Survey (JAS), specified for a British working population. The major modifications have been the inclusion of questions related to leisure and group activities, as well as to work activities. Factor analytical studies of the NAS have indicated the existence of three factors, drive (D), social orientation (O) and irritability/time urgency (Ir).

RESULTS

Table 2 shows the norm and standard deviation scores of the experimental and control groups, and of a large general population. No significant differences were demonstrated between the two groups on any of the factors, although the difference for drive (D) approached significance. Coronary patients, however, can be very concerned not to appear different from other individuals (Arlow, 1945). This may be reflected in responses to questions concerning the patients image of himself. A search was therefore conducted in an attempt to reveal some more obscure differences which might be contained in the questionnaire responses.

Table 2. Norms and Standard Deviations of the NAS Scales

Scale	General Population N = 489		Myocardial Infarction N = 22		Matched Control N = 22	
	Norm	SD	Norm	SD	Norm	SD
Drive	20.7	5.8	22.5	7.0	19.7	7.0
Social Orientaion	10.7	4.3	10.8	4.2	11.7	3.1
Irritability	13.1	4.1	14.4	3.0	13.5	4.3
Myocardial Infarction	——	——	13.0	2.4	9.5	2.7

As a preliminary strategy the questionnaire items were rank ordered on the basis of differences in score between the two groups. The ten items with the largest discrepancies in scores were used to form a fourth myocardial infarction (MI) scale. This scale, as expected, discriminated significantly between the two groups. The items were then studied individually and a conceptual analysis undertaken in order to develop a profile of a high scorer on the MI scale. It became apparent that this prototype MI scale revealed a combination of two types of behavioural description, as shown below.

> (a) Is easily annoyed.
>   Takes work seriously.
>   Sets work deadlines.
>   Volunteers for added responsibilities.
>   Does not take time over jobs.
>   Does not slow down when tired.
>   Tries to do more than one job at once.

> (b) Enjoys sitting and chatting to fellow worker at
>   tea breaks and lunch times.
>   Regards free time as a chance for a complete rest.
>   Does not like doing anything else when watching
>   television.

CONCLUSION

The items under (b) are not normally included in descriptions of the type A pattern. The high scorer on the MI scale is dedicated to leisure as well as work. This feature has emerged because of the inclusion in the NAS of items relating to leisure and group activity. The addition of a particular profile of non work behaviour could provide an important advance in the diagnosis of the type A pattern, and in the efficacy of intervention programmes.

FURTHER WORK

Work is now proceeding to investigate relationships between the NAS scales and responsiveness of the circulatory system. Various forms of stress are being applied to volunteer students at Ulster Polytechnic, these students having previously completed a student version of the NAS. As yet the MI scale has not correlated with any circulatory parameters, but the scales for Drive and Irritability are showing some significant correlations with heart rate and blood pressure.

Table 3. Correlations of NAS Scales

Scales	Correlation	p
MI : Ir	0.42	< 0.01
MI : O	−0.35	< 0.01
O : Ir	0.10	------

It is also worth noting that the MI scale correlates significantly and positively with irritability (Ir), and significantly and
negatively with orientation (O) in this student group. This would
tend to confirm the view that the items of maximum difference between
the original patients and controls were not random occurrences. (See
Table 3).

The NAS is still in the development phase, but we believe in
its usefulness as a measuring instrument.[*]

* _____
Thanks are extended to Mrs. G. Dick for her valuable secretarial
   help.

REFERENCES

Arlow, J. A., 1945, Identification mechanisms in coronary occlusion,
      Psychosomatic Medicine, 7:195–209.
Brand, R. J., Rosenman, R. H., Sholtz, R. I., and Friedman, M., 1976,
      Multivariate predictors of coronary heart disease in the
      Western Collaborative Group Study compared to the findings of
      the Framingham Study, Circulation, 53:348–355.
Cox, T., Watts, C., and Barnett, C. A., 1981, "The experience and
      effects of task–inherent demand. Final report to USARI
      (European Office)," University of Nottingham, Nottingham.
Friedman, M., and Rosenman, R. H., 1959, Association of specific overt
      behaviour pattern with blood and cardiovascular findings,
      Journal Am. Med. Assoc., 159:1286–1296.
Hill, D. A., 1979, Physical Activity, Mood and Anxiety in Normal and
      Post–Coronary Males, in: "Stress and Tension Control,"
      F. J. McGuigan, W. E. Sime, and J. M. Wallace, eds., Plenum,
      New York.
Jenkins, C. D., 1976, Recent evidence supporting psychologic and
      social risk factors for coronary disease, New Engl. Journal
      Med., Part I, 294:987–994, Part 2, 294:1033–1038.

Jenkins, C. D., Jyzanski, S. J., and Rosenman, R. H., 1976, Risk of
        new myocardial infarction in middle-aged men with manifest
        coronary heart disease, Circulation, 53:342-347.
Rosenman, R. H., Friedman, M., and Straus, R., 1964, A predictive
        study of coronary heart disease: The Western Collaborative
        Group Study, Journal Am. Med. Assoc., 189:15-22.
Rosenman, R. H., Brand, R. J., and Jenkins, C. D., 1975, Coronary
        heart disease in the Western Collaborative Group Study: Final
        follow up of 8.5 years, JAMA, 233:872-877.
Roskies, E., Spevak, M., Surkis, A., Cohen, C., and Gilman, S., 1978,
        Changing the coronary-prone (Type A) behaviour pattern in a
        nonclinical population, Journal Behav. Med., 1.2:201-216.

TENSION CONTROL IN SPEECH PATHOLOGY

Robert L. Casteel

Speech-Language Pathologist
Portland State University
Portland, Oregon

TENSE WORDS - TENSE MUSCLES

The significance of word symbols is that much of our own behavior
is governed by our internal speech, how we talk to ourselves. Our own
words are our Pavlovian bells. In the delightful words of Fink
(1953), "The uninformed stress the voluntary aspects of behavior. But
among those who know psychology best, it is habits two to one." A
vocabulary of tension teaches tension. The person with a vocabulary
of tension is in trouble if it is true that words are our map in time
and place. Let us analyze some common maps drawn by the overly tense.

> "I only need to get a grip on myself."
> "On my job, I am involved in a physical struggle for existence."
> "Boy, I'm scared stiff."
> "You give me a pain in the neck."
> "He is a headache."
> "Get off my back."
> "I hope I don't get cold feet."

Thus, with Pavlovian bells going off to the left of us, to the right
of us and in our very midst, we maintain a stiff upper lip (to mention
one muscle group) and charge after the truism that anyone can do
anything if he only tries hard enough. Panic sets in only when we
find this "physically" impossible. Brodnitz (1967) in his article on
"Semantics of Voice" wrote, "Thus language may become a source of
misunderstanding in human communication and may lead to inappropriate
behavior."

Fortunately, relaxation may be taught as well. When we learn that all we have to fear is tension itself, then we are ready to loosen our safety belt. If we have been a "sit tight" personality that is unchanging and unbending, it is time to reevaluate and to let go. To permit the body to function as it was designed to function with fluidity and ease.

It is therefore the purpose of this paper to address the themes (language) of physiological responsibility and how semantic reorientation relates to modifying stuttering and functional voice disorders. By way of amplification, how we talk about stress and tension will explain how intervention in stuttering and functional voice disorders can be achieved through a tension reduction model.

THEMATICS

Inappropriate tension as the result of stress frequently is a key element in maintaining communicative disorders. As a speech-language pathologist who has spent most of a professional career working with people who stutter and those with voice disorders, I can say that over-tensing is focal to the disorders.

If the client is to modify his/her behavior, it is beneficial to talk about responsibility and choice. Often the client perceives himself as a helpless victim of events over which he has no control. He sees himself as in a confrontation with words or sounds which (he believes) requires him to attack with great energy in order to overcome his impediment. In truth, much of the problem is the very energy he brings to the speaking task.

To facilitate language change, Casteel and Stone (1982) developed a model of Parameters of Behavioral Intervention Thematics which was greatly influenced by the works of Sanders (1971) and Williams (1957). The model was made up of four language categories with antithetical descriptors: Undesirable Language Responses (ULR) and Desirable Language Responses (DLR). The categories dealt with the individual's 1) description of behavior, 2) responsibility for behavior, 3) attitudes about performance and 4) awareness of behaviors used in the speaking task. When analyzing what the person was saying, we declared as ULR's those which were obscure, negative, passive, denying self regulation, pessimistic or demonstrating ignorance of appropriate response patterns. By contrast those which were judged to be DLR's were composed of appropriate description of physiologically based behavior (I pushed too hard with my tongue) which was stated in the active voice in a positive construct (I can touch lightly and move my tongue sooner). Additionally, we looked for language which reflected awareness of the behavior being brought to the task which is the product of the individual's choice (I tightened my throat muscles). Finally we looked for optimism (I can do the things I need to do to

talk) and knowledge of specific behaviors which result in success
(permitting the air to flow) or failure (over-tensing my vocal folds).

With this model it was possible for us to assess both the instructions of the clinician and the self instruction of the client.
Did they recognize what they were doing with their muscles? Were they
aware that more could be achieved if they brought less struggle between agonistic and antagonistic muscle groups? Were they trying to
make air flow or were they permitting air to flow? Could they pin-
point where they were bringing inappropriate tension which interfered
with vocal production? Did their attitude reflect knowledge of coping
behaviors they could use? Clues to these questions and others were
reflected in how they talked about a communication disorder. If they
were to become self managers of their communication acts, they would
need DLR's to guide them in making appropriate choices and to help
them take responsibility for what they were doing. For, what they
were doing determined success or failure.

VOICE DISORDERS

Hyperfunctional dysphonia is the single largest group among all
voice disorders (Brodnitz, 1967). The culprit is how we use the
muscles of the larynx. We over-constrict. Perkins (1968) has written
that appropriate constriction in voice production is no constriction.
Many of these hyperfunctional voice disorders end in lesions of the
vocal folds; however, a more subtle hyperfunctional disorder is functional aphonia or dysphonia. In this latter disorder there is no
visable lesion and the mechanism appears to be functioning normally
upon visual inspection.

In developing a program to reinstate voice in the functional
dysphonia population, Stone and Casteel (1982) noted that tension was
present in the vocal track even when not using nor attempting to use
full voice. The strain could be heard in sustained exhalation or in a
voiceless sigh. The first task was to reduce inappropriate tension in
noncommunicative vocal production which was preceded by relaxation.
Basic training in progressive relaxation was paired with modelling and
instruction in the language of responsibility which laid a backdrop
for reinstating "neuromuscular coordination of the larynx and associated structures" (Stone and Casteel, 1982).

A two-way grid was laid out with a vocal skills progression from
least to greatest degree of muscle activity on the abscissa and a
speaking situtation hierarchy for stress on the ordinate. At the
bottom of each column was a "speaking" situation with least likelihood
of a hyperfunctional response. The hierarchy varied from "sustaining
sound" through eleven intervening steps culminating in "talking on the
phone". The degrees of muscle activity varied through sigh, whisper,
breathy, full voice and vocal variety.

Through this systematic approach the clients were able to reinstate a normal "set" (Boone, 1983) for phonation. It was as if at
some time in the past the client used greater effort to talk and this
effort became the way to talk. For example, the client might say "I
have to tighten my throat to make sound" or "If I don't push hard
nothing will come out for it gets stuck". Here we see a language of
tension to justify struggle. There is no evidence that this is physiologically based information nor does it suggest a choice on the part
of the speaker. It is as if something "has happened" to the person.
What has "happened" to the person is that he has adopted a counterproductive speaking set.

STUTTERING

In stuttering, the more the person struggles the more severe the
disorder is considered to be. Stuttering consists of breath holding,
overtightening of the vocal folds, pushing too hard with the tongue,
holding the lips together in a tight, pursed manner or <u>in other ways</u>
<u>investing too much energy in what is normally an effortless, coordin-</u>
<u>ated exhalation of air</u>. The coordination demands voicing, then reso-
nation and finally articulation. When over-tensing is brought to the
speech act, there is a breakdown in the synergy due to over-
constriction. It is very much the same as driving with the
brakes on in a car.

Once again, for the person who has stuttered all of his life, it
is as if something is happening to her and she feels she must struggle
to talk. When analyzing the language of people who stutter, one can
note the responses are vague which suggests a poor understanding of
the behavior known as stuttering. In fact there is a denial of responsibility for speech behavior in general. Casteel and McMahon
(1978) developed a program which teaches finer and finer discriminations between what one does to be fluent and what one does to be
dysfluent. It takes into consideration the motoric difference between
stuttering and permitting air to flow for speech. Since fluent speech
is a finely coordinated act which can be disrupted by stress and
tension, the program begins with progressive relaxation. This is
followed by a four stage management program which removes stress and
tension from the speaking situation. Stage I takes place with the
clinician in a quiet room. The clients slow their speech rate, while
maintaining a breathy quality and reducing articulation pressure. The
clients soon discover that by such sacrifices they have a choice which
permits them to be fluent. They not only learn to describe what they
do with their speech mechanism when they stutter, but also what they
do to be fluent. This is a discrimination task which they refine over
time in four stages.

In Stage II the clients demonstrate to themselves that when they do enough things correctly, they can increase their speaking rate to normal. By Stage III they are capable of more subtle discriminations which permit them to be less breathy and more precise in their artic- ulation while maintaining fluency. By the time they are proficient in Stage III, they are able to maintain fluency in medium and high stress situations outside the clinic. Ultimately, in Stage IV, they talk in all situations easily and effortlessly when they make the appropriate choices and take responsibility for their speaking behavior.

In 1981, Mathew completed a thesis on subjects seen in the Portland State University stuttering program. She looked at the language of responsibility used by the clients in a follow-up ques- tionnaire several years after dismissal. DLR's of clients who did maintain fluency significantly above baseline severity were compared with DLR's of clients who relapsed toward baseline severity. The subjects' increasingly greater gain (rho of +.64) was moderately related to overall desirable language usage. That is to say they used more descriptive language, took more responsibility for their behavior and were more optimistic as to the choices available to them for promoting fluent speech.

SUMMARY

In this brief paper, we have looked at the interaction of tension with the language of responsibility, voice disorders and stuttering. It appears that when change needs to take place, if we commit our- selves to a scientific analysis of what we are doing, we can make changes, and when we persist in making appropriate, reality based changes, the task becomes easier through practice.

REFERENCES

Boone, D. R., 1983, "The Voice and Voice Therapy, 3rd Edition," Prentice Hall Inc., Englewood Cliffs.

Brodnitz, F. S., 1967, Semantics of the voice, Journal of Speech and Hearing Disorders, 32:325-330.

Casteel, R. L., and McMahon, J., 1978, The modification of stuttering in a public school setting, Journal of Childhood Communication Disorders, 2:6-17.

Casteel, R. L., and Stone, R. E. Jr., 1982, Intervention in childhood voice problems: intervention thematics, in: "Phonatory Voice Disorders in Children," M. Filter, ed., Charles C. Thomas, Springfield, 166-180.

Fink, D. H., 1953, "Release from Nervous Tension," Simon and Schuster, New York.

Mathew, K. F., 1981, An analysis of the relationship between the
        degree of maintained fluency improvement of former Portland
        State University stuttering clients and the overall language
        themes they used. Unpublished master's thesis, Portland State
        University.
Perkins, W. H., 1968, Optimal vocal functioning, <u>Journal of
        California Speech and Hearing Association</u>, 48-51.
Sander, E. K., 1970, Talking plainly about stuttering: guidelines for
        the beginning clinician, <u>Central States Speech Journal</u>,
        21:248-255.
Stone, R. S., Jr. and Casteel, R. L., 1982, Restoration of voice in
        nonorganically based dysphonias, <u>in</u>: "Phonatory Voice Disorders
        in Children," M. Filter ed., Charles C. Thomas, Springfield,
        132-165.
Williams, D. E., 1957, A point of view about 'stuttering'. <u>Journal of
        Speech and Hearing Disorders</u>, 22:390-397.

SELF-HELP GROUPS FOR AGORAPHOBICS:

THEIR ROLE IN COPING WITH ANXIETY AND DEPRESSION

David Hodgson and Keith Oatley

Laboratory of Experimental Psychology
University of Sussex, Brighton, England

"A group is the beginning of everything," wrote Peter Uspenskii, "one man can do nothing, attain nothing... a group of people can do what one man can never do" (Uspenskii, 1950).[*]

Over the past decade Uspenskii's belief in the value of self-help groups, expecially in the area of health care, has attracted increasing popular support – a recent study (Webb, 1982) identified more than 2,000 such groups in the London area alone. At the same time, research into the structure, dynamics and effectiveness of such groups remains limited. However, as Robinson (1980) commented: "Self-help has received surprisingly little systematic analysis either at a broad general level or at the level of particular groups and activities." We have been studying the role and therapeutic efficacy of self-help groups concerned with helping agoraphobics and comparing their outcomes with the results achieved by using a recently developed, home based, treatment programme administered under professional guidance.

In this paper we discuss the part played by self-help groups in treating stress related health problems and consider the types of mutual aid available to agoraphobics. Preliminary data from our study of these groups will be presented.

---

[*] Editor's note: This is open to a variety of interpretations, depending on the context. For instance it has also been written about groups that "you can search all the parks in each and every city but you'll never find a statue for a committee". Self-help groups, though, may be another thing.

THE FUNCTION AND STRUCTURE OF SELF-HELP

Self-help health groups largely attract their members from people who consider themselves ignored, neglected, forgotten or rebuffed by conventional systems of care. In many instances they arise out of a disillusionment with established helping services and from changing ideas concerning what medicine is able to achieve. Prior to joining such groups, members usually consider themselves to be in some way "defective" and therefore "inferior" to other people - a view frequently shared by the wider society. They seek to achieve their goals by establishing fellowships of individuals united by their "differences," whether these are mental, physical, psychological or social.

Marie Killilea (1976) notes seven features which characterise the self-help group:

(1) Common experience of the members.
    Groups aid individuals to cope with long-term defects, deprivations and life cycle transitions (Robinson and Henry, 1977).

(2) Mutual aid and support.

(3) The helper principle.
    Members commonly help one another by "doing" but it also widely recognised that the helper usually benefits the most from the exchange. He, or she, is less dependent, has the opportunity of observing his, or her, own feelings at a distance and benefits from the sense of social usefulness and comfort drawn from the thought "I must be well if I can help" (Robinson and Farrell, 1980).

(4) Differential association.
    All self-help groups have a tendency to destigmatize their members by mutual reinforcement of self-concepts of normality.

(5) Collective willpower and belief.
    Members validate one another's feelings and attitudes.

(6) Information resources.
    The group collects and passes on expertise, both at the level of practical coping skills and general knowledge.

(7) Constructive action towards shared goals.
    The concept of learning and change through doing. Effective groups will be action-orientated.

THE VALUE OF SELF-HELP

Many writers such as Hurwitz (1970), and Cole (1983), are un-stinted in their praise for self-help groups but we lack much data on their functioning.

SELF-HELP AND PERSONAL CHANGE

The factors influencing individual change within a group can be grouped under three headings; emotional, cognitive and behavioural (Barish, 1971; Borkman, 1976).

Emotional

Members are offered unconditional care and concern, understand-ing, warmth and empathy. They help members make the transition from the margins of society toward more normative social roles or toward a more normal definiton of a deviant social status (Cole, 1983).

Cognitive

They provide a new reference group which brings about a reduction of alienation and anomie by validating the members sense of self-worth.

Behavioral

They can give advice, share equipment, exchange coping skills and become an arena for social experimentation.

It is important to keep these three points in mind when consider-ing the role, and the effectiveness, of the three different types of self-help groups for agoraphobics described below.

THE NATURE OF AGORAPHOBIA

Contrary to popular belief, and the definition offered by several standard medical references, ("Psychiatric Dictionary," Fifth Edition; "Dorland's Medical Dictionary," 22nd Edition) agoraphobia means a fear not of open spaces but of places where people gather (Gr: agora – meeting place), and was so named by the German neurologist Carl Westphal in 1871.

The "American Diagnostic and Statistical Manual of Mental Disorders," Third edition, (1980) (DSM-III) provides the following diagnostic criteria:

A. The individual has marked fear of and thus avoids
    being alone or in public places from which escape
    might be difficult or help not available in the
    case of sudden incapacitation, e.g., crowds, tun-
    nels, bridges, public transportation.

B. There is increasing constriction of normal activi-
    ties until the fears or avoidance behaviour domin-
    ate the individual's life.

C. Not due to major depressive episode, obsessive
    compulsive disorder, paranoid personality disorder
    or schizophrenia.

The prevalence has been estimated at 0.5% of the population
(Agras et al, 1969). Terhune (1949), Errera and Coleman (1963) and
Marks (1969) suggest that the proportion of psychiatric patients
diagnosed as phobic is between 2-3%, the majority being agoraphobic.
About two thirds of these are women (Marks, 1970; Terhune, 1949).

The age of onset among the clinical population is variously
reported as 24 years (Marks and Gelder, 1965); 28 years (Burns and
Thorpe, 1977); 29 years (Marks and Herst, 1970); and 31 years (Buglass
et al, 1977).

It is clear, however, that the disability usually starts early in
adult life, onset being very uncommon in childhood and increasingly
rare after the age of 40 (Burns and Thorpe, 1977). Marks and Gelder
(1966) point out that it contrasts with the childhood origin of most
simple phobias and the onset of most social phobias in later adolescence. The starting point is often reported as a panic attack,
arising seemingly spontaneously and without warning. "The onset may
be so acute that the patient returns home from the first attack of
panic, stays indoors - sometimes in bed - for days, and thereafter has
great difficulty leaving the house" (Mathews, 1981, p. 13).

In time, fear may be aroused not only by being away from home but
even when alone in the house, or in certain rooms of the house. Some
sufferers can, however, travel almost any distance - with little or no
anxiety - provided they are accompanied by a trusted companion,
spouse, teenage son or daughter, parent, close relative or a friend
who knows about their condition.

The majority of patients see their General Practitioners and many
seek help from other specialists both within and outside medicine. In
a survey of 1,200 agoraphobics belonging to the Open Door Association,
Marks and Herst (1970) found that of the 95% who had seen their GP's,
67% had been referred for psychiatric help. A further 15% had sought
help from ministers of religion or spirit healers. These findings are
supported by the results of our own survey of three different types of
self-help organisation.

AGORAPHOBICS AND SELF-HELP

Agoraphobics make special demands on both conventional medicine and self-help organisations as a result of their inability to travel far from home on their own. Attending a meeting, even when all those present are fellow sufferers, can prove an insurmountable obstacle for many.

Given these difficulties, and the geographical isolation of many agoraphobics, it is not surprising that the only two national self-help organisations currently active in the UK – The Open Door Association and the Phobics Society – provide their services almost exclusively via the mail. We have termed such organisations "Postal" groups.

The remaining self-help groups come into two main categories. There are those which offer a clearly structured programme and regard their primary function as teaching members such well-established therapeutic procedures as relaxation, goal setting, and in vivo desensitisation. We have termed these "Active Training" groups. Finally there are those offering mostly companionship, social support and mutual reassurance. We have called these "Social" groups.

Where training is offered, it consists largely of relaxation sessions – often by means of commercially recorded cassette material – and a general encouragement to go out more. The social nature of the groups is emphasised by visiting speakers, who often talk on subjects far removed from agoraphobia, and expeditions to places of local interest. Such groups tend to have a relatively limited membership, around 100 – 200 in total with some 15 – 25 members attending meetings and to be locally based.

THE PRESENT STUDY

One purpose of this research is to determine whether any of these approaches – Postal, Active Training or Social – is more or less effective than any other in reducing anxiety, stress and depression while increasing the sufferer's mobility. New members of self-help groups in each of the categories listed above were contacted immediately on joining and asked to cooperate by providing the following information:

(i) A fairly detailed description of their agoraphobia, including information about its duration and nature, mode of onset, help received both within and outside the medical services prior to joining the group, attitude of their spouse and families and so on.

(ii)   A completed anxiety/depression questionnaire - The
       Leeds Scales for the Self-Assessment of Anxiety and
       Depression (Snaith et al, 1976).

(iii) A day-to-day diary in which they list any trips
       away from home which caused anxiety. Details here
       include duration of visit, whether alone or accom-
       panied, distance travelled and mode or transport,
       and feelings generated by the experience. Clients
       are instructed to keep records for a period of one
       week prior to embarking on any training programme,
       and these entries constitute a baseline of activi-
       ty. A method for scoring these diaries which
       relies, as far as possible, only on what they have
       written, has been developed. The intention is to
       make the diaries of people providing postal data
       comparable with those who are in therapy. A two
       part score is generated by the method. One part is
       a score of new items achieved on a quasi-hierarchy.
       This is intended to be approximately comparable
       with Mathews et al's (1981) 15 point scale of
       behavioral items gained. The second part is a
       score reflecting situational anxiety, derived from
       both the changes in excursions from safe places and
       the anxiety ratings in the diary for these
       excursions.

(iv)   In addition, those attending Active Training groups
       were asked to complete both the Leeds Scale and a
       specially designed Repertory Grid form (Kelly,
       1955). Their spouses were also requested to com-
       plete these forms.

Information from self-help group members is compared with similar data
obtained from female agoraphobics and their partners seen by the
authors on a one-to-one basis. These clients are being treated via a
home based programme developed by Mathews, Gelder and Johnston (1981).
In this approach the partner takes an active role in bringing about a
reduction of anxiety and encouraging progressive exposure to the
feared situations. Follow-up assessments are being carried out at
two, six and twelve months. The following groups are involved:

Postal Group

    The Phobic Society, founded in 1971, has around 2,500 members,
90% of them women. It was set up by a former agoraphobic and offers
members encouragement, advice and support by means of regular
"Progress Reports". These offer practical advice - for example de-

scriptions of relaxation procedures and tips for coping with specific
difficulties - and print letters from readers describing their experiences and successes. It does not organise official self-help groups
at a local level, although members sometimes meet informally, usually
in one another's homes, to discuss their difficulties.

Active Training Groups

     There is currently only one organisation offering a structured
training programme, Stresswatch, which has 25 local self-help groups
in different parts of the country. It was set up as a non-profit
making organisation in 1976. Unlike the majority of self-help organisations in this field, Stresswatch was established by two psychologists (David Hodgson and Robert Sharpe) rather than agoraphobics.
All the group leaders are, however, former agoraphobics. Leaders
receive training in behavioral methods, including relaxation and progressive desensitisation in vivo, but each group is largely autonomous
and members belong to their local group rather than a centralised
organisation. Training involves regular practice between meetings,
recording progress on charts and learning both general and specific
procedures for coping with anxiety. For example deep muscle, quick
and differential relaxation methods are taught, carefully graded hier-
archies of feared situations created and successes monitored. In some
groups progress charts are kept and members are awarded gold "stars"
which reinforce each attainment.

     In addition to the specific training programme, members are
encouraged to enjoy one another's company socially, and meetings end
with general conversation over coffee. Guest speakers are invited,
but these are almost always specialists in the relevant area - -
psychologists, psychiatrists and family doctors.

Social Group

     There is a great number of locally based self-help groups which
offer a variety of mutual support and skill training programmes.
Members meet, usually once a week, to discuss their progress, offer
mutual encouragement and occasionally learn a specific procedure such
as relaxation.

     Typical of such groups is the Way Out Club, based in Norwich,
England. Founded in 1976 by Nora Patchett, a former agoraphobic, it
has a membership of 143, 94% of whom are women. Meetings, which are
held once a fortnight, attract some 15 members. The term "Club" is
appropriate, since there is emphasis on social activities, including
talks from guest speakers on a wide range of topics, organised walks
and coach trips. Members are also encouraged to learn relaxation from
printed leaflets.

COMPARISONS BETWEEN GROUP MEMBERS AND CLINICAL PATIENTS

As a number of authors (e.g. Mathews et al, 1981; Thorpe and Burns, 1983) have observed, and as can be seen from Table 1, agoraphobics experience a very high level of generalized anxiety. This is indicated by the scores on the Leeds Specific Scale for Anxiety (SAA) in the right hand column. Although agoraphobic and anxiety symptoms fluctuate, it seems likely that the sufferers whose statistics appear in Table 1 have been highly anxious for an average of more than ten years. Many, but not all, are also either moderately or severely depressed - they are not merely situationally phobic. The great majority of members in all three types of groups under study (96%) had seen their family doctors, a figure close to that (95%) found by Marks and Herst (1970) in their survey of Open Door members. A further 65% had seen either a psychiatrist or a psychologist, or both. Again this figure is close to that (67%) found in the Open Door study.

Although 51% expressed themselves as satisfied with their family doctor's attitude and help, only 33% felt the same about the psychologists or psychiatrists they had been. This is not, perhaps, very surprising since self-help groups of this type are likely to become a repository for those patients who have not been helped by psychiatry.

Table 1. Mean age of reported onset of agoraphobic symptoms, age when entering psychological therapy or joining self-help group, and severity of generalised depression and anxiety. (Leeds Self-Assessment of Depression (SAD) and Anxiety (SAA) Specific Scales - maximum score on each 18).

Source	Number of class	Age of onset	Age entering therapy or club	Leeds SAD	SAA
Hospital Out-Patients (i)	30	31	38	–	–
Open Door (ii)	1200	29	42	–	–
Phobic Society	35	27	38	11	15
Way Out Club	14	24	38	7	12
Stresswatch	27	27	41	10	14
Ind. Therapy	22	28	45	9	13

(i) Buglass et al, 1977.   (ii) Marks and Herst, 1970.

Psychiatric pharmacotherapy is not successful in treating this group of patients, although some symptomatic relief may occur (cf. Mathews et al, 1981) and patients naturally perceive that lack of success. Complaints also included a lack of sympathy on the part of, especially, psychiatrists (74%), the infrequency of therapy sessions (46%) and the anxiety aroused by having to attend a hospital (65%).

A sense of shame at having to visit a mental hospital was also mentioned by a number of agoraophobics (32%). The higher level of satisfaction with family physicians is interesting in view of the fact that most did no more than prescribe benzodiazepines and anti-depressants. Expectations are that the physician will provide symptomatic relief, but not be able to offer any long term solutions.

How successful are voluntary organisations in helping people deal with the anxiety, stress and depression associated with agoraphobia? Our longitudinal data are preliminary at present, since not all the members within each group have completed their first reassessment. The results from some members of the Phobic Society and some of the clients receiving the home based treatment programme are shown in Table 2.

Evidently joining the Phobic Society is associated with a significant decrease in depression which is not accompanied by any decrease in anxiety. The suggestion that such a reduction is due to the sense of "fellowship" noted by Robinson (1980) and the hope that, through membership, the agoraphobia will be alleviated, is supported by the views of members at this stage; 90% reported feeling "reasonably satisfied" with their membership of the society while 65% commented they no longer felt so isolated. It may also be, of course, that new enterprises such as joining a society, only become possible at a time then the depression is starting to lift. In any case depressions are known to resolve while anxiety symptoms fluctuate but are more chronic.

Table 2. Mean decreases in the scores on the Leeds Self-Assessment of Depression (SAD) and the Leeds Self-Assessment of Anxiety Specific Scales (SAA) for Phobic Society members and individual clients.

Group	Number of Cases	SAD	SAA
Phobic Society	10	3.7 **	1.4 n.s.
Ind. Therapy	18	0.8 n.s.	2.3 **

** Difference significant p < .005
Wilcoxon Matched Pairs Signed-Rank Test. One tailed.

A further point is that joining such a Society offers the hope of change without making any specific demands on the individual. She, or he, is encouraged to make more trips away from home but not given any specific instructions to do so, nor is behavior monitored in any way. Thus they are unlikely to feel under any real pressure to make changes in her or his lifestyle and can enjoy the hope of change without experiencing any of the inevitable frustrations and set-backs associated with a more active and directive programme. This explanation is strengthened by an analysis of diaries kept by Phobic Society members which show no significant increases in journies away from home over the baseline measure.

With clients seen on the one-to-one home based treatment programme there was no significant decrease in depression two months after the start of therapy (See Table 2). The anxiety of these clients, although still high, shows a significant decrease over baseline measure (p < .005, one-tailed) while an analysis of diary entry data indicates that the mean number of items gained on a behavioural hierarchy was comparable to that found by Mathews et al (1977).

The conclusion suggested by these preliminary findings is that by offering fellowship and a measure of hope to agoraphobics, without putting them under any obligation to venture out more, it is possible to reduce - at least in the short term - their depression. This approach does not, however, lead to any change in either behaviour or anxiety.

In self-help groups which provide a more structured training programme and require their members to work actively towards reducing agoraphobic restrictions, one might expect a pattern of response similar to that found with clients receiving treatment on a one-to-one basis.

Depression, at least in the early months of such programmes, is likely to remain high as clients meet the challenges imposed by the need to bring about major changes in their way of life. Increased mobility will be accompanied by a high, if declining, level of generalised anxiety. Under these conditions the need to successfully motivate members, so that they continue with training despite the associated negative affect, will clearly be crucial. Whether such motivation can be provided most effectively by the spouse, as in a home-based treatment programme, or by fellow sufferers within a self-help group remains to be determined.

REFERENCES

Agras, W.S., Sylvester, D., and Oliveau, D., 1969, The epidemiology of common fears and phobias, Comprehensive Psychiatry, 10:151-156.

Barish, H., 1971, Self-help groups, Encycl. Soc. Work, 2:1163.

Borkman, T., 1976, Experiential knowledge: A new concept for the analysis of self-help groups, Social Science Review, 50:445.

Buglass, D., Clarke, J., Henderson, A. S., Kreitman, N., and Presley, A. S., 1977, A study of agoraphobic housewives, Psychological Medicine, 7:73-86.

Burns, L. E., and Thorpe, G. L., 1977, The epidemiology of fears and phobias with particular reference to the national survey of agoraphobics, Journal of International Medical Research, Supplement (5), 1-7(a).

Cole, S. A., 1983, Self-help groups, in:"Comprehensive Group Psychotherapy", H. I. Kaplan, and B. J. Sadock, eds., Williams and Wilkins, Baltimore.

Errera, P., and Coleman, J. V., 1963, A long-term follow-up study of neurotic phobic patients in a psychiatric clinic, Journal of Nervous and Mental Disorders, 136:267-271.

Hurwitz, N. T., 1970, Peer self-help psychotherapy groups and their implications for psychotherapy, Psychotherapy: Theory Research and Practice, 7(1):41-49.

Killilea, M., 1976, Mutual help organisations: interpretation in the literature, in: "Support Systems and Mutual Help: Multidisciplinary Explanations," G. Caplan, and M. Killilea, eds., Grune and Stratton, New York, p. 37.

Kelly, G. A., 1955, "The Psychology of Personal Constructs," Norton, New York.

Marks, I. M., 1969, "Fears and Phobias," Heinemann, London.

Marks, I. M., 1970, Agoraphobic syndrome (phobic anxiety state), Achives of General Psychiatry, 23:538-553.

Marks, I. M., and Gelder, M. G., 1965, A controlled retrospective study of behaviour therapy in phobic patients, British Journal of Psychiatry, 111:561-573.

Marks, I. M., and Gelder, M. G., 1966, Different ages of onset in varieties of phobias, American Journal of Psychiatry, 123:218-221.

Marks, I. M., and Herst, E. R., 1970, A survey of 1,200 agoraphobics in Britain. Features associated with treatment and ability to work, Social Psychiatry, 5:16-24.

Mathews, A. M., Gelder, M. G., and Johnston, D. W., 1981, "Agoraphobia Nature and Treatment," Tavistock Publications, London.

Mathews, A. M., Teasdale, J., Munby, M., Johnston, D., and Shaw, P. A., 1977, A home-based treatment program for agoraphobia, Behaviour Therapy, 8:915-924.

Robinson, D., 1980, Self-help groups, in: "Small Groups and Personal Change," P. B. Smith, ed., Methuen, London.

Robinson, D., and Farrell, C., 1980, Kings Fund Project Paper No. RC8.

Robinson, D., and Henry, D., 1977, "Self-help and Health," Martin Robinson and Co. Ltd., London.

Snaith, R. P., Bridge, G. W. K., and Hamilton, M., 1976, The Leeds scale for the assessment of anxiety and depression, British Jounal of Psychaiatry, 128:156-165.

Terhune, W. B., 1949, The phobic syndrome: A study of eighty-six
        patients with phobic reactions, <u>Archives of Neurology and
        Psychiatry</u>, 62:162-172.
Uspenskii, P. D., 1950, "In Search of the Miraculous," Routledge and
        Kegan Paul, London.
Webb, P., 1983, "Ready, Willing But Able? The Self-Help Group,"
        (In press).
Webb, P., 1982, "Back to Self-Help," (privately published),
        Merton, Sutton and Wandsworth Area Health Education Service.
Wolfenden Committee, 1975, "The Future of Voluntary Organisation,"
        Croom Helm, London.

THE PERCEPTION BY DENTISTS AND PATIENTS OF

FEAR IN DENTAL TREATMENT

S. J. E. Lindsay

Psychology Department, Institute of Psychiatry
   and
Royal Dental Hospital
University of London, London

The fear of dental treatment is widespread. In the United Kingdom over 40 per cent of adults delay visits to dentists because of the fear of treatment (Todd, Walker and Dodd, 1982). Fifty-four percent of adults visit a dentist only when in trouble (Todd and Walker, 1980), 29 percent of these being too afraid to seek attention more frequently (Todd et al, 1982). Thus the fear of routine dental care is one of the most significant factors in preventing the efficient delivery of dental care. This is partly responsible for chronic dental pain suffered by such people; two to five million working days and one million nights of sleep are interrupted by this problem (Sheiham and Croog, 1982).

Although fear of treatment deters many people from seeking dental care, this source of distress deserves solution in its own right. The terror experienced by many patients in the dental chair is itself a serious problem. Behavioural scientists have now established many procedures for treating and managing fears, by behaviour therapy and medication (Mavissakalian and Barlow, 1981) and these with some limitations (Lindsay and Woolgrove, 1982), should be making their impact on dental fear.

The fear of dental treatment is interesting also from a theoretical point of view. The factors which encourage the persistence of fear are especially evident in dentistry. The persistence of fear being characteristic of anxiety about dental treatment. People remain afraid in spite of their experiencing, time after time, no pain or other distress in dentistry (Lindsay, 1983). The fear of dental treatment, being so persistent, provides opportunities for investigating issues which may be relevant to their fears.

213

The techniques which have been developed for treating fears have mainly required that the patient be referred to a specialist qualified in behaviour therapies. Unfortunately most dentists are unaware of sources of behavioral specialists help for frightened patients. It would thus be appropriate to consider those procedures which can be practised by dentists themselves and which do not add significantly to dentists' work load or to the time spent in giving dental treatment. There is already much pressure on dentists to complete treatment quickly.

It is evident that the strain which is involved in managing nervous patients is very high in general dental practice which is one of the most stressful professional occupations (Cooper, 1981). Hence any procedure which can reduce this source of the dentist's workload is highly desirable. Unfortunately dentists receive very little education as undergraduates or postgraduates in the management of fear in patients.

The techniques which have become most popular among dentists are those which are:

1) easy to administer and

2) require minimal time to apply.

Unfortunately these have been pharmacological treatments such as relative analgesia (oxygen – nitrous oxide sedation) and oral diazepam (Lindsay and Roberts, 1980; Yates and Lindsay, 1983).

Although relative analgesia often acts quickly in reducing the distress of very nervous patients, it does not help all patients in this way (Lindsay, 1982). In addition, there is very little scientific evidence that relative analgesia can help very anxious adults accept dental treatment (Lindsay and Roberts, 1979).

There is, therefore, good reason for further examination of these treatments and for establishing additional ones which can be implemented successfully by dentists. Towards this end, the following studies sought to identify those expectations in adult patients which contribute to their continuing to be afraid of treatment. The investigations also sought to determine how accurately dentists can anticipate the experience of patients during treatment and how well dentists' predictions match those of their patients.

In this study one hundred adults were asked to describe what they would usually expect to experience during a number of dental procedures; some common ones (e.g., drilling and injections), others less familiar (e.g., wisdom tooth extractions). The subjects were given a number of rating scales referring to specific sensations which they might experience "drilling, sharp, pinching, tugging, hot, tingling, sore, tender, tiring, wretched and miserable." These had been derived following pilot studies with the McGill Pain Questionnaire. Scales were also provided to determine the subjects' expectations about the certainty with which they would anticipate these sensations. They were also asked to estimate the apprehension and discomfort which they could expect.

The dental procedures were the following: polishing, wisdom extraction, incisor extraction, infiltration injection, nerve block injection, impressions, scaling, fast drill, slow drill, matrix band and molar extraction.

Forty dentists in one dental teaching hospital in London completed one similar questionnaire to estimate the sensations, apprehension and discomfort which they expected the average patient would experience during dental treatment with the same eleven procedures.

It was found that dentists and patients anticipated similar patterns of sensations for dental treatments but for some, the dentists expected less intense sensations than did the patients. For the most alarming and most intensely felt procedures such as extractions and drilling the patients attributed stronger sensations than did the dentists. This was not evident for less threatening and milder dental operations (See Figure 1).

One notable difference between dentists and patients was observed in the pattern of sensations attributed to similar procedures. Patients expected similar sensations for infiltration injections and

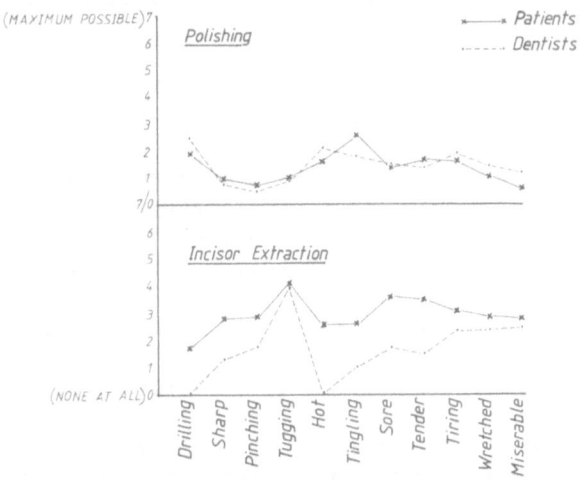

Figure 1. Sensations in Dental Treatment Anticipated by
Dentists and Patients.

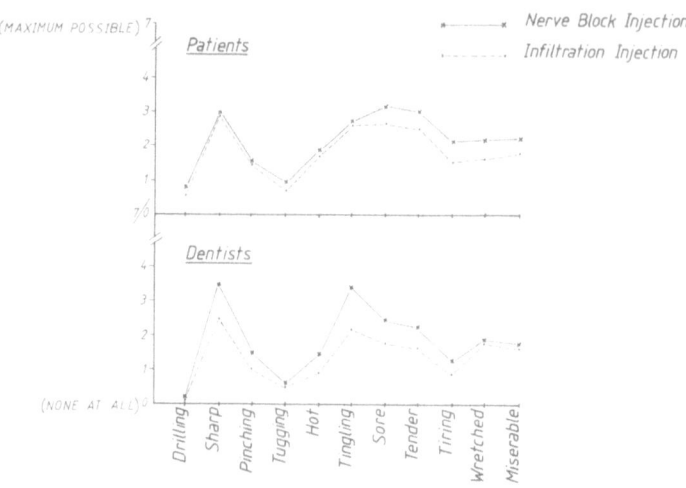

Figure 2. Sensations in Nerve Block and Infiltration Injections
Anticipated by Patients and Dentists.

mandibular nerve block injections as shown in Figure 2. Dentists on
the other hand associate stronger tingling and sharp sensations with
nerve block injections than with infiltrations. This may reflect
dentists' knowledge of the anatomy and physiology of these injections.

For the other procedures it was not clear whether dentists or
patients gave the more realistic estimates. It was possible that
dentists, for the most unpleasant procedures, are insensitive to the
intensity of the experience undergone by their patients. On the other
hand, patients may not have a realistic idea of the sensations involved in these treatments.

To check on the accuracy of the patients' ratings for injections
and fast drilling a second sample (30 patients) was asked to rate
immediately before treatment the sensations expected for these treatments. Immediately afterwards they described what they had experienced on identical scales. Almost all subjects experienced less
intense sensations than they expected (See Figure 3). Similar discrepancies were noticed for apprehension and discomfort also.

It was possible that the patients, experiencing the relief of
emerging from treatment, underestimated the intensity of the experience. However, their ratings before treatment, completed when they
would have been anxious about their experience, were very similar to
those anticipated by the earlier sample who gave their ratings at
least one day before the time of a dental appointment. It is probable, therefore, that dentists have more realistic ideas about the
intensity of sensations in dental treatment than do patients.

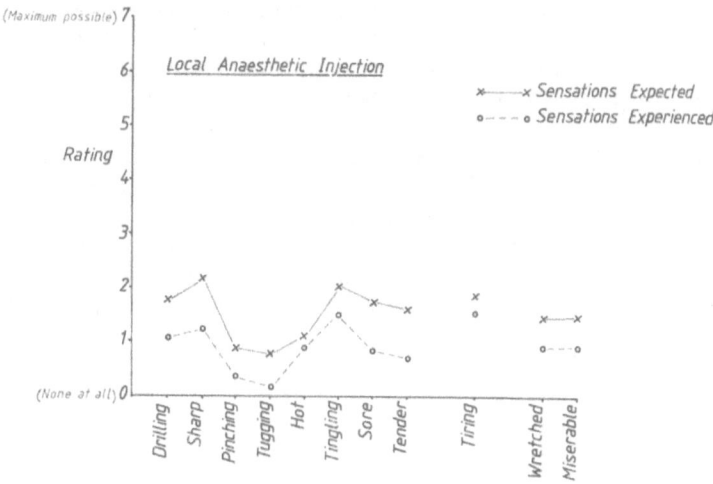

Figure 3. Sensations Expected and Experienced by
Patients in Treatment.

When the dentists estimated the apprehension which they assumed
patients experience, they appeared to anticipate greater fear than did
the patients themselves. However, the differences were not statistically different (Figure 4). The dentists and patients agreed also
about the amount of discomfort anticipated for dental treatment.

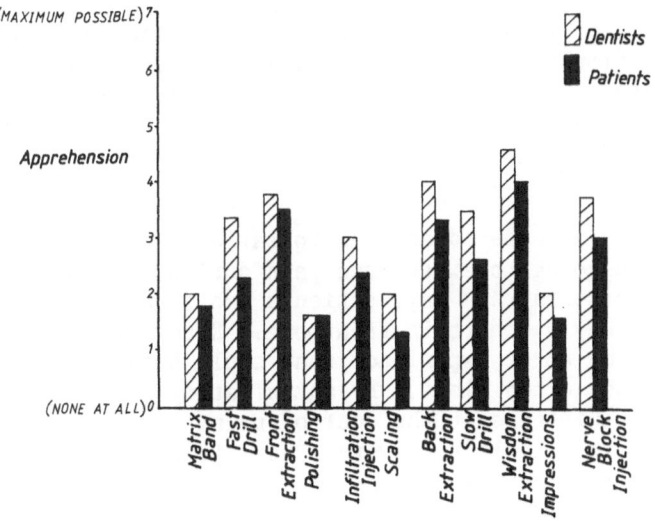

Figure 4. Apprehension Expected by Dentists and Patients.

In summary, it would appear that dentists have a more realistic idea about the level of sensations involved in dental treatment but both dentists and patients overestimate the fear and discomfort which is expected for most dental treatment.

It is clear that the anticipation of treatment is more distressing than the experience. This is surprising because the subjects in this investigation would have had much experience of routine treatment. Thus it appears that patients continue to expect more discomfort, more fear and more intense sensations than they experience in spite of many positive experiences, i.e., less discomfort than expected.

It is hardly surprising that dentistry is stressful for dentists when they anticipate along with their patients that treatment will be more alarming and more uncomfortable than it turns out to be.

For the patients, the more intense are the sensations expected, and the greater the discomfort anticipated, the more apprehensive they become. In addition, the more uncertain the subjects were about the sensations they expected, the more apprehensive they would be about undergoing treatment.

There are probably several other reasons why fear of treatment persists unrealistically in patients (Lindsay, 1983). Laboratory research has shown that people are more apprehensive when they are anticipating unpleasant events which rarely happen than when they are waiting for such stimuli which occur more frequently. Electric shocks have been the stimuli used in this research. In addition, subjects in such experiments overestimate the frequency of strong electric shocks. With weaker electric shocks subjects make more realistic predictions. Overestimates of likelihood are greatest for the most infrequent, most unpleasant stimuli. These demonstrations offer clear parallels with dental treatment where most adults can recall having undergone at least one painful experience in treatment (Lindsay, unpublished data) but where such events are very infrequent, a conclusion confirmed in this study.

There may even be advantages to patients in their anticipating that they are going to experience pain during treatment. By anticipating the worst in this way, patients avoid receiving an unpleasant surprise if they are hurt in treatment. Receiving an unexpected pain is probably the worst experience that could befall a nervous dental patient as they often claim. By anticipating pain these patients may reduce impact distress at the cost of being continually highly apprehensive. It is not clear, however, how widespread this protective but maladaptive reaction is.

It must be emphasized here that, although nervous patients very frequently expect to be hurt in treatment, it would be misleading to

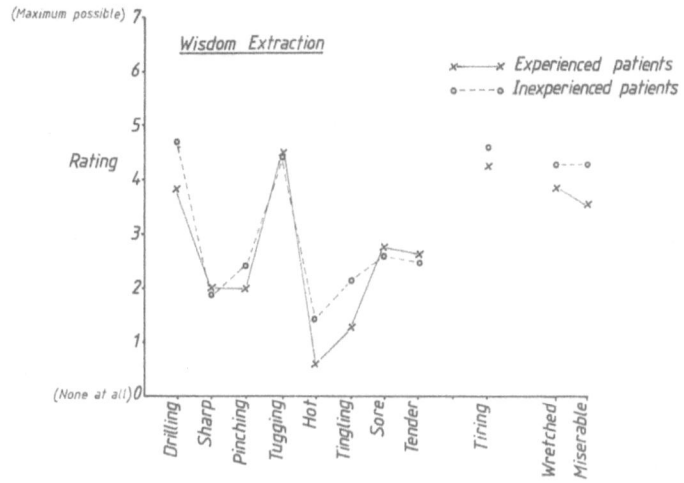

Figure 5. Sensations Expected by Experienced and Inexperienced
Patients for Wisdom Extractions.

conclude that they are nervous only because they have experienced pain
in the past. There are other factors which influence the fear of
treatment (Lindsay, 1983).

Patients can, for example, anticipate the pattern of sensations
which are involved even in procedures which they have not experienced
such as extractions, as shown in Figure 5, and they are correspondingly afraid.

What can be done to reduce patients' inappropriate expectations
about treatment and their uncertainties about these expectations?

A number of laboratory studies (Lindsay, 1983) have indicated
that if subjects are prepared beforehand by being told what sensations
they can expect during a pain tolerance task, they will experience
less distress during the task itself than they would without that
information. This may not be helpful, however, to the most apprehensive subjects who may become even more apprehensive by this procedure.
Continuous information from the dentist during treatment is helpful.
Thirty patients were offered here information during local anaesthetic
injections and drilling (Lindsay, 1973). the patients were warned
that the injections would produce a "sharp" sensation followed by
"tingling" and "numb, cold" feelings. During drilling they were
warned to expect "drilling and tingling sensations". Four patients
did not want to be given that information. Nevertheless, they were no
more apprehensive than the remaining patients. All but two of them

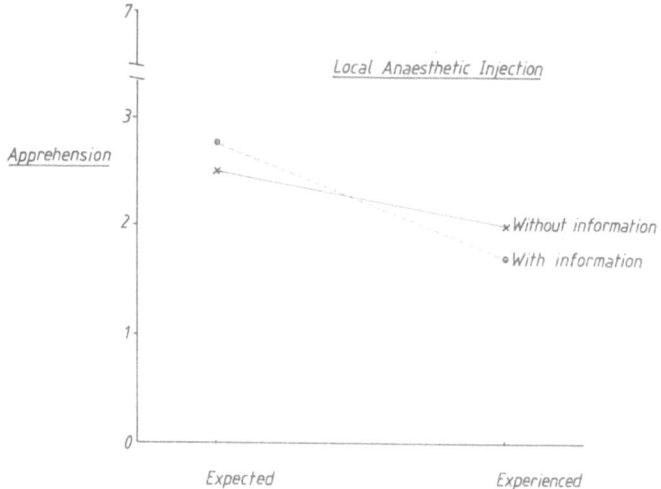

Figure 6. Effects of Sensation-Information on the Reduction
of Apprehension in Treatment.

said immediately after treatment, that the information had been helpful. However, they experienced significantly no less apprehension or
discomfort than a control group who did not receive such information,
as can be seen in Figures 6 and 7. Ratings of sensations were also
not affected.

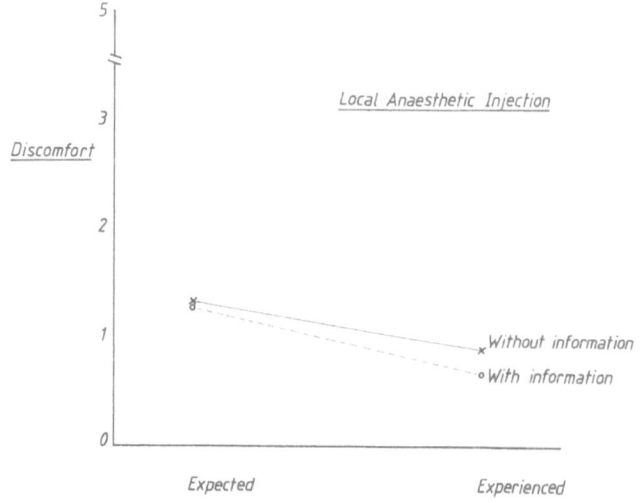

Figure 7. Effects of Sensation-Information on the Reduction
of Discomfort in Treatment.

It was encouraging to note that even the most nervous patients experienced less apprehension and discomfort than they expected. Indeed, this discrepancy was a great or even greater than that reported by the moderately anxious subjects. Studies are now being conducted to determine what other cues may have occurred in the control condition which may have limited the distress there also.

It is possible that the sensation information reduced expectations about sensations which were not covered by the assessment procedure used here. A pilot study indicated that besides the sensations which were identified by our questionnaire, some subjects expected many others such as throbbing, stinging and shooting sensations during nerve block injections.

It is possible that the patients' knowledge that the dentist was to inform them continuosly about treatment, made them less apprehensive beforehand. That would be very desirable since it is in the pretreatment period where patients are most apprehensive.

The present studies do, however, indicate that adult patients are fairly certain about what sensations to expect in treatment. Even for unfamiliar procedures they attribute over 70 percent likelihood to the sensations which they correctly anticipate. Regarding future studies, the exact time of the onset and withdrawal of syringes and drills may be even more important than the information provided in this study.

REFERENCES

Cooper, C. L., 1980, Dentists under pressure, in: "White Collar and
    Professional Stress", C. L. Cooper and J. Marshall, eds.,
    Wiley, London.
Lindsay, S. J. E., 1982, Is Relative Analgesia effective?, Presented
    to: British Paededontic Society, Bath, United Kingdom.
Lindsay, S. J. E., 1983, The Fear of Dental Treatment: a critical and
    theoretical analysis, in: "Contributions to Medical Psychology,
    III," S. Rachman, ed., Pergamon, Oxford.
Lindsay, S. J. E., and Roberts, G. J., 1979, Does relative analgesics
    work? British Dental Journal, 147:206.
Lindsay, S. J. E., and Roberts, G. J., 1980, Methodology for research
    on dentally anxious children: the example of relative analgesia,
    British Dental Journal, 149:175.
Lindsay, S. J. E., and Woolgrove, J., 1982, Fear and pain in
    dentistry, Bulletin of the British Psychological Society,
    35:275.
Mavissakalian, M., and Barlow, D. H., eds., 1981, "Phobia: Psycho-
    logical and Pharmacological Treatment," Guilford Press,
    New York.

Sheiham, A., and Croog, S. H., 1981, The psychosocial impact of dental
        diseases on individuals and communities, Journal of Behavioural
        Medicine, 4:257.
Todd, J. E., and Walker, J., 1980, Adult Dental Health in England and
        Wales, H.M.S.O., London.
Todd, J. E., Walker, J., and Dodd, P., 1982, Adult Dental Health 2,
        H.M.S.O., London.
Yates, J., and Lindsay, S.J.E., 1983, The efficacy of oral diazepam in
        nervous child dental patients (unpublished paper).

STUDIES OF STRESS AND TENSION IN OCCUPATIONAL SETTINGS

OCCUPATIONAL STRESS TESTING IN THE REAL WORLD

Wesley E. Sime, Bronston T. Mayes, Hermann Witte,
Daniel Ganster and Gerald Tharp

University of Nebraska
Lincoln and Omaha, Nebraska

INTRODUCTION

The role of emotional stress in the etiology of numerous chronic diseases has been clearly established (Hoiberg, 1982). Coronary heart disease, in particular, has the most profound accumulation of literature supporting the causal effect of emotional stress upon atherosclerotic changes, as well as signs and symptoms in the form of angina pectoris hypertension and coronary vasospasm (Eliot, 1979; Eliot and Sime, 1980). Numerous other studies have shown a relationship between emotional stress (measured in several different ways) with hypertension, irritable bowel syndrome (including ulcers and colitis), and Raynaud's syndrome (Ford, 1982). Further evidence of the pathological consequences of emotional stress stems from the literature showing that quite a number of functional disorders are treated successfully with a variety of stress management interventions including biofeedback, progressive relaxation, and autogenic training.

Until recently there was very little specific information regarding the exact nature of the provocative factors that make up the overall milieu of stress. The most useful information in this regard comes from experimental evidence in laboratory investigations or from the contrasting field research done in large populations usually associated with some form of overwhelming disaster (Taylor and Frazer, 1982). The literature dealing with stress following disasters such as fire, flood, tornado, and wartime is frought with numerous methodological problems. The biggest concern from the design standpoint is the fact that populations who experience and survive a disaster can in no way be compared to a randomized control population (after the fact) with respect ot occurrence of stress-related disorders. Thus, in such an arbitrarily selected group, one has no apparent control over the

variables which might determine the susceptibility or vulnerability
due to genetic resistance or weakness. In addition, the disaster
population may experience varying levels of stress based upon differing perceptions of loss. For example, a person who has experienced
significant loss of material goods (home, business, etc.) may perceive
this to be a terrible misfortune if material goods seem to be highly
valued items. On the other hand, a person who does not care much
about material possessions (when experiencing the same significant
loss of those possessions) might not feel it to be any great misfortune at all. Similarly, the life change events which include death of
a spouse, might have variable interpretation ranging from: 1) a relief
for the victim who is suffering to 2) a tremendous loss of a cherished
and much needed victim. Another problem with disaster or life change
event studies is the fact that they must be studied in retrospect. In
no way can the researcher make arrangements ahead of time and be
prepared to analyze such populations and most assuredly the researcher
has no experimental influence over how and when the disaster experience is to be invoked. The disaster research to date, represents
pre-experimental learning of an empirical nature which provides suggestive evidence laying the groundwork for more carefully controlled
and properly designed studies.

Laboratory Stress Research

The use of laboratory studies is obviously a far more controlled
environment for dealing with stress measures. The population is well
controlled and can be randomized. The stress or stimuli is well-
controlled and can be invoked in a temporal sequence so as to evaluate
the resultant strain effects of varying intensities and time duration
(Sime, Buell and Eliot, 1980a). In addition, laboratory studies
afford the researcher the luxury of multi-faceted observation by
monitoring behavioral, physiological and biochemical changes which are
occurring simultaneously during the stressful experience. In carrying
out such studies there are relative advantages and the disadvantages
of human versus animal research. With animals there is much greater
freedom from ethical issues regarding the actual stress induction or
the use of deception by implication or with the anticipation of threat
or danger. Similarly the use of noxious and stress stimuli, such as
severe shock, trauma and chemical disruption is necessarily and appropriately frowned upon by human subjects protection committees. Thus,
the limitation inherent in laboratory research is the fact that the
"relevance" of the stress or experience is usually greatly impaired.
It is extremely difficult to simulate a "real-life" stressor.
Attempts have been made to use laboratory simulation to create frustration, anger or hostility in situations of extreme competition or
performance against a standard. These have proven fairly productive
with respect to physiological change and personality behavior pattern
characteristics in identifying patients at risk of re-infarction
(Sime, Eliot and Buell, 1980b).

Field Studies on Stress

In the systematic experimental investigation of occupational stress phenomena, it is neither appropriate to utilize occasional disaster situations (job loss or traumatic change in job responsibilities) nor to use the laboratory controlled environment. In effect, the job setting becomes the best possible laboratory for experimental investigation. Thus, taking laboratory equipment into the field setting and testing employees at the job site seems to be the most useful and relevant manner of investigation for this purpose. Baseline measurements can be obtained during the periods of the day when employees would be likely to experience peak moments of situational stress. Additional information can also be gained from the use of simulated task demands during which time physiological parameters can be monitored. The most useful measurement for baseline purposes during the occupational working hours is obtained in an unobtrusive manner through the use of urine sampling for the observation of catecholamines (epinephrine and norepinephrine).

Other important measures to be obtained in the occupational work setting would include personality variables, attitudes towards the working conditions, and the employee behavior pattern (Type A) and self-reported perceptions of occupational strain. These self-report measures may then be correlated with more objective measures of metabolic strain as provided with the physiological and biochemical measures.

In summary, it would appear as though the most useful measures of occupational stress testing in the real world would include field testing on the actual job site with employees who have the consent and approval of the employer for anonymity to provide personal information and documented physiological and biochemical characteristics. These measurements taken at the job site would appear to have greater relevance than actual laboratory studies while maintaining somewhat better control than the disaster studies described above.

The outcomes of two separate occupational studies are presented herein. These represent two very different occupational subgroups with varying job characteristics, one with predominantly female employees and the other with predominantly male employees.

STUDY ONE

Subjects

The study population consisted of a group of 79 employees, predominantly female (92%) with an average age of 39 years and a mean education level of 14.5 years. The participants were recruited from a larger population of 230 employees of a public agency charged with delivering social services and benefits to welfare clients in the community. The average length of employment among the participants was four years and the participant job classification ranged from secretary to the agency director with representation throughout all levels and functions.

The work task of the participants generally included determining eligibility for benefits, counseling and handling emergency cases. Workers typically cited heavy workloads, inadequate resources, and frequent crises as the primary sources of stress. In addition, employees were faced with an arbitrary delay in salary raises which fueled further aggravation with the administrators and worsened the work climate.

Noting the employee concerns, the administrators welcomed an opportunity for research in sources of occupational stress and found also that employees were very interested in assistance with their work-related stress problems. The researchers recruited participants through an informational seminar which outlined job stress versus health outcomes. Following the assessments described herein the employees were provided with an experimental stress management program, the results of which has been described previously (Ganster, Mayes, Sime and Tharp, 1982).

Criterion Measures

Data collection for dependent variables to represent indicators of occupational strain were itemized according to the classification components: psychological, physiological and somatic.

The psychological strains consisted of self-reported anxiety, depression and irritation. A 20-item scale by Spielberger, Gorsuch and Lushene (1970) was used for assessment of chronic (trait) anxiety. Depression and irritability were assessed by scales from Caplan, Cobb, French, Van Harrison and Pinneau (1975) using six-item and three-item measures, respectively.

The physiological strains were assessed according to urine catecholamine values and several physiological (vital signs) variables. Catecholamine assessment focused upon epinephrine and norepinephrine, which are known metabolic indicators of elevated sympathetic nervous activity in the face of some form of physical or emotional (perceived) demand or threat. While urine catecholamine measures provide a rough assessment of prior (2-6 hour) accumulation of sympathetic activity, the physiological measure including heart rate, blood pressure, electrodermal response (palmar sweating) peripheral blood flow (skin temperature) and pulse wave volume provide an acute assessment of the current state reflecting the degree to which laboratory assessment, close personal interaction and cognitive challenge are interpreted to be threatening or strain producing. By conducting the physiological assessment on premises of the job site and presenting typical personal interaction with some cognitive challenge, we presume that relevance to occupational strain assessment is achieved. During the personal interview, each employee was assessed for Type A behavior pattern characteristics. This quasi-psychological (behavioral) variable was used together with self-report measures of the same for Type A according to the Jenkins Activity Survey (Jenkins, Rosenman and Zyzanski, 1974).

The somatic complaints as indicators of strain included such items as : headaches, dizziness, nausea, muscular discomfort and pain, sweaty palms, cold hands and flushing of the face. These are some of the typical signs and symptoms commonly associated with emotional stress.

Data Analyses

Data analyses in this preliminary study were focussed upon the association between Type A behavior pattern (TABP) and the physiological measures of individual strain. TABP is currently of great interest regarding its predictive value in coronary heart disease and related physical symptoms thereof, i.e., angina, hypertension. Thus the validity coefficients (correlations) between TABP and physiological indicators of acute sympathetic nervous system would also carry meaning with extrapolation to the heart disease symptoms.

Table 1. Mean and Standard Deviation for Heart Rate, Systolic and Diastolic Blood Pressure, Skin Temperature, Skin Conductance (sweat), Pulse Wave Volume, Adrenalin, Noradrenalin Taken at Baseline Rest and at Peak During Stress in 79 Female Employees.

	Base	Peak	Change
Heart Rate (bts/min)	75 ± 13	91 ± 15	21% ± 15
Systolic Pressure (mm Hg)	117 ± 11	139 ± 14	19% ± 10
Diastolic Pressure (mm Hg)	74 ± 8	87 ± 9	18% ± 8
Skin Temperature (°F)	82 ± 6	80 ± 6	-3% ± 0.4
Skin Conductance (mmhos)	15 ± 3	40 ± 7	286% ± 62
Pulse Wave Volume (mm)	54 ± 21	16 ± 10	-68%± 22
Adrenalin (micrograms/100 ml urine)	1.3 ± 0.2		
Noradrenalin (micrograms/100 ml urine)	7.8 ± 1.5		

Results

The mean, standard deviation and calculated percent change at baseline rest and at peak during stress for all physiological and biochemical values are shown in Table 1. Heart rate and blood pressure at rest for this population are well within normal limits for the group-at-large, 75 bts/min, 117/76 mm Hg, respectively. In this predominantly female population, even the vital signs (heart rate, blood pressure) were well within normal limits, 91 bts/min, 139/84 mm Hg, respectively. The percent change values of approximately 20% among these variables are similar to that reported in other similar groups under stress conditions (Dembroski, McDougall and Lushene, 1979).

Skin temperature, as an indicator of peripheral blood flow, dropped during stress in appropriate comparison with the pulse wave volume (also an indicator of peripheral blood flow) which dropped in much greater magnitude (-68% versus -3%, see Table 1). However because of the close association between these two variables and because of the relative advantage of skin temperature assessment (absolute digital readings using portable equipment requiring no special interpretation) pulse wave volume is not used for future field investigations (Study Two) because it requires cumbersome equipment which has no digital output thus involving some subjective interpretation.

Skin conductance (palmar sweating response) is the variable which shows the most dramatic response to the stressor experience. Normal values at rest in subjects who are most relaxed and comfortable should be close to zero. Thus the average rest value of 15 micromhos indicates that the subjects were quite stressed just sitting quietly in the test environment. Then, with the mental arithmetic stressor the average increase was 25 micromhos (268%).

Urine catecholamine values for adrenalin and noradrenalin are reported also in Table 1. There is, however, no value during stress since urine output is not obtainable at frequent intervals.

Physiology and Behavior Pattern

Of greatest importance in this study is the identification of physiological mechanisms by which behavior pattern is associated with coronary heart disease. Previous studies have established the Type A heart disease link (Rosenman and Chesney, 1980).

In the present study several physiological variables (percent change) were found to be associated with behavior pattern. Behavior pattern, in this study, was assessed both by interview (Rosenman and Chesney, 1980) and by the Jenkins Activity Survey, (JAS), (Jenkins, Rosenman, Zyzanski, 1974). Heart rate was strongly associated with the content portion of the interview as well as with the accumulated

sum of all content and stylistic values (Continuous A). Similarly
skin temperature was strongly associated with these variables (r = -
0.27) as well as with the JAS and the 2-Category A rating. Blood
pressure was only modestly associated with behavior pattern (r = -0.26
with JAS, speed and impatience) and this was an unexpected inverse
relationship. On the other hand, skin conductance was strongly associated with 3 or the 5 interview measures of behavior pattern, perhaps
indicative of the acute changes that occur in that interpersonal
interaction during the interview. Noradrenalin (but not adrenalin)
was significantly associated with several of Type A measure from both
the interview and questionnaire (JAS) technique.

Table 2. Construct Validity Coefficients for Type A Behavior
Pattern Variable Together With Percent Change During Cognitive
Challenge for Physiological Reactivity in 63 Female Employees

Type A measures	Acute/Physiological Reactivity (Percent Change)					
	heart rate	skin temp.	syst. BP	dia. BP	skin cond.	Noradrenalin
** Interview **						
Stylistics A	12	-19	-15	-17	20	-17
Content A	30^{***}	-27^{**}	-01	03	38^{***}	-21
Continuous A	26^{*}	-27^{**}	-08	-07	35^{***}	-23^{*}
4-Category A	12	-22	-13	-24	31^{***}	-15
2-Category A	18	-23^{*}	-16	-07	16	-24^{*}
** Jenkins Activity Survey **						
Type A	04	-36^{***}	-17	17	00	-6
Speed & Impatience	-12	-14	-26^{*}	03	05	-18
Job Involvement	07	-13	-13	-12	15	04
Hard-Driving Competitive	07	-27	01	05	06	22

*p < 0.10 **p < 0.05 ***p < 0.01

STUDY TWO

Subjects

The study population consisted of employees from three different occupational settings. Two of these were public services agencies (police and fire department in a midwestern urban location, population = 200,000) that housed the state capitol and predominantly small industry corporations. The third occupational grouping was employees from an electrical contracting firm. This group made up slightly more than one half of the total population of 692 employees in this study. Most were unionized electricians of either apprentice or journeyman status located in an isolated rural community 800 miles away from the corporate headquarters wherein the remainder of this group including management and office staff were based.

Of the 692 employees in the cohort, complete data was available on 411 (59%). Of the total group and the sample analyzed, the majority were males (88%) whose average age was 34 years and length of employment was 4.2 years with an average of 14 years of formal education (2 years of college). Further demographic and lifestyle characteristics are provided in Table 3.

Table 3. Demographic and Lifestyle Characteristics of
411 Employees (88% male) from Three Organizations
Representing a Cross-Section of Job Levels.

Attribute	Mean	Dev.	Range	Units
Height	67.0	3.4	60-80	inches
Weight	179.2	32.1	95-340	pounds
Age	34.6	9.4	18-63	years
Education	13.7	1.9	8-22	years

Recently had a health exam	40%
Drink Alcoholic Beverages	82%
Smoke cigarettes, pipe or cigar	45%
Recreational drug usage	12%
Engage in some sport activity	86%

Criterion Measures

A far more sophisticated and diversified data assessment package was utilized in study two. In addition to the self-report questionnaires, physiological monitoring and biochemical assessment of urine utilized in Study One, there were also numerous additional self-report measures of personal strain and personal observation of the strain-producing elements of the job environment, as well as two separate interviews. One interview was with the employee to amplify and verify self-report measures and the other, more importantly, was with the employee's immediate supervisor thus providing a cross matching of information in an unprecedented two-dimensional approach to occupational assessment.

Procedures

Only about one-half of the employees were located in the nearby vicinity of the university-based research. Thus a large portion of the data collection was obtained in a field experiment setting right on the job site. This was particularly difficult to arrange for the group because the job site was a power plant construction site in a remote rural area with only temporary physical facilities. Thus, extensive efforts were made to create a testing environment with a large mobile home and trailors at one site and in a converted warehouse at the second construction site.

Questionnaire measures were presented in professionally-designed booklets on two separate 45' sessions. Interviews were obtained in a private non-threatening environment. Urine samples were obtained before work and again after three hours on the job. The samples were immediately frozen and stored securely for later analysis. Physiological measures during personal interview and cognitive challenge were obtained in a closed private field laboratory station with assurance of confidentiality.

Data Analyses

A multivariate analytical model was used to handle the numerous and diverse measures obtained in this study. In this analysis each dependent variable was regressed upon a set of independent variables. A hierarchial model was used in all regressions involving physiological strains. In some cases where more than one measure was obtained for a single parameter, the first sampling was used as a covariate. A significant change in R^2 was required for the statistical test in order to inspect the betas for multiple independent variables to further explain the R^2 change. The traditional simultaneous regression model was used for the remaining analyses wherein all independent variables were entered into the equation at the same step.

Results

    Demographic and lifestyle characteristics are presented in Table 3. This predominantly male population (55%) is fairly representative in age (34 years), height (5 ft 10 in) weight (179 lbs) and education (2 years beyond high school). The majority (86%) engage in some form of sport activity for health or recreational purposes, but only 40% have recently had a medical check-up (health exam) for preventive health purposes. A very large portion (82%) consume some form of alcoholic beverage (beer, wine, liquor) on an occasional or regular basis. Surprisingly, the incidence of smoking behavior was only 45% and by contrast the reported use of recreational drugs (e.g., marijuana) was fairly high (12%).

    The physiological measures at rest and in response to cognitive challenge during the interview are recorded in Table 4. Heart rate and blood pressure values at rest were, on the average, well within normal limits (HR = 72 bts/min, BP = 129/80 mm Hg), though noting the range of values there were some individuals exhibiting tachycardia and hypertension. This observation was particularly obvious at peak stress wherein even the mean values rose to borderline hypertensive criteria (147/90 mm Hg).

Table 4. Physiological measures at rest and in response to peak cognitive challenge for 411 employees.

	Resting Base		Peak Stress	
	Mean ± Dev.	Range	Mean ± Dev.	Range
Heart Rate (bts/min)	72 ± 13	43–126	81 ± 15	45–132
Systolic Pressure (mm Hg)	129 ± 13	97–172	147 ± 8	98–244
Diastolic Pressure (mm Hg)	80 ± 13	54–114	90 ± 13	47–138
Skin Temperature (°F)	87 ± 6	81–96	86 ± 6	72–95
Skin Conductance (micromhos)	5.4 ± 5	1–78	9.9 ± 8	1–75
Adrenalin[*] (micrograms/100 ml urine)	1.08 ± 1.5	0–16	1.05 ± 1.3	0–14
Noradrenalin[*] (micrograms/100 ml urine)	6.67 ± 8.3	0–82	5.91 ± 7.0	0–50

[*] Rest values for adrenalin and noradrenalin were taken at the start of the work shift, while peak stress values were taken after at least 3 hours of work on the shift.

Skin temperature, as an indicator of level of peripheral blood flow (influenced by sympathetic nervous activity), was unremarkable in these preliminary analyses. On the other hand, skin conductance (palmar sweating) was quite variable among subjects and differentially reactive to the cognitive challenge. Catecholamine measures of adrenalin and noradrenalin in this population were in the same range as values obtained on a previous population (Ganster et al, 1982), but were somewhat more variable, subject to concern over reliability.

Regression analysis of the relationship between job stressors and strain, as well as between outcomes (personally and organizationally) and acute physiological reactions are presented in Table 5A. There was a very significant relationship between the degree of role ambiguity and conflict versus the self-report of worry and arousal. In addition, the levels of noradrenalin and arousal were associated with the degree of control and responsibility of other subordinates. A combined interpretation would suggest that the compound effect of responsibility, control and discord (ambiguity and conflict) yield remarkable strains, both self-reported and metabolic through urine catecholamines. In separate analyses, adrenalin was correlated significantly with irritation ($r = 0.24$) and with somatic complaints ($r = 0.30$) further supporting the hypothesized link between job stressor, strain and health outcome.

Personal and organizational outcomes are reported in Table 5B. Adverse health symptoms are clearly associated with the level of personal satisfaction and self-esteem, as well as with commitment to the organization. Acute physiological measures of heart rate and systolic pressure were found to be associated with self-esteem and the resultant evaluation of good performance in the organization. The link between personal/organizational attributes and physiological signs is apparent, particularly as it relates to future health problems.

DISCUSSION

The adverse health consequences associated with psychological job stress is well documented (Hurrell and Colligan, 1982; Hoiberg, 1982). Recognition of this fact is further substantiated by the growing number of corporations that are offering stress management to their employees (Parkinson, 1982). A Yale-NIOSH study of 80 organizations and/or labor unions revealed that a high percentage of these (50-80%) offered some form of stress management either in the form of "education about stress" or specific programs in biofeedback relaxation, coping or assertiveness (Neale, Singer, Schwartz, and Schwartz, 1983). In spite of this growing interest, until recently there has been a relative paucity of scientific studies evaluating either the mechanisms and personality interactions that cause job strain or the effectiveness of specific intervention programs (Newman and Beehr, 1979).

Table 5. Standardized betas and R^2's for the regressions of personal strain on job stressors and of personal/organizational outcomes on acute reactions (decimals and nonsignificant betas omitted).

Table 5A.

| | PERSONAL STRAIN | | | |
Job Stressors and Task–Role Expectations	Worry	Arousal	Adrenalin	Nonadrenalin
Ambiguity	32***	–	–	–
Conflict	12*	12*	–	–
Workload	–	–	–	–
Skill Usage	–	–	–	–
Responsibility	–	–	–	08*
Scope of Job	–	–	–	–
Control	–	–17**	–	–09*
R^2	14***	06***	01	02*

Table 5B.

| | OUTCOMES | | | |
| | PERSONAL | | ORGANIZATIONAL | |
Acute Physiological Reactions	Esteem	Satisfaction	Performance	Commitment
Health Symptoms	–37***	–28***	–	–12**
Systolic Pressure	12*	–	–	–
Diastolic Pressure	–	–	–	–
Heart Rate	–	–	–17***	–
R^2	15***	09***	03**	02

*p < 0.05   **p < 0.01   ***p < 0.001

The purpose of the present investigation and the report herein was to deal with the specific sources of job stress, the physiological and psycholgoical indicators of strain and the early signs and symptoms of adverse health reports.

No attempt was made to evaluate specific interventions.

The populations under investigation in these two studies were remarkably diverse just in the sex-role dominance of each. The welfare agency had predominantly female employees and the larger group of construction workers, firefighters and police were predominantly male. This difference can probably be accounted for by sex-role stereotyping and by great difference in salary scale which has in the past favored male-oriented positions, particularly in blue collar unionized, heavy physical labor related jobs. It is tempting to try to compare job dissatisfaction, personal strain and health outcomes in these two groups, but the obvious difference in sex and salary would tend to confound any meaningful interpretation.

The demographic characteristics and baseline physiological parameters for these two groups (see Tables 1, 3 and 4) were well within the normal range for employees in this age group (Roskies, Spevack, Surkis, Cohen, and Gilman, 1978). However, the lifestyle characteristics noted in Table 3 are likely quite representative of the conservative, blue-collar population at large, i.e., fairly high proportion of alcohol usage, low report of recreational drug usage, high proportion involved in sport activity and a low report of recent health exam. These are not value judgements of character, but rather trends in lifestyle common to this segment of the population.

The relationship between behavior pattern and the physiological responses noted in Table 4 was particularly important because it was obtained on a female population of adult employees across the entire age spectrum (19-63 years). Other previous studies have documented this meaningful relationship (linking personality to heart disease through a logical physiological mechanism) but most have dealt with either adult male high-risk (or post-infarct) groups or with very young college students (MacDougall, Dembroski, and Musante, 1979). Thus, to demonstrate that physiological reactivity (percent change heart rate, etc.) is linked to Type A measures in working women is particularly useful in observing and evaluating the impact of increased dual career families in our society.

The results in the larger study reported in Table 5A and B are most intriguing because they serve to introduce a general model of job stress. We have reported on the results of this theoretical model previously (Mayes, Ganster, Sime, and Thart, 1984). Basically it states that stressors (organizational structure, task-role conflict and personality) are linked to some very specific acute and chronic strains (psychological, physiological and behavioral) which can have some very dramatic outcomes in either adverse health symptoms for the employee or poor performance for the organization.

In summary, we believe strongly that occupational stress investigations are most effective when they are conducted by an interdisciplinary team using multiple physiological and psychological assessments carried out in the actual job site on 'real, live' human employees (not laboratory animals). Our results have demonstrated the feasibility of such large scale studies and have established a large body of evidence showing the link between job strain, personality characteristics and personal health and/or organizatonal performance outcomes.

REFERENCES

Caplan, R. D., Cobb, S., French, J. R. P., Jr., VanHarrison, R. V., and Pinneau, S. R., 1974, "Job Demands and Worker Health (HEW Publication No. N105H)," U.S. Government Printing Office, Washington.

Dembroski, T., MacDougall, J., and Lushene, R., 1979, Interpersonal interaction and cardiovascular response in Type A subjects and coronary patients, Journal of Human Stress, 5:28-36.

Eliot, R. S., 1979, "Stress and the major cardiovascular disorders," Futura Publications, Mount Kisco, New York.

Eliot, R. S., and Sime, W. E., 1980, Stress and the heart: environmental, behavior patterns and management in cardiac patients, in: "Advances in Heart Disease," Volume 3, D. Mason, ed., Grune and Stratton, New York.

Ford, M., 1982, Biofeedback treatment for headaches, Raynaud's disease, essential hypertension and irritable syndrome, a review of longterm follow-up literature, Biofeedback and Self-Regulation, 7:521-536.

Ganster, D., Mayes, B. T., Sime, W., and Tharp, G., 1982, Managing organizational stress: A field experiment, Journal of Applied Psychology, 67:533-542.

Hoiberg, A., 1982, Occupational stress and disease, Journal of Occupational Medicine, 24:445-451.

Hurrell, J. J., and Colligan, J. M. J., 1982, Psychological Job Stress, in: "Environmental and Occupational Medicine," W. N. Rom, ed., Little, Brown and Company, Boston.

Jenkins, C. D., Rosenman, R. H., and Zyzanski, S. J., 1974, Prediction of clinical coronary heart disease by a test for the coronary-prone behavior pattern, New England Journal of Medicine, 290(23):1271-1275.

MacDougall, J., Dembroski, T., and Musante, L., 1979, The structured interview and questionnaire methods of assessing coronary-prone behavior in male and female college students, Journal of Behavioral Medicine, 2:71-84.

Mayes, B. T., Ganster, D. C., Sime, W. E., and Tharp, G. D., A multivariate test of a general model of job stress.

Neale, M., Singer, J., Schwartz, J., and Schwartz, G., 1983, Yale-NIOSH occupational stress project, <u>presented at</u>: Society of Behavioral Medicine Meeting, Baltimore, Maryland.

Newman, J. D., and Beehr, T., 1979, Personal and organizational strategies for handling job stress: A review of research and opinion, <u>Personal Psychology</u>, 32:1-43.

Parkinson, R., ed., 1982, "Managing health promotion in the workplace: Guidelines for implementation and evaluation," Mayfield Publishing Company, Palo Alto, California.

Rosenman, R. H., and Chesney, M. A., 1980, The relationship of Type A behavior pattern to coronary heart disease, <u>Activity Nerv. Super.</u>, 22:1-45.

Roskies, E., Spevack, M., Surkis, A., Cohen, C., and Gilman, S., 1978, Changing the coronary-prone (Type A) behavior pattern in a nonclinical population, <u>Journal of Behavioral Medicine</u>, 1:201-216.

Spielberger, C. D., Gorsuch, R. L., and Lushene, R. E., 1970, "Manual for the state-trait anxiety inventory," Consulting Psychology Press, Palo Alto, California.

Taylor, A., and Frazer, A., 1982, Post disaster body handling and victim identification work, <u>Journal of Human Stress</u>, 8:4-12.

STRESS AT WORK: A REVIEW OF AUSTRALIAN RESEARCH

Robert Spillane

Management Studies Centre
Marquarie University
North Ryde, Australia

If we consider stress to refer to a broad class of problems
concerned with demands which tax the human system - the physiological
and psychological systems - then stress research has a long history in
Australia. Bernard Muscio (1971), for example, lectured under the
auspices of the Workers' Educational Association in 1916 on occupational selection, scientific management and work fatigue. The study
of fatigue at work is a recurring theme in business publications and
research papers between 1910 and 1930. During the 1930s psychologists
studied the effects of time and motion techniques, training, rest
pauses, changes in layout or work and executive stress (Marshall and
Trahair, 1981). Since 1945 psychologists and sociologists have been
particularly active in the field of occupational stress to the extent
that in the 1980s stress rates as a prominent factor in industrial
relations.

Research into occupational stress has followed two predictable
paths. Studies initiated by management are generally concerned with
the assessment of job satisfaction and attitudes to work. Research
initiated by trade unions is more likely to be concerned with ergo-
nomic studies and the identification of stressors in specific work
environments. Similarly, theoretical assumptions reflect managerial
and labor perspectives where the phenomenology of the Right confronts
the determinism of the Left. Management tends to focus on the personalities and coping styles of employees whilst union officials emphasise environmental determinants of stress and ill-health. Management
is inclined to avoid stress research or to support stress management
programs which emphasise counselling, psychotherapy, exercise or relaxation training. Labor representatives favor job redesign, worker
participation or penalty payments for stressful, dangerous work
(Lansbury and Spillane, 1983).

In recent years occupational stress researchers have adopted more sophisticated methodologies to respond to the demands of the various industrial parties for more 'objective' assessment techniques. Cross-sectional, questionnaire studies have little action potentiality in the volatile field of Australian industrial relations. A small group of researchers responded to this need by founding the Brain Behaviour Research Institute an aim of which was the psychobiological study of occupational stress (Campbell and Singer, 1983). Building on the work of Swedish researchers (Frankenhaeuser and Gardell, 1976) studies investigated relationships between work environments (characterised by understimulation, overstimulation, lack of control over job), psycho-biological indices of stress (adrenaline, noradrenaline, cortisol secretion patterns) and subjective reports of job satisfaction, coping styles and states of well-being. The application of this psychobio-logical approach has added a new dimension to stress research and has significantly influenced the local industrial scene (Bartley, 1981).

It is the aim of this paper to review new developments in the field and their implications for stress management programs, work reform, workers' compensation and industrial relations generally.

As the action potentiality of stress research is influenced by the context of organizational life in Australia it is necessary to embed this review in relevant local conditions.

THE CONTEXT OF ORGANIZATIONAL LIFE IN AUSTRALIA

The characteristic talent of Australians is not for improvisa-tion, nor even republican manners; it is for bureaucracy (Davis, 1964). This view of the Australian contrasts with the conventional image of the rugged individualist who scorns authority and is aggres-sively egalitarian. Both views, however, have validity. According to Encel (1970) the ambiguity of Australian attitudes towards authority is a reflection of the paradox that the quest for equality has been satisfied to a large extent by the establishment of bureaucratic institutions.

The bureaucratic quality of life in Australia has deep historical origins and arises in part from the special problems of early settle-ment and nation building. Early Australians saw no contradiction between bureaucracy and egalitarianism since they looked upon the state as a vast public utility, whose duty it (was) to provide the greatest happiness for the greatest number (Hancock, 1930).

Employing organisations in Australia tend to fall into two main categories: large-scale, predominantly foreign-owned corporations and small-scale, Australian-owned enterprises. Some three hundred comp-anies dominate the rest and of these two-fifths are overseas-control-led. Not surprisingly, the quality of management which has emerged

under these conditions has been subject to considerable criticism.
Pym's (1971) comments still hold true. Namely, that Australian managers are highly dependent on ideas and practices imported from
abroad, that they are unwilling to accept the need for change, that
they fear anyone who might threaten the established order through the
introduction of radical alternatives and that they are dominated by
'sleepers' rather than 'thrusters'.

Arbitration tribunals (both federal and state) have shaped the
industrial scene since the turn of the century. The Australian Con-
ciliation and Arbitration Act of 1904 required employers to recognise
trade unions which were registered under the Act and empowered unions
to make claims on behalf of all their members within an industry.
However, the tribunals are not empowered to hear matters which in-
fringe upon the 'rights' or 'prerogatives' of management.

An unanticipated consequence of the arbitration system was the
rapid growth of a large number of relatively small and weak trade
unions. There are approximately 300 trade unions which cover more
than 55 percent of the labor force. However, the thirty largest
unions account for almost two-thirds of the total union membership.
A similar problem exists on the employers' side where the number of
employers' bodies is even greater than that of trade unions. Industrial relations continues to be a major source of problems with
Australia one of the more 'strike-prone' countries in the industrialised world.

The social context of Australia has been radically changed since
1945 by the addition of migrants, particularly from non-English speaking countries. During the past twenty-five years more than 3 million
migrants have arrived giving Australia the highest proportion of
overseas-born workers in any country except Israel. Migrants predominate in the less skilled, lower paid, least attractive jobs. While
there is some occupational mobility, especially among those who obtained skills-training prior to arrival in Australia, the main upward
movement is from unskilled to semi-skilled jobs. According to Ford
(1975), management and trade unions have neglected or made only token
efforts to aid the migrant worker: "in some areas there has been a
deliberate effort to create a class of industrial serfs".

A large number of studies have highlighted the problems faced by
migrant women in the workplace (Storer, 1976). Approximately one-
quarter of migrant women are employed in unskilled or semi-skilled
industrial work compared with only 8 percent of Australian women. The
concentration of migrant women in some of the lowest paid jobs with
the most primitive working conditions has caused Ford (1975), to call
them "the newest and last of the industrial cannon fodder". Ford and
his research team found that it was not uncommon for migrant women to
work the day shift in a factory while their husbands took care of pre-
school children. At night, the men worked while the women took over

the child care or vice versa. Not surprisingly, the rates of accidents and sickness among migrant women in the labor force are considerably greater than average.

The tendency of employers to rely on migrants to fill the 'dirty jobs' in Australian industry has also retarded the reform of working conditions in many industries (Birrell, 1981). Employers have often been willing to tolerate high levels of labor turnover, absenteeism and even expensive Workers' Compensation premiums rather than restructure or improve working conditions, because they perceived that there was a sufficient pool of recent immigrants desperate to obtain employment. Although Indo-Chinese refugees have now replaced Southern European migrants in their willingness to suffer inadequate pay and working conditions, there is a growing awareness among trade unions, governments and certain employers that the long-term interests of Australian society will be better served by raising the standards of health and safety in the workplace. Stress research has an important part to play in this respect.

AUSTRALIAN ATTITUDES AND VALUES

Hofstede (1980) in a study of social attitudes across forty countries, found that Australians scored extremely high on individualism, higher than average on masculinity, were average for power distance and lower than average on uncertainty avoidance. Australians adopted a "calculative" attitude to work and employment, based on self-interest, were easy-going in their attitudes to work rules, but gave little support to reforms that would require more involvement by employees in the decision-making process. Most Australians were willing to allow management to exercise authority in the workplace in return for economic and job security.

In another international study, England (1975), compared value orientations among private sector managers in five industrialised societies. He reported that Australian managers placed considerable emphasis on humanistic values, such as loyalty, tolerance and employee welfare, had a relatively low level of achievement orientation and were not strongly motivated towards high levels of growth or profit. England's study portrays a consistent pattern of benevolent paternalism by managers towards their employees but little concern or vision about the future. When transferred to the industrial relations arena, these attitudes result in the strong defence of managerial prerogatives and bitter resistance to any expansion in the influence of employees over decisions about the organization of work. These authoritarian and conservative attitudes are not confined to private sector managers but are found within government and trade union organizations as well (Spillane, 1980a).

Walker (1959) asked executives and trade union officials for
their views on industrial conflict in Australia. Both groups tended
to focus their attention on legal and economic factors as the major
contributors to industrial conflict. In his follow-up study, more
than twenty years later, Spillane (1980b) reported that although
economic issues remained important the emphasis had shifted to psychological factors. Managers were inclined to attribute industrial
conflict to greed, lack of cooperation and poor team spirit among
their employees; union leaders blamed the autocratic, selfish and
uncooperative attitudes of management. Spillane claimed that these
findings were consistent with a dominant individualist ethos in
Australian society in which conflict is attributed to individual
personalities, or more specifically, to personality disorders. The
tendency to "psychologise" industrial conflict was strongest among
executives and reflected a prevailing managerial view of the importance of maintaining a strong defence of managerial prerogatives.

A national survey of 2000 non-managerial employees was commis-
sioned by the federal Labor government in the mid 1970s. Emery and
Phillips (1976) reported that although the majority of Australians
appeared to be satisfied with their work in general, less than one-
third indicated high job involvement. People in highly bureaucratised
jobs were four times as likely to be dissatisfied as those whose jobs
were low on this dimension. The term "bureaucratization" was used to
describe a work situation in which employees felt isolated on the job,
distant from their colleagues and supervisors and feared that management could easily replace them. The greatest polarisation between
labor and management occurred in highly bureaucratized work environments. The researchers also identified a group of "disadvantaged
workers". They tended to be foreign-born, unskilled, poorly educated
and saw very limited prospects for improvement in life, were involved
in a higher than average number of accidents, took more sick leave,
and tended to withdraw from social contacts in the workplace. By
contrast, those people who were most satisfied with their work, had
higher quality jobs which permitted greater vaiety of experience,
opportunities for learning, freedom to organize, mental challenge and
a desirable future. People in these jobs tended to enjoy good health,
had a belief in their capacity to adapt to change and possessed a
strong sense of hope for the future. One of the main implications of
the survey is the importance of job and organizational design to
create the preconditions for work motivation, satisfaction and
involvement.

Modern research on managers and employees in Australia continues
to monitor the individual's continuing adjustment to work. Bordow
(1977) believes that research over the last 25 years has shown that
the individual worker may be experiencing a growing sense of power-
lessness, meaninglessness and isolation and that the work itself may
be declining in quality, humanization and adaptability. In Australia
we have yet to develop the consciousness of the need to change our

system of industrial relations, redesign our systems of work and organization and upgrade the skills of our nation's labor force in order to cope with future changes.

STUDIES IN OCCUPATIONAL STRESS AND HEALTH

Australian statistics reveal familiar relationships between mortality and occupational status. Professionals record the highest life-expectancies followed by executives/managers, farmers, sales and clerical personnel, military personnel, tradespeople, process workers, laborers, service, sport and recreation workers, and finally, miners, quarry workers (McGlashan, 1977).

The contribution stressors make to disease and mortality has attracted considerable research interest. Andrews and Tennant (1978) argue that stress makes a modest contribution to both physical and mental illness. In a study of 863 people in a Sydney suburb they attributed about 37 percent of mental illness and 20 percent of physical illness to stress and social factors including life-event stress, history of adverse childhood, poor social support and low occupational status (Andrews, Tennat, Hewson and Schonell, 1978). Whilst they believe there is no substantial evidence to suggest that psychological stress leads directly to physical illness they argue that stress may complicate existing physical illness and may lead to role changes which alter health habits.

Glaser, Darby and Wilkinson, 1981, identified by cluster analysis five groups with different health profiles. These were:

- Worriers (younger blue-collar suburbanites, stressed, fatalistic, powerless, hypochondriacal smokers);

- Dependents (old blue-collar, stressed, fatalistic, poor health);

- Modern Millies (young, educated, female, independent, optimistic, average health, preventive orientation, good health knowledge);

- Independents (young, educated, self-determined, optimistic, skeptical, but emphasize importance of diet, healthy);

- Young Executives (male, affluent, conventional, low health knowledge, average health).

Otto (1970) investigated the distribution of negative and positive life experience among 799 men and women in selected occupations and examined relationships between quality of life experience, symptom awareness and medical help-seeking. Semi-skilled women reported a predominance of negative over positive experiences while male managers reported more positive experiences. Quality of life experience was significantly related to symptom awareness and to medical helpseeking. Women reported high symptom levels more frequently than men with comparable life experience and they sought medical help more often than men.

O'Brien, Dowling and Kabanoff (1978) found little relationship between somatic conditions (asthma, heart trouble) and the gap between desired and perceived skill utilization. However, items relating to mental well-being were significantly affected by perceived deficiencies in skill-utilization. The most frustrated employees were those whose jobs were perceived to be mentally undemanding. Feelings of stress, anxiety and fatigue were a consequence of working in low-quality, externally-controlled jobs. The importance of skill-utilization and control over job was demonstrated by Otto (1980) who found that car plant workers were under greater pressure and had less control than workers from a government factory. The car workers experienced more stress and reported higher frequencies of psychological and physiological symptoms. Such problems are not confined to blue-collar employees. In a study of high school teachers Otto (1981) found the most prominent stressors were concerned with the alienating relationship between teachers and their employers. Ninety-six percent of those questioned said they lacked influence over decisions which affected them. Lack of influence over organizatinal decisions, lack of consultation and appreciation by superiors were particular problems.

The literature on occupational stress and health is growing rapidly. However, many studies consist of uncontrolled observations with poor study design and a lack of critical evaluation in assessing results. It is rare to see studies which include personality and coping styles as mediating variables in the work environment - reported stress/ill-health relationship. Thus, it is not surprising that researchers should turn to psychobiological approaches to develop different indices of work stress (Bartley, 1980).

THE PSYCHOBIOLOGY OF OCCUPATIONAL STRESS

Several studies have been undertaken by researchers associated with the Brain Behaviour Research Institute which adopt a psychobiological approach to occupational stress (Wallace, 1983). Research has proceeded on the following assumptions: there is a physiological cost attached to all work situations - in some cases the cost is large, in other cases it is small; this cost varies with the type of work environment and the personality of the worker; this cost can be assessed independently of survey-type questionnaires by urinary analysis of catecholamines and cortisol.

Many jobs require workers to maintain performance under difficult

conditions. Those who achieve this goal often do so at a cost.
Adrenaline, noradrenaline, cortisol levels, heart rate and blood
pressure have been seen to increase as a result of attempting to
maintain job performance under conditions of job overload and job
underload. Where performance is maintained under difficult conditions, physiological cost increases. High physiological cost leads to
stress reactions, lack of feelings of well-being and possibly, in the
long term, to ill-health. Although there is no direct evidence for a
causal relationship between hormonal excretion patterns and disease in
humans, data from several sources suggest that if excretion is prolonged, damage to various organs may occur (Henry and Stephens, 1977).

Bassett (1982) has shown that there are two stages in the stress
induced development of ischemic heart disease and that the hormones
adrenaline, noradrenaline and cortisol have a mediating role in this
process. In the first stage the interaction of increased levels of
all these hormones leads to greater sensitivity of the heart. However, the sensitivity is not matched by the usual increase in dilation
of the coronary arteries. At this point a functional but not a structural change has occurred. In the second stage there is a structural
alteration of the arteries (coronary occlusion) which occurs when
there have been repeated stress induced increases in hormones. These
repeated hormone increases lead to a number of changes in the arteries
in both structure and function, culminating in coronary arterio-
sclerosis. Bassett has demonstrated the sequence of this development
with histological, biochemical and electron microscopy techniques.

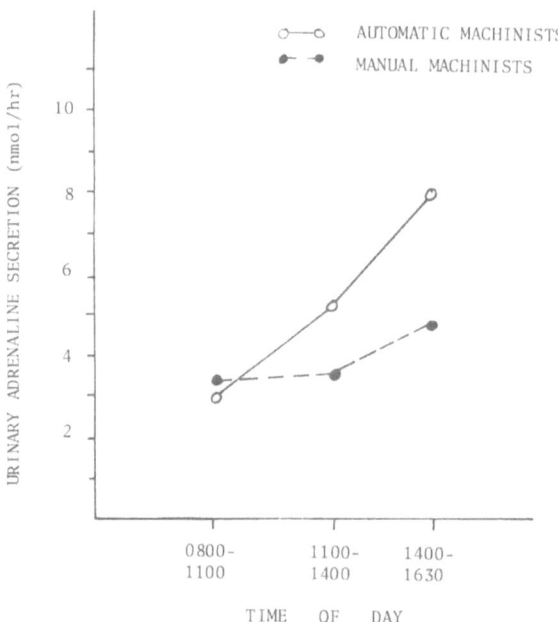

Figure 1. Mean Urinary Adrenaline levels for automated and manual
operators in a clothing factory during a working day.

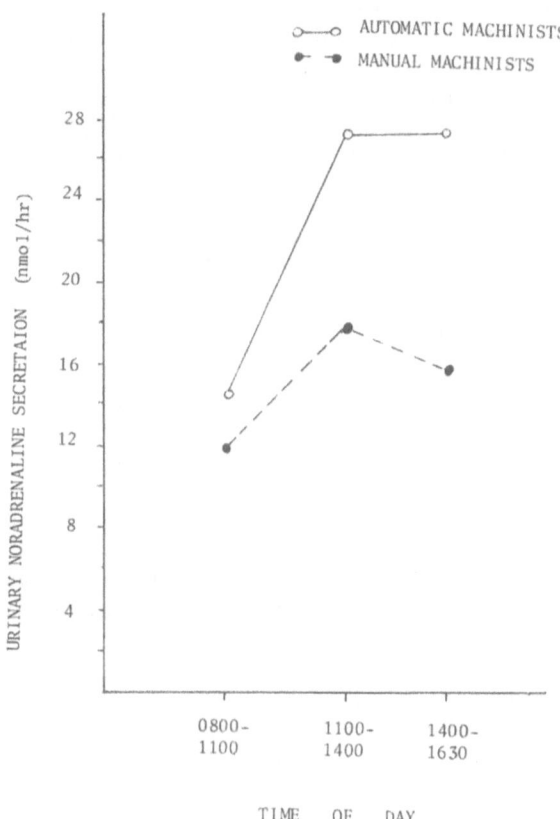

Figure 2. Mean Urinary noradrenaline levels for automated and manual operators in a clothing factory during a working day.

THE CASE OF THE CLOTHING WORKERS

Female operators in a clothing factory were studied and it was found that women working on machine-controlled production processes had higher catecholamine levels than did workers on manually-operated machines which allowed more control over the work pace and greater utilization of skills (Singer, Spillane and Romas, 1982). The results, however, showed a hormonal increase for <u>both</u> manual and machine-controlled operators at the end of the working day (see Figures 1-2). This is contrary to the normal 24 hour change pattern of the catecholamines in which levels should decline towards the end of the afternoon in preparation for rest and sleep during the evening. The hormonal patterns were reflected in the operators' questionnaire reports which revealed that "ability to unwind" after work was the best predictor of job satisfaction and health. The end of day arousal was much greater for the mechanised workers than for manual workers, although this difference could only be detected by hormonal analysis. None of the questionnaire responses revealed a difference which suggests that workers cannot explain to themselves or to others the origin of these feelings. In fact, a feature of this study was the high levels of job satisfaction reported. The high end of day hormonal levels in the clothing workers are probably a result of the fast pace originating in the incentive payment system (piece-rates). These results may constitute an early warning signal of the effects of mechanisation and payment by results systems on health.

In a second local study the hormonal patterns of the clothing operators were compared with those of women working in a munitions factory (Romas, Beeby, Coleman, Spillane and Singer, 1983). Two similar levels of mechanisation were studied in both the clothing and munitions environments. However, whilst the clothing workers were paid by the piece, the munitions workers were paid a fixed wage. Results confirmed the prediction that the stress response as measured by urinary hormones would be greater under the piecework system than under the fixed wage system. It is likely that working on a mechanised production system results in qualitative underload while the piece-rates may produce an increased stress response due to quantitative overload. Stress may be the result of working in mechanised work because, relative to manual work, operators would have difficulty in increasing the production pace to take advantage of incentive payments. Workers would therefore have less control over production rates since they could modify only the input part of the production cycle. For this reason the higher stress levels were found among those workers who worked on the more highly mechanised systems on piece-rates.

THE SHIFTWORK STUDIES

Several studies of shiftworkers have assessed the possible effects of shiftwork on the health, and well-being of employees (Wallace, 1983b).

A large scale study of nearly 900 electricity workers was undertaken to establish the existence and nature of differences between shiftworkers and dayworkers on variables related to quality of life. Shiftworkers reported more frequent occurrence of almost all health symptoms and experienced more 'nervous' problems. More shiftworkers reported receiving medical attention for asthma, lung and breathing problems, stomach ulcers, arthritis and high blood pressure but not for diabetes, cancer, hernia or heart disease. These health differences were not due to severe effects on a small proportion of shiftworkers but rather moderate effects on a large proportion of the group studied. Shiftworkers reported a more frequent use of laxatives, sleeping pills, pain-killers and cough medicines. No differences were found between the groups in alcohol intake although shiftworkers smoke

more and drink more coffee and tea than dayworkers. Shiftworkers
report more dissatisfaction with daily amount of sleep, have less
sleep and more trouble getting to sleep. In general, shiftworkers
perceived their health to be worse than others of the same age, worry
more about their health and believe that work affects their health
more than do dayworkers. Shiftworkers reported more interference to
their family lives especially in terms of the time available to spend
with their wives and children. However, shiftworkers were not more
likely to be divorced or separated from their families.

An unexpected finding was the response pattern of the shiftworkers aged over 50 years. This group generally reported fewer
problems than the other age groups and rarely differed significantly
from the equivalent dayworker group. This is in marked contrast to
the 40-49 year group where the largest differences between shiftworkers and dayworkers were found. The researchers interpreted the 50
year old group as a survivor population, a group of workers relatively
well adjusted to shiftwork.

The researchers also compared dayworkers with previous and current shiftworkers. The results showed that on health items previous
shiftworkers were intermediate between dayworkers and shiftworkers
suggesting the possible influence of lasting health deficits due to
exposure to shiftwork. If shiftworkers are a survivor population then
the previous shiftwork group includes those who failed to survive and
the effects of shiftwork may best be understood by an in-depth study
of this group. Of particular interest here would be the personalities
of those who chose to leave shiftwork and the reasons for so doing.

Wallace (1983) studied the relationship between shiftwork and
catecholamine levels in a service industry. She argues that the
problem for shiftworkers is that they attempt to work with inappropriate levels of hormones which may have a more immediate effect on
health than prolonged exposure to stressful events. When adrenaline
and noradrenaline levels are elevated the person is aroused and alert.
If this peak occurs when a shiftworker is trying to rest, he may well
experience sleep problems and related nervous disorders. Enzymes
involved in the digestive processes also have rhythms and digestive
and gastric problems can occur if these are out of phase.

Wallace found that: (1) high adrenaline and noradrenaline levels
suggest that afternoon shiftworkers are irritable and tense when they
begin their shift; (2) night shiftworkers show a marked drop in adrenaline in the middle of their shift, indicating a sluggish state not
conducive to optimal mental performance; (3) both night and afternoon
shiftworkers leave work with adrenaline levels as high as day-workers
- they may be too alert for sleep shortly after leaving work; (4) the
flat noradrenaline pattern of nightworkers suggests this hormone
rhythm has not adjusted - the levels are as low as those typically
found during the sleep period of dayworkers.

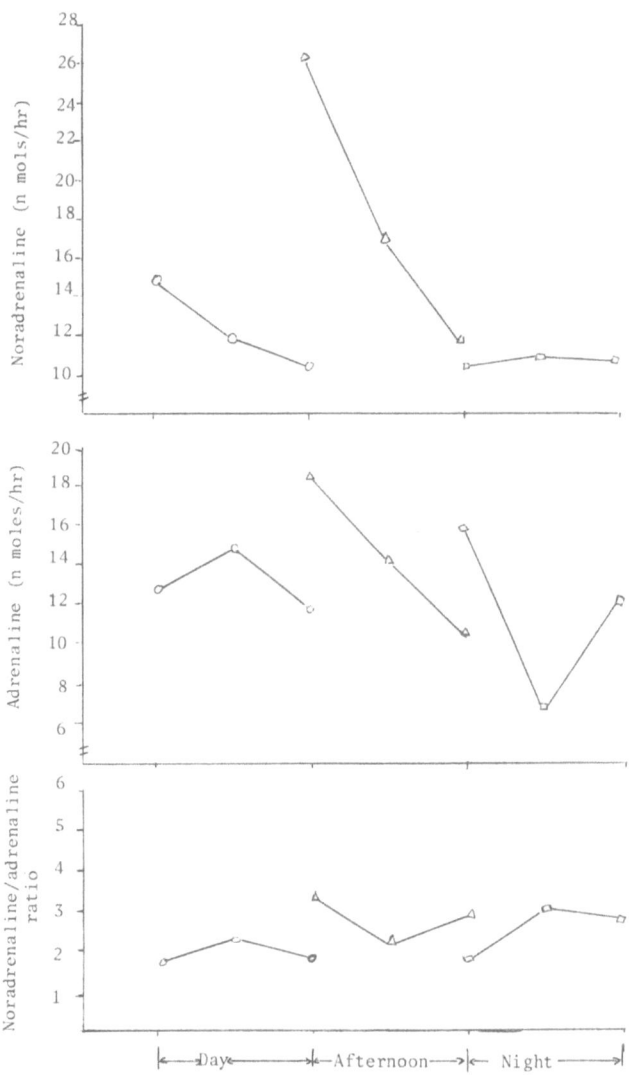

Figure 3. Adrenaline and noradrenaline levels for three groups of shiftworkers obtained from urine samples given at beginning, middle and end of each shift. Bottom graph shows ratio of noradrenaline to adrenaline.

Catecholamine levels and the ratio of noradrenaline to adrenaline are plotted in Figure 3. In the studies (Fibiger, Singer and Miller, 1983) which have explored the implications of various ratio values, lowest ratios were found there there was the highest mental involvement. The highest ratios were found following physical activity to

the point of exhaustion, and during sleep when periods of mental
activity are virtually excluded. The higher ratio values in the
afternoon shiftworkers tend to support the view that they are in a
state of irritability and tenseness, rather than in a mentally alert
state. It also agrees with Wallace's interpretation of physical
fatigue during the night shift. Both shifts finish work with rather
high ratios, suggesting workers are physically tired rather than
mentally fatigued.

Wallace and her colleagues are currently investigating appropriate patterns of hormone levels for well-adjusted shiftworkers.
They argue that if workers maintain the pattern that is normal for
afternoon and night, their hormonal responses are tuned to rest when
they need to be alert. If the job they are doing is the same as
dayworkers, then perhaps the most appropriate patterns of hormonal
response is the same as is found in dayworkers who have no health or
work-related problems. This requires a complete 12 hour shift in
rhythms. Research is continuing in this area.

Singer (1983) argues that an essential first premise from local
research into shiftwork is that greater affluence of shiftworkers may
be a necessary but not a sufficient condition for a relative improvement in quality of life. More money may be essential to allow a
higher material living standard, but it alone will not improve health
and well-being. Shiftwork is a problem which must be solved by workers and management through critical discussion, planning and change.
Changes will be of two types - redesign of jobs and redesign of shiftworkers' lifestyles. Traditionally, management opposes the first
approach and labor the second approach.

Singer offers the following suggestions: reassess the need for
shiftwork; design rosters so that their direction of rotation incorp-
orates circadian principles; redesign certain jobs to include alarm
systems to allow workers to sleep on the job as practiced by firemen,
doctors; arrange working hours and leave to increase opportunities for
weekend interaction with families; allow flexibility to exchange
rosters for special events, such as educational activities. With
respect to changes in lifestyles Singer suggests that night shiftworkers should avoid eating heavy meals and retain normal breakfast
and dinner times where they are not in opposition to digestive
rhythms. Sleep can be improved by noise, light and temperature reduction in the home (heavy drapes, sound-proofing, redeployment of children's play areas, air-conditioning).

Research teams associated with the Brain Behaviour Research
Institute are currently conducting studies on a wide range of employees in both private and public sectors. They aim to establish a
national databank for the identification of stressful occupations.
Use of urinary hormonal analysis adds a new dimension to local research which has relied on cross-sectional studies, crude determin-

istic models and inadequate controls. Singer argues that the research
program provides a basis for the prediction of long term consequences
of stress in relation to health as well as an assessment of the
effectiveness of stress management programs.

STRESS MANAGEMENT PROGRAMS AT THE WORKPLACE

Australian managers and union officials have 'rediscovered' occupational stress and stress management programs. More significantly,
occupational stress has had an important impact in the fields of
industrial relations and workers' compensation.

Bennett and Hely (1982) quote the case of Anderson Meat Packing
Company Pty. Ltd. versus Giacomantonio, a worker in the packing plant
who "sees God in the cool room of meat works - falls in faint -
terrified and confused - nervous shock arises wholly in course of
employment ... incapacitated for work whilst demoralised". In this
case the worker obtained an award despite an appeal by the company to
a higher court.

An increasing number of cases are being brought to industrial
courts where 'stress' is thought to be a major cause of ill-health.
Claims for compensation for pathological conditions exacerbated by
occupational stress have also increased. Consequently, management is
expected to have knowledge of the symptoms of stress, the conditions
which are likely to provide such symptoms, and methods for ameliorat-
ing these conditions.

To date management's response has been either to ignore the issue
altogether or to focus attention on the individual rather than the
work environment. Programs which emphasize employee fitness through
exercise, relaxation and attention to health are favored. An example
is the G.J. Coles company which introduced an executive fitness program at a cost of $300,000 (Bishop, 1981). However, job redesign or
worker participation programs aimed to reduce job stress do not exist.
Where such programs have been introduced economic and more general
industrial factors have inspired them (Lansbury, 1980). Even in these
cases there are few examples of the challenges to organizational power
structures and traditional managerial practices which may be necessary
if work stressors are to be reduced (Anreatta and Rumbold, 1975).

A stress management program with police cadets was conducted by
Singer (1983). He used relaxation training, exercise and cognitive
restructuring and monitored the results over ten week training period.
Although there were few differences between the stress management
group and a control group (who received lecture material only), the
type of coping responses used by the former group showed a healthy
trend towards more social modes of stress abreaction. Deep breathing
and cognitive restructuring were more frequently used by the cadets.

Winkler (1981) employed stress seminars and assertion training with union officials. He argues that stress is a function of work conditions and economic processes and that assistance in the development of effective union shop stewards is an important strategy for the prevention of stress.

Spillane draws on the Rational-Emotive Therapy perspective (Ellis and Harper, 1977) to draw a distinction between stress (practical problems) and distress (emotional problems about practical problems). He argues for the need to study stress at the organizational level and distress at the individual level. His research uses urinary hormonal analysis to monitor stress together with questionnaire data to assess stress and distress reactions. Stress management programs are designed to achieve more control for employees over their jobs and cognitive restructuring to cope with distress (Landsbury and Spillane, 1983).

Kirk et al, (1983) studied the hormonal responses of executives to relaxation training. They reported a differential reduction in hormonal secretion following the relaxation with both cortisol and adrenaline levels decreasing below basal levels, while noradrenaline levels remained unchanged. In a related study they found that the hormonal response of meditators to a stressful situation was significantly lower than controls, thereby confirming the therapeutic value of meditation in stress management. During the study Kirk found significant correlations between basal hormonal levels, a measure of irrational thinking (Jones, 1968) and hormonal change after relaxation. The mediating role of irrational thinking in the stress/distress response is under intensive investigation (Kirk and Spillane, 1983).

CONCLUSION

The study of occupational stress and employees health and well-being has become increasingly important in Australia. New techniques in stress research have overcome many of the limitations of questionnaire studies. By measuring the excretion of specific hormones, the physiological cost of performing work may be indexed and a national databank developed. Stress research has made a significant contribution to the work reform movement, workers' compensation and industrial relations. Stress management programs which emphasize the importance of personal controllability are playing a crucial role in the prevention and amelioration of occupational stress.

REFERENCES

Andrews, G., Tennant, C., Hewson, D., and Schonell, M., 1978, The relation of social factors to physical and psychiatric illness, American Journal of Epidemiology, 108:27.

Andrews, C., and Tennant, C., 1978, Being upset and becoming ill: an appraisal of the relation between life events and physical illness, Medical Journal of Australia, 1:324.

Anreatta, H., and Rumbold, B., 1975, "Organisation Development in Action," Productivity Promotion Council of Australia, Melbourne.

Bartley, H., ed., 1980, "Stress at Work," Brain Behaviour Research Institute, La Trobe University, Bundoora.

Bartley, H., ed., 1981, "Work Effectiveness," Brain Behaviour Research Institute, La Trobe University, Bundoora.

Bassett, J. R., 1982, Psychological stress and the coronary artery in ischemic heart disease, in: "The Coronary Artery," S. Kalsner, ed., Croom Helm, London.

Bennett, M., and Hely, K., 1982, Stress as an Occupational Health Hazard, in: "Advances in Behavioural Medicine," J. L. Sheppard, ed., Cumberland College of Health Sciences, Sydney.

Birrell, R., and Birrell, T., 1981, "An Issue of People: Population and Australian Society," Longman Cheshire, Melbourne.

Bishop, W. H., 1981, Employee Fitness: A new trend in the 1980s, in: "Work and Health: Issues in Occupational Health," M. Hatton, ed., Australian National University, Canberra.

Bordow, A., ed., 1977, "The Worker in Australia," University of Queensland Press, St. Lucia.

Campbell, F., and Singer, G., 1983, "Stress, Drugs and Health: Recent Brain Behaviour Research," Pergamon Press, Sydney.

Davis, A. F., 1964, "Australian Democracy," Cheshire, Melbourne.

Ellis, A., and Harper, R. A., 1977, "A New Guide to Rational Living," Wiltshire, Hollywood.

Emery, F. E., and Phillips, C. R., 1976, "Living at Work," Australian Government Publishing Service, Canberra.

Encel, S., 1970, "Equality and Authority: A Study of Class, Status and Power in Australia," Cheshire, Melbourne.

England, G. W., 1975, "The Manager and his Values: An International Perspective," Ballinger Press, Cambridge, Mass.

Fibiger, W., Singer, G., and Miller, A. J., 1983, Relationships between catecholamines in urine and physical and mental effort, (unpublished), La Trobe University, Bundoora.

Ford, G. W., 1975, A study of human resources and industrial relations at the plant level in seven selected industries, in: "Policies for Development of Manufacturing Industry: A Green Paper," The Jackson Report, Vol. 4, Australian Government Publishing Service, Canberra.

Frankenhaeuser, M., and Gardell, B., 1976, Underload and overload in working life: outline of a multidisciplinary approach, Journal of Human Stress, 2:35.

Glaser, S., Darby, D. N., and Wilkinson, I. F., 1981, "Stress, control
        and health care attitudes and behaviour," quoted by J. Powles
        in: "Work and Health: Issues in Occupational Health," M.
        Hatton, ed., Australian National University, Centre for
        Continuing Education, Canberra.
Hancock, W. K., 1930, "Australia," Benn, London.
Henry, J. P., and Stephens, P. M., 1977, "Stress, Health and the
        Social Environment: A Sociobiologic Approach to Medicine,"
        Springer Verlag, New York.
Hofstede, G., 1980, "Culture's Consequences: International Differences
        in Work Related Values," Sage Publications, London.
Jones, R. G., 1968, "A Factored Measure of Ellis' Irrational
        Belief System, with Personality and Maladjustment Correlates,"
        Unpublished Doctoral Dissertation, Texas Technological College.
Kirk, A., Coleman, G., and Singer, G., 1983, "Changes in
        neuroendocrine responses following the relaxation response in
        trained and untrained subjects," Unpublished paper, Brain
        Behaviour Research Institute, La Trobe University, Bundoora.
Kirk, A., and Spillane, R., 1983, "Adrenal hormone excretion and
        irrational thinking," (unpublished), Macquarie University,
        Sydney.
Lansbury, R. D., ed., 1980, "Democracy in the Workplace," Longman
        Cheshire, Melbourne.
Lansbury, R. D., and Spillane, R., 1983, "Organisational
        Behaviour: The Australian Context," Longman Cheshire,
        Melbourne.
Marshall, J. G., and Trahair, R. C., 1981, "Industrial Psychology
        in Australia to 1950: An Annotated Bibliography," La Trobe
        University Library, Bundoora.
McGlashan, N. D., ed., 1977, "Studies in Australian Mortality,"
        University of Tasmania, Hobart.
Muscio, B., 1971, "Lectures on Industrial Psychology," Angus and
        Robertson, Sydney.
O'Brien, G. E., Dowling, P., and Kabanoff, B., 1978, "Work, Health and
        Leisure," National Institute of Labour Studies, Adelaide.
Otto, R., 1979, Negative and positive life experience among men
        and women in selected occupations, symptom awareness and visits
        to the doctor, Soc. Sci. & Med. 13:151.
Otto, R., 1980, "Occupational Stress among Factory Workers," La
        Trobe University Working Papers in Sociology No. 58, Bundoora.
Otto, R., 1981, "Occupational Stress among High School Teachers,"
        (unpublished), La Trobe University, Bundoora.
Pym. D. L., 1971, Social change in the business firm, in: "Australian
        Management and Society," D. Mills, eds., Penguin, Ringwood.
Romas, N., Beeby, M., Coleman, G., Spillane, R., and Singer, G., 1983,
        "The effects of mechanisation and piece work on urinary
        catecholamine responses," (unpublished), La Trobe University,
        Bundoora.

Singer, G., Spillane, R., and Romas, N., 1982, "Report on Automated
        Work in a Clothing Factory," (unpublished), La Trobe
        University, Bundoora.
Singer, G., 1983, Future Options in the Shiftwork Arena, in:
        "Shiftwork in Australia," M. Wallace, ed., La Trobe University,
        Bundoora.
Singer, G., 1983, "Police Academy Report," (unpublished), La Trobe
        University, Bundoora.
Spillane, R., 1980, Attitudes to Australian industrial relations:
        the influence of political affiliation, Human Resources
        Management Australia, 18:17.
Spillane, R., 1980, Attitudes of business executives and union leaders
        to industrial relations: twenty-three years later, Journal of
        Industrial Relations, 22:317.
Storer, D., 1976, "But I wouldn't Want my Wife to Work Here," Centre
        for Urban Research and Action, Melbourne.
Walker, K. F., 1959, Attitudes of union leaders and business execu-
        tives to industrial relations, Occupational Psychology, 33:157.
Wallace, M., ed., 1983, "Shiftwork in Australia," La Trobe University,
        Bundoora.
Wallace, M., ed., 1983, "Shiftwork in Australia," La Trobe University,
        Bundoora.
Wallace, M., 1983, Shiftwork Case Studies, in: "Shiftwork in
        Australia," M. Wallace, ed., La Trobe University, Bundoora.
Winkler, R. C., 1977, Stress Management at Work, read at:
        Australian Psychological Society Annual Conference, Sydney.

AUTOGENIC TRAINING AS A STRESS MANAGEMENT

TOOL IN AIR TRANSPORT OPERATIONS

F. H. Hawkins

Aviation Human Factors Consultant
P. O. Box 75577, Schiphol Airport (C)
1118 ZP Amsterdam, Netherlands

STRESS AND THE AVIATION INDUSTRY

More than a century ago people were writing of "the stresses of modern living". At that time aviation was confined to balloons, so we must be cautious in identifying stress with aviation, or any other particular industry or occupation for that matter. Or even with the nature of modern society.

Nevertheless, the aviation industry does involve a cocktail of stressors which is unique when combined with a critical need for a high level of human performance. Those in the industry responsible for safety and efficiency are often reminded of the problem when trying to explain dramatic accidents resulting from less than optimum human performance. But profound discussion of human performance in accident investigation reports is regrettably rare, as illustrated in the traditional use of the term "pilot error" as a common manner in which to close the investigation file.

Air Traffic Control has been the subject of numerous stress studies (Hopkin, 1982) and while the problem appears from recent research to be somewhat less severe than was thought earlier, it nevertheless merits attention. ATC officers are engaged in a very critical activity in which, in spite of a generally high degree of automation, small human errors can have catastrophic consequences. A high level of vigilance is required continuously and a momentary weakening of this vigilance can bring disaster. Furthermore, air traffic control is normally a round-the-clock activity involving shift work. This disruption of the natural circadian rhythms of the body introduces unfavorable influences on human performance with loss of motivation and sleep disturbance as well as the inevitability of sometimes having to work during the low phase of the circadian performance curve (Klein et al, 1980).

Flight crews — in particular, long-range flight crews — also face a unique combination of stressors. Early examination of flying stress goes back a long way (Flack, 1918), though stress was then interpreted in physiological terms only. Perhaps the earliest stressors recognised were those created by the immediate environment — noise, vibration, temperature and humidity extremes and acceleration forces. Many of these now have less significance than formerly, and they have been replaced by more complex factors, the implications of which are not always yet fully understood; not least, their impact on motivation and other behavioural characteristics. Working/resting patterns for crews while away from base in long-range flying are now totally irregular; not even the degree of regularity involved in shift work is maintained. Upon this irregular working/resting pattern is superimposed the disturbing influence of transmeridian flying on the circadian rhythms. Cases of pilots — sometimes both pilots — dropping off to sleep during cruise flight at night are certainly not unknown, as revealed from confidential incident reporting programs. To these physiological circadian rhythm disturbances is also added the affective stress associated with family separation for approaching two thirds of the working life. This can be particularly unpalatable during the years when the children are growing up and a wife needs support, and may appear to her at times as putting career before family. The effect of domestic stress on flying efficiency has been examined elsewhere (Haward, 1974). Other affective psychological stresses are also present, such as the feelings of insecurity induced by 6-monthly medical and proficiency checks; four times each year the pilot's licence, and thus his career, are put under challenge. Unlike most other professional people, he is routinely required to perform, in perhaps hostile conditions, with a company check pilot or state aviation inspector looking over his shoulder, constantly posing a psychological threat to his security. Role overload and other role pressures are certainly not confined to flight crews but are additional elements with some crew members, and occasionally, a fear of flying may also be present (Preston, 1968). Even with the young and enthusiastic pilots in military aviation, effective stress is still present. One study cited 71% of pilots as admitting to being worried by personal and domestic problems during the previous year (Aitken, 1969).

It has long been known that affective stress can produce certain characteristic types of pilot error (Davis, 1949). While earlier studies were concerned with a wartime environment, long-range commercial aviation has introduced its own stressors. More recent studies have also shown impairment of flying skill with affective stress (Haward, 1983).

At certain times the pilot may suffer from cognitive stress. This is most likely to occur at the more critical and high workload phases of flight, such as take-off and landing — particularly in

adverse weather conditions. It may also arise during emergency
situations.

It has been said that a pilot is a type who typically denies his
internal emotional life. Under stress he is likely to seek a constructive solution and acts out his frustrations if he does not succeed in achieving a particular objective. If he should be unavoidably
confronted with his emotional life he seems to possess inadequate
strategies for coping with the situation (Ursano, 1980).

The pilot lives a life of deadlines. He is under constant pressure to maintain a public relations image. He is exhorted endlessly
to be professional, responsible, vigilant, dedicated, avoid complac-
ency, and be economically conscious. He works under the threat of
immediate media spotlight in the event of a deviation from routine
operational circumstances which could influence safety - or at least
which could make a good story. Yet to admit to suffering from the
pressures upon him may be seen, in a society which extols achievement
and competitiveness, as an admission of failure. And so, too often
the existence of the related symptoms are denied by the individual and
ignored by his company. Until, that is, they become apparent through
behavioural changes, sickness or reduced performance. This might be
too late to avoid an accident and maintain operational and commercial
efficiency. It may also be too late to save the individual's career.

An analysis of safety reports (Lyman and Orlady, 1981) has shown
significant human performance decrements to exist associated with what
was described as fatigue. The majority of the incidents concerned
involve such unsafe events as altitude deviations, take-offs and
landings without clearance and the like. Monitoring and vigilance
tasks suffered severely. Fatigue has been the subject of much study,
but its origins have not been clearly established. The relationship
between stress and fatigue, while warranting serious discussion, is
beyond the scope of this paper.

In organized industry, the limiting of fatigue has generally been
accomplished by means of work rules — usually restrictions applied to
the number of hours worked — though attention to chrono-biological
factors has usually been inadequate. Even government regulatory agencies are sometimes reluctant to become involved due to the controversial nature of the problem and the possible implication of commercial
cost (FAA, 1981). It is unrealistic to imagine that in the aviation
environment the stressors can be totally removed, though with proper
chronohygiene their severity can be reduced. Neither is it realistic
to suggest that "those who cannot stand the heat should get out of the
kitchen." Even though we may have techniques which would enable us to
determine those who are better able to adapt to circadian rhythm
disturbances, there will surely be sociological and economic problems
in being able to select or schedule staff taking such criteria into
account. Adaptation to occupational stressors is thus likely to
remain a problem for a significant part of the relevant population.

In discussing occupational stress it is not possible to ignore the interaction of work and non-work factors. It is delusory to instruct staff to leave their domestic problems at home; the brain and body suffering from the effects of stress at home cannot simply become healed on arriving at work. Family separation due to work can be expected to induce stress at home. Frustrations at home can be expected to be reflected in attitudes at work. Insomnia, often a symptom of stress, may be reflected in reduced performance at work. Studies of occupational stress often fail to take this interaction into account.

CURRENT STRESS MANAGEMENT TECHNIQUES IN AIR TRANSPORT

Alcohol

As stress may be seen as a source of adaptive energy, which becomes a problem only if it becomes excessive, its management, rather than its total avoidance, may be the appropriate objective. Perhaps the most widely used technique for managing stress today — and for centuries past -- is alcohol. The availability of tax-free liquor on international flights does not encourage resistance to this traditional approach. It is now believed that physiological damage from alcohol occurs at daily consumption levels considered as moderate and far lower than those leading to neuro-psychiatric symptoms (Pequignot, 1979). Recent scientific research shows that in a given country alcohol pathology is linked to the average per capita alcohol consumption and that the proportion of excessive drinkers is growing faster than the per capita consumption. For the 20 years up to 1972 this consumption rose by 276% in the Netherlands, by 182% in Germany and by 133% in Denmark (Godard, 1979).

But in a skilled occupation where optimum performance is required for the protection of life, the use of alcohol has dimensions beyond purely medical ones. Drinking drivers are estimated to cause one fifth of all deaths on the road in the United Kingdon, and two thirds of the drivers killed at night have a blood alcohol concentration (BAC) above the legal limit. Such irresponsibility could never be permitted in commercial aviation; in fact, alcoholism can be the basis for permanent loss of a pilot's licence. Significantly, in the context of this paper, alcohol impairs performance -- for example, reaction times increase, tracking performance is lowered, distance judgement is degraded, and visual acuity is decreased. A second effect, highly dangerous in the flying environment, is that it creates an illusion of improved performance; "you think you are doing fine".

Dedicated research has been carried out to demonstrate the adverse effect of alcohol on the performance of specific flying tasks, such as instrument approaches (Billings, Wick, Gerke, and Chase, 1973).

In spite of all that is known in research circles of the impairment of performance due to alcohol, some 50% of general aviation pilots in the USA responding to a survey stated it would be safe to fly within 4 hours after drinking alcohol. Analysing the answers further it was estimated that 27-32% of the respondents considered flying after drinking, within a time period that would result in a 15 mg % BAC or higher, to be a safe behaviour. The relationship between these attitudes and actual behaviour can only be hypothesised. However, it is noted that about 20% of general aviation pilots killed in accidents in the USA are found to have a BAC of 15 mg % or higher (Damkot and Osga, 1978).

In commercial aviation, evidence of working under the influence of alcohol is usually difficult to obtain but in any case it is far less than in general aviation. Nevertheless, dramatic reminders occur from time to time to suggest that the problem cannot be totally ignored (e.g., Anchorage, DC8, Jan. 1977, all aboard killed, pilot under the influence of alcohol).

Drugs

The 20th century's tool for the management of stress is drugs — tranquilisers, sedatives, hypnotics. And then, sometimes, amphetamines to counteract the resulting drowsiness. In road transport a highly significant statistical correlation has already been shown to exist between the use of tranquilisers and accidents (Skegg, Richards, and Doll, 1979). More recently it was demonstrated that car-driving behaviour appeared to be changed in such a way as to increase the risk of an accident on the road 12 hours after (the morning after) taking a single dose of temazepam (Normison, Euhypnos) or fluranzepam (Dalmane, Dalmadorm) (Betts and Birtle, 1982). This is particularly interesting in the case of temazepam as it has a relatively short half-life (less than 6 hours) and has previously been shown to have little effect in psychomotor test the following morning (Bond and Lader, 1981), and the authors called for replication of the study. Flurazepam has a half-life of 47-100 hours and its influence on a tracking task was included in the study noted immediately below.

Commercial airlines, however, are perhaps understandably reluctant for various reasons to speak out on either the existence of a stress problem or the techniques currently used to manage it. Even after research showing the effect of certain benzodiazepines and barbiturates on an adaptive tracking task (Nicholson, Borland, Clark and Stone, 1976) one of the world's major airlines announced to its crews, in response to concern expressed by some of its captains, that the use of "one nitrazepam (Modagon) a day" was not harmful (the half-life of nitrazepam is 25-30 hours). A survey carried out amongst flight crew members of a European airline indicated that almost half the crew members interviewed occasionally used pharmacological aids to sleep when away from home on long-range flying duties. At home the

figure dropped to 13% (Table 1). The drug most commonly used at that
time was nitrazepam, though as a result of some increase in awareness
of various aspects of Human Factors in the airline concerned since
this survey, some change in preference may now have taken place. This
survey was carried out independently of company and staff union
(Hawkins, 1980). In a few cases crew members admitted to taking such
drugs as little as 6 hours before departure. There are also occasions
when crew members will use both alcohol and drugs within one 24-hour
period, contrary to advice and with unpredictable effects on performance. One must be cautious, however, in too hastily criticising crews
for their actions in these respects. Each individual crew member is
faced with the difficult task of making a differential assessment of
the loss of performance — and motivation — resulting from sleep
deprivation compared with that resulting from the use of drugs. They
generally have nothing upon which to base this assessment apart from
subjective feeling, which can, of course, be very misleading.

Table 1. Use of pharmacological aids to sleep by
airline flight crew members while away from home and at home.

a) Within the working environment (i.e., on the line, away from home)

Crew category	Ss	Av.age	Never	Rarely	Sometimes	Frequently
Cockpit crew	81	48	54%	16%	27%	3%
Cabin crew (M)	41	45	56%	12%	25%	7%
Cabin crew (F)	236	26	54%	24%	18%	4%
Total crew	358	33	54%	21%	21%	4%

b) Outside the working environment (i.e., at home)

Crew category	Ss	Av.age	Never	Rarely	Sometimes	Frequently
Cockpit crew	83	48	90%	6%	3%	1%
Cabin crew (M)	40	45	75%	15%	3%	7%
Cabin crew (F)	225	26	89%	7%	3%	1%
Total crew	348	33	87%	8%	3%	2%

The incidence of drug taking may be expected to vary widely among different ethnic groups, and staff from different countries are likely to favour different drugs. In the USA, for example, where nitrazepam is not marketed, flurazepam was likely to be favoured at the time of the survey.

Considerable progress has been made in recent years in the development of hypnotics for those engaged in skilled occupations and it is likely that there will continue to be an evolution in the type of drug favoured by flight crews.

Diazepam is currently often favoured for both night-time and day-time sleep though it has a long half-life and, according to one source, should not be ingested more than once every 48 hours and more than twice in 7 days. Temazepam has a short half-life but appears to be less effective when used to achieve sleep outside the normal body sleeping phase. Midalozam may also have potential in the context of sleep in the long-range aviation environment of disturbed biological rhythms (Nicholson and Stone, 1982). Recommendations from other sources may be more restrictive, however, with flight duties being excluded until, for example, 10 days after using diazepam (Mohler, 1983).

Even if it were possible to produce a drug with a hypnotic or chronobiological effect which has no side effects to influence performance for more than a short period, the problem still remains of how to police any safeguards placed on its use. And the wisdom of relying on any drugs for a lifetime management of occupational stress is debatable.

Non-Pharmacological Approaches

It is likely that both genetic and cultural backgrounds will influence the stress control strategy which a person finds most suitable. Individual strategies may include physical activities such as exercise, particularly sport, and diet. They may involve psychophysiological methods such as meditation, Yoga, biofeedback, hypnosis as well as other techniques for "relaxation". They may implicate outside intervention through counselling, guidance or psychotherapy.

Little systematic work appears to have been done by organizations in the air transport environment to examine the different individual strategies available, and to make recommendations and offer facilities to staff, though some thought has been given to the problem (Hawkins, 1980). Nevertheless, staff of 3 national airlines in Europe have had Autogenic Training courses made available to them. In 2 cases the courses were sponsored by the airline and in the other case by the staff themselves, without active company participation. The rest of this paper is devoted to this technique and its application in the airline environment.

AUTOGENIC TRAINING PROGRAMMES FOR AVIATION STAFF

The Technique

Autogenic Training (AT) is a psychophysiological technique prescribed by a trained therapist and carried out by the subject himself.
It is designed to promote the subject's own homeostatic mechanisms.
In an industrial application such as this it has the objectives of
improving the quality of working and domestic life and of providing a
personal tool for the long-term management of stress. It can be
expected to have a favourable influence on working efficiency as well
as personal well-being. In air transport a relationship with safety
can be clearly inferred.

AT appears suitable for this particular industrial application
for several reasons. Once the technique is learned it is in the
subject's own hands; he needs no further booster or assistance from
the therapist for lifetime use of the tool which he has acquired. He
becomes self-reliant, participates in the maintenance of his own well-
being and does not develop dependency upon a therapist. No equipment
or drugs are needed in either the acquisition of the technique or its
routine application. It enhances the performance of the body's own
homeostatic, self-regulating, mechanisms rather than interfering with
them. It is easy to learn and has no ritualistic, religious, mystical
or folkloric connotations. It can be utilised effectively by the
majority of people. Finally, its application, particularly in the
clinical field, has been very well documented over more than half a
century and these reports also include special applications in highly
skilled occupations such as that of surgeons (Labhart, 1980). It
must, however, be taught by a skilled practitioner and for optimum
results must be industry-adapted and case-adapted. Furthermore it has
physiological implications which must be well understood by the
therapist.

The vast majority of literature available on AT is concerned with
its clinical application, with a relatively small amount devoted to
application in sports and education and very little to industrial use.
Yet increasingly, large organisations are introducing AT as a facility
for their staff. It has been used, amongst others, by Renault and
SNECMA. By I.G. Farben. By IBM, Marks & Spencer and the British
National Coal Board. By railway workers, airline crews and telephone
operators. The observations reported from several industrial programmes may be summarised as:

. improvement of health and efficiency

. long-term protection against industrial/professional stressors

. decrease of absenteeism, errors and accidents

. improvement in interpersonal relations

Establishing a Programme

For an effective AT programme to be established in an airline or other industrial environment, proper planning and organisation are essential:

- Staff Requirements: A fully trained therapist or instructor is required and he may be expected to hold degree level or equivalent qualifications in psychology or medicine. He should have undertaken an AT therapist's course in a centre approved by the International Committee of Autogenic Therapy (ICAT). He must be fully aware of the physiological and psychological implications (contra-indications, etc.) of AT. There must be a routine consulting line of communication with a physician very familiar with AT, in connection with the screening of trainees and the differential diagnoses which may be required during the course. For various reasons it has been found preferable that the therapist not be an employee of the organisation.

- Training Location: A quiet, comfortable, correctly furnished room must be available. This could be in the medical department of the company, but a case can be made for choosing a restful location remote from the working environment.

- Consultation: Full cooperation can only be assured by adequate consultation with management, the medical department and relevant staff associations. It is likely to be necessary to arrange informative talks to groups to staff to dispel prejudices and misunderstandings and to publish exploratory articles in staff magazines. Some scheduling cooperation may be required from the company.

- Training Logistics: AT lends itself well to group instruction. However, groups should be small, ideally about 5, though some flexibility is possible with careful screening. The group should be homogeneously constituted taking into account such factors as status, and psychological and medical condition. The course consists of 8-10 weekly sessions of 1-1 1/2 hours each.

- Motivation: Benefits of AT often do not appear to the trainee until after some 3 or 4 weeks of training, although the internal therapeutic process begins much sooner. It is therefore necessary to utilise various techniques to enhance motivation so as to avoid drop-outs during these early weeks.

Effectiveness

The programmes which have been used in airlines were practical, industrial applications of AT and not controlled experimental research studies. It would be extremely difficult in the commercial aviation environment to carry out such controlled studies in the field. The variables are so extensive — flight schedules, crew functions, domestic situations and so on — that establishing matched control groups would seem an unrealistic objective. Not least, because commercial airlines do not readily accept the role of research establishments. However, controlled studies of many therapeutic applications of AT have already been carried out, and so, after more than 60 years of clinical use, AT is not without justification (Luthe, 1970). There are, nevertheless, certain specific laboratory research studies which could usefully be done on aspects of AT directly relevant to long-range air transport operations.

Analyses of results of 2 of the airline programmes have been made. The first Swedish programme (Sommer and Byron, 1978) analysis was particularly related to the effects of AT on sleep in the disturbed circadian rhythms of long-range flying. Improvement in sleep was reported for all the different categories of flying staff (Figure 1). A more extensive programme in 1981, controlled from the same Swedish source, involved AT application for flying personnel in Stockholm, Malmo, Copenhagen and Oslo. This is the most ambitious systematic programme of AT applied in a civil airline (Byren, Unesthal and Yanis, 1981).

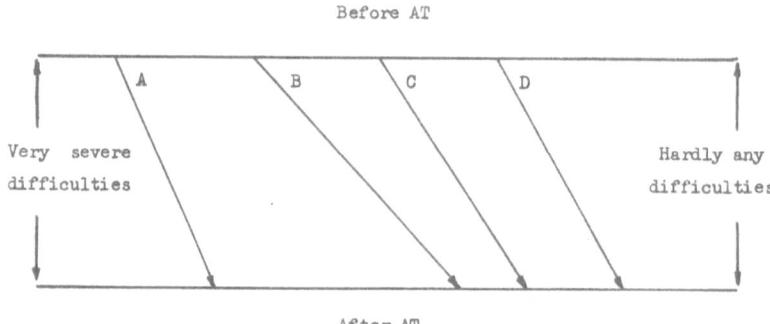

Figure 1. Capacity to fall asleep in conditions of disturbed biological rhythms before and after an Autogenic Training course. Subjects were all airline flying staff and for purposes of this analysis were divided into 4 equal groups (A, B, C, D) based on their initial sleeping difficulty (Sommer and Byron, 1978).

The first programme in Holland analysed several behavioural and performance aspects as reported by the trainee and, independently, by the peer (wife, husband, etc.) as well as recording blood pressure changes (Hawkins, 1982).

In these Dutch 8-week programmes, too, special reference was made to the subjective reporting of sleep both in the working environment and in the home environment. The most common cause of complaint of long-range flight crews relates to the disturbance of sleep and body rhythms (Hawkins, 1980) and this problem can be directly related to human performance (Klein and Wegman, 1980). In the first programme in Holland, 43% of the trainees had indicated sleep problems as one reason for wanting to participate in the programme. After the course, 70% of trainees reported an improvement in sleep in the disturbed waking/sleeping pattern of the working environment (Figure 2). Interestingly, 77% of peers reported that the trainee slept better in the normal, home environment after the course.

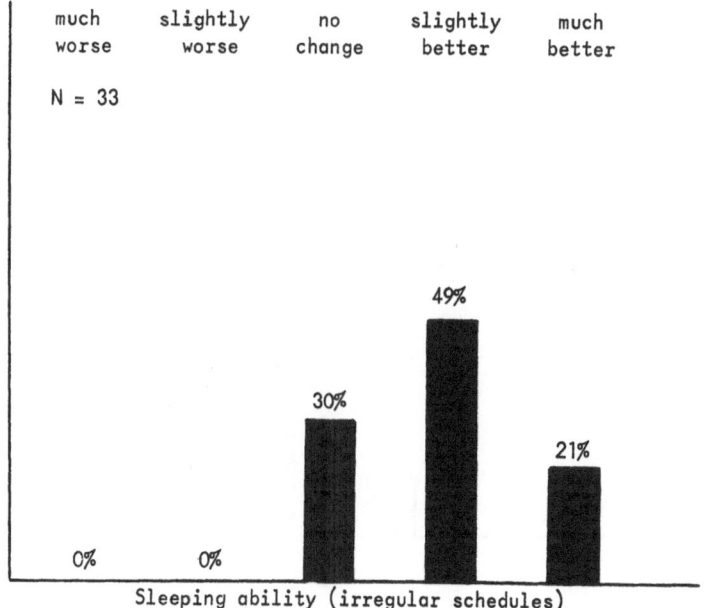

Figure 2. Questionnaire responses from flight crew members on their ability to sleep in conditions of disturbed biological rhythms in long-range flying following an Autogenic Training course (Hawkins, 1982).

The mean of the sleep latency times reported by all participants was reduced from 26 minutes before the course to 14 minutes after the course. Before the course, 13 participants reported a sleep latency time of more than 30 minutes in the normal regular waking/sleeping environment. After the course, only 2 reported more than 30 minutes.

While all of these sources reflect only subjective reporting of sleep improvement, there is a consistency in the information which suggests confirmation of previous conclusions that sleep tends to improve following the use of AT.

As might have been expected, the ability to relax in this stress-ful environment improved, with 89% of trainees reporting a favourable change (Figure 3). This was largely confirmed by the independent report of peers quoting 83% of trainees experiencing an improvement in relaxation capability.

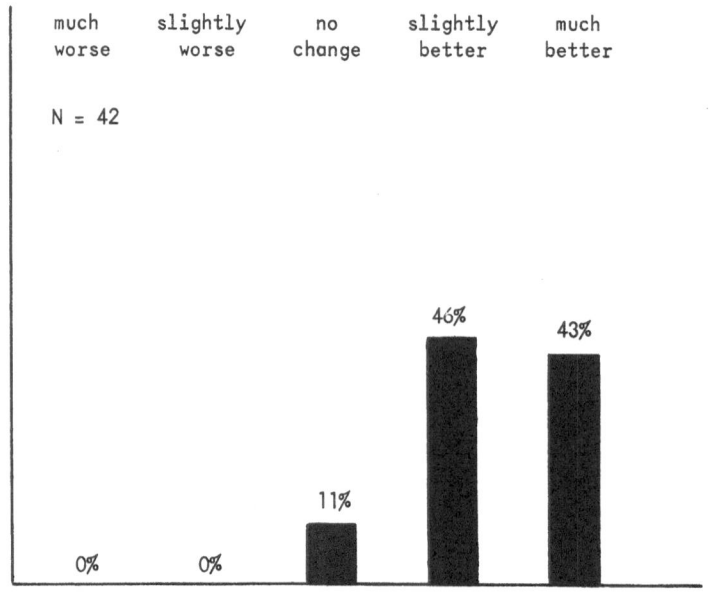

Figure 3. Questionnaire responses from trainees on their ability to relax, following an Autogenic Training course (Hawkins, 1982).

AT has normally been shown in clinical application to have favourable influence on hypertension. Analysis of the first course in Holland revealed that the blood pressures of the 10 participants (all flying staff) having the highest blood pressure of the group at the beginning of the course, had all fallen by the end of the 8 weeks (Figure 4). All were originally mild hypertensives.

Amongst other improvements reported by trainees were better performance in sport (54%), improvement in the ability to concentrate (44%), and improved efficiency at work (31%).

These programmes did not, of course, establish a causal relationship between AT and the therapeutic benefits achieved, as they were not designed as controlled experimental studies. The programmes in Holland, involving an industry – and case-adapted application of AT did, however, suggest that:

a)   a technique such as AT can be made acceptable to a sophisticated, technically oriented, population of this kind.

b)   motivation in such a population can be maintained at a level which results in minimal drop-out during the course.

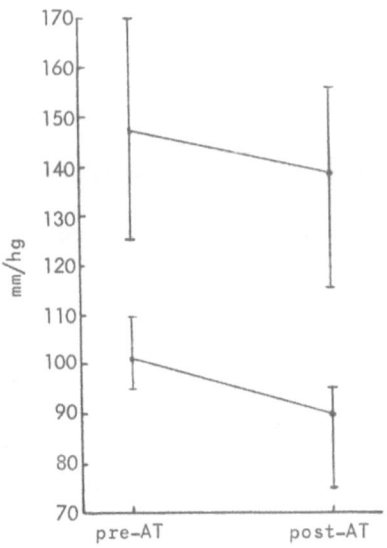

Figure 4. Mean systolic and diastolic blood pressure for a group
        of 10 flight crew subjects with hypertensive tendencies,
before and after 2 months of Autogenic Training (Hawkins, 1982).

c)   the majority of participants can be expected to report benefits
      in several behavioural and performance aspects, reflecting an
      apparent improvement in personal stress management.

d)   highly individual personalities, often represented in flight
      crews, can be integrated homogeneously enough for group training.

e)   an effective programme can be established in a transportation
      organisation, involving scheduled travel away from base, without
      disruption of those schedules, and with only a small degree of
      logistical cooperation.

f)   a "fixed package" course with a cut-off date and without an
      openended clinical therapeutic facility, appears acceptable, in
      spite of the rather less than clear distinction between indus-
      trial and clinical application. Nevertheless, a referral facil-
      ity should be available for use in isolated cases.

g)   the most common complaint reported as being amenable to such a
      programme appears to be sleep disturbance, but many other aspects
      such as concentration, memory, anxiety and personal relation-
      ships, often variable with stress, were also reported as changing
      favourably.

h)   various disorders of a psychosomatic nature, not specially
      related to the occupation, may be expected to show improvement
      during the programme.

i)   most participants believe that their company or organisation
      should support such stress management schemes for staff, although
      confidentiality is considered important.

     While these programmes were not considered as clinical applica-
tions of AT, clinical symptoms, in addition to insomnia, may be expected to appear occasionally. In the Dutch programmes, for example,
such cases included tinnitus, bulimia nervosa, bruxism, amnesia, and
several cases of hypertension; all reportedly responded favourable to
AT. In addition, case-adaptation may be required, even in an industrial programme, for conditions such as stomach disorders, migraine,
pregnancy, and individual disturbing responses.

CONCLUSION

     In air transport operations, as in other activities where safety
depends upon a high level of human performance, the use of alcohol and
drugs as tools of stress management entails risks which make their use
questionable. Necessary safeguards are difficult to apply
effectively.

There are, however, non-pharmacological methods available which appear to be effective in many cases when carefully planned, industry- and case-adapted and taught with skill.

In the case of one such technique, Autogenic Training, the bene-fits accruing have been demonstrated to involve improved efficiency in addition to improved well-being. It would therefore seem to be in the interest of organisations to facilitate the acquisition of such techniques by interested staff members. Such facilitation involves disseminating information, and preferably some logistical and finan-cial support.

REFERENCES

Aitken, R. C. B., 1969, Prevalence of worry in normal aircrew, Br. J. Med. Psychol., 42:283.
Betts, T. A., and Birtle, J., 1982, Effect of two hypnotic drugs on actual driving performance next day, Br. Med. J., 285:852.
Billings, C. E., Wick, R. L., Gerke, R. J., and Chase, R. C., 1973, Effects of ethyl alcohol on pilot performance, Aerospace Med., 44(4):379-382.
Bond, A., and Lader, M., 1981, After-effects of sleeping drugs, in: "Psychopharmacology of Sleep", D. Wheatly, ed., Raven Press, New York.
Byren, S., Unestal, L., and Yams, M., 1981, "Autogenic and mental training for flying personnel," (Swedish language), SAS, Stockholm, Sweden.
Damkot, D. K., and Osga, G. A., 1978, Survey of pilots' attitudes and opinions about drinking and flying, Aviat. Space and Environm. Med., 49(2):390-394.
Davis, D. R., 1949, "Pilot Error", HMSO, London.
FAA, 1981, Operation review program notice no. 7. Flight crew member flight and duty time limitations and rest requirements. Notice of withdrawal of supplemental notice 78-38, Fed. Reg, 46:119.
Flack, M., 1918, "Flying Stress," Medical Research Committee, London.
Godard, J., 1979, Alcohol consumption in Europe and its consequences, in: "The medico-social risks of alcohol consumption", Working party report, Commission of European Communities 1979, ISBN 92-825-1007-7.
Haward, L. R. C., 1974, Effects of domestic stress on flying efficiency, Rev. Med. Aeronaut. Spat., 13:29-31.
Haward, L. R. C., 1983, Effect of Autogenic Training upon flying skill in pilots under stress, presented at: International Interdisciplinary Conference on Stress and Tension Control, Brighton, England.
Hawkins, F. H., 1980, "Sleep and body rhythm disturbance amongst flight crew in long-range aviation," MPhil Thesis, Univeristy of Aston, Birmingham, England.

Hawkins, F. H., 1982, Autogenic Training programme for aviation staff:
    and analysis of results, (unpublished report).
Hopkin, D., 1982, "Human factors in air traffic control," NATO-AGARD-AG-
    275.
Klein, K. E., and Wegmann, H. M., 1980, "Significance of circadian
    rhythms in aerospace operations," NATO-AGARD-AG-247.
Labhart, F., 1980, The influence of Autogenic Training or betablockade
    on stress in surgeons — a preliminary report, in:
    "Psychosomatic cardiovascular disorders — when and how to
    treat," Kielholz, Siegenthaler, Taggart, and Zanchetti, eds.,
    Hans Huber, Berne/Stuttgart/Vienna.
Luthe, W., 1970, "Autogenic Therapy: Vol. IV; Research and Theory,"
    Grune and Stratton Inc., New York.
Lyman, E. G., and Orlady, H., 1980, Fatigue and associated performance
    decrement in air transport operations, NASA report CR 1666167.
Mohler, S. R., 1983, Medicines and the pilot, FSF, Human Factors
    Bulletin, Vol. 29, No.3.
Nicholson, A. N., Borland, R. G., Clarke, C. H., and Stone, B. M.,
    1976, "Experimental basis for the use of hypnotics by aerospace
    crews," NATO-AGARD-CP-203.
Nicholson, A. N., and Stone, B. M., 1982, "Sleep and wakefulness
    handbook for flight medical officers," NATO-AGARD-AG-270(E).
Pequignot, G., 1979, General assessments of the risks, in: "The
    medico-social risks of alcohol consumption," working party
    report, Commission of European Communities, ISBN 92-825-1007-7.
Preston, F. S., 1968, Twelve year survey of airline pilots, Aerospace
    Med., 39:312-314.
Skegg, D. C. G., Richards, S. M., and Doll, R., 1979, Minor
    tranquilisers and road accidents, Br. Med. J., 1:917-919.
Sommer, H., and Byren, S., 1978, "Autogenic Training for flight
    personnel in Scandinavian Airlines," presented at: The 1st
    European Congress of Hypnosis in Psychotherapy and
    Psychosomatic Medicine, Malmo, Sweden.
Ursano, R. J., 1980, Stress and adaptation: the interaction of the
    pilot personality and disease, Aviat. Space and Environm. Med.,
    51(11):1245-1249.

THE EFFECT OF AUTOGENIC TRAINING UPON

THE FLYING SKILL OF PILOTS UNDER STRESS

L. R. C. Haward

Professor of Clinical Psychology
University of Surrey
Guildford, Surrey, England

It was Descartes, the great French scientist, who wrote in 1637 that the only proper way to conduct a scientific discourse was to define all terms and prove all propositions. This paper examines the definition of the word 'stress', and reviews briefly a series of studies supporting the proposition that autogenic training has a beneficial effect on pilots under stress.

Stress entered the English language nearly six centuries ago, being recorded as early as 1410 AD. It seems to owe its origins to an aphatic form of 'distress', that is, stress is related to distress. At this time stress was conceived as a purely physical force applied to an object or person, and by 1655 it was used with particular reference to a force applied to a person to extort a confession or compel a desired action - a grim reminder of the prevalence of torture in Tudor times. Stress was still conceived as a physical force; identical to that studied by the engineer. Even the types of stress were the same. Compression stress was applied by the thumbscrews, tensile stress by the rack, and shear stress by the Iron Maiden. The definition of stress used in the 15th century is still in use by the engineer and physicist of today.

In contrast to the five centuries of consistent usage in the natural sciences, the concept of stress in the social sciences is comparatively recent. Pronko and Leith (1956) studied the psychological literature for the pentade preceding World War II and were unable to find one single mention of the term stress. Yet within twenty years the Annual Reports on Stress listed 25,000 titles. Contradictions abound. Sometimes stress is used as a stimulus, sometimes as a response. Other anomalies are manifest. Selye (1951) refers to stress as an internal condition, while to Basowitz (1954) it is external. Lacey (1952) regards stress as specific to the individual

275

while Funkenstein (1953) says it is specific to the conditions.

The term stressor has also been introduced, neologistically, inconsistently, and quite superfluously. Some writers use it to denote a single stress or subclass of stress (Mefford & Wieland, 1966); others use stressor to refer to the stimulus, and stress to refer to the response, or consequences (Fisher & Agnew, 1955). There is, however, a suitable and existing term for the latter. It is called strain. When Selye talks of people living under strain and exhibiting stress, he reverses completely the original meanings of these two terms. Nearly a quarter-century ago, the present writer put forward a special plea (Haward, 1965), that the definitions of stress employed by the social sciences should be consistent, if not synonymous, with those of the physical sciences. It may be argued that every human is unique, and unlike inanimate matter, will perceive and react to stress differently. While this is true, the alloys of metallurgical science are capable of existing in an infinite number of combinations and solid states, and are changed by experiences (e.g., metal fatigue) so that variation is common in both animate and inanimate material. The differences in respect of stress and strain are those of degree, not of kind, and probably only the simplicity of structures makes the strains of metals more predictable than those of people.

In this paper, the term stress refers to the stimuli or antecedent condition, and the response or consequence is termed strain.[*] In this context, stress control and management are seen as a separate and different processes to the management and control of strain. The subjective experience of stress, and the neuro-physiological mediators between imposed stress and perceived strain pose special problems, in which the concept of anxiety plays an important role. Many so called measures of stress are measures of anxiety, and contemporary investigators such as Bridges (1974) regard the physiological concomitants of stress and anxiety as identical. Yet stress is not always accompanied by anxiety. When stress is defined objectively in terms of the stimulus this confusion disappears. The physiological and behavioural concomitants of anxiety are then legitimately part of the response we call strain, and need no differentiation.

Flying stress was first investigated in 1916 and reported on at the close of World War I, (Flack, 1918). In the following half century, the physical forms of stress, such as hypothermia, and hypoxia, more latterly vibration, gravitation and biorhythm displacement, received virtually all the attention by Government agencies. The physical stresses of flying have dominated the literature and are well-summarized by Gillies (1965)

---

[*] Editors Note: In the field of relaxation, stress, (as a stimulus) evokes a state of (muscular) tension, as in Edmund Jacobson's classic work.

During World War II, combat anxiety led to the study of psychological stress chiefly by Davis (1948, 1949) and his co-workers and
Hall et al (1965), and the effects of this type of stress upon flying
performance has since become an object of study in various aeromedical
laboratories, (e.g., Haward, 1973, 1975, 1978, 1979a).

The research under review proceeded from this point and studied
more subtle forms of stress, both physical and psychological. Since
the methods of inducing stress were common throughout the series,
these are briefly described first to avoid later repetition. Studies
of forced respiration caused by wearing an oxygen face mask (Haward,
1959, 1960, 1968) led to the development of a potent and practical
method of inducing physical stress, which has been employed in subsequent research. Interference with breathing disturbs the respiratory cycle which becomes the strain being measured. By building
positive feedback into the system, the strain augments the stress and
so leads to sharper discrimination. The emotional stresses of combat
flying have no parallel in peace time. There are, however, other
forms of psychological stress. These included personal incompati-
bilities on the flight deck, illicit emotional relationships with
other crew members (Haward, 1974), marital disharmony, and other
domestic problems (Haward, 1977a). It has been well established that
stress can lead to flying accidents and more than one hundred aircraft
accidents occur every week on average (Haward, 1979b). Although
psychological stress is common enough in all occupations, in flying
there can be little tolerance of error, and insignificant mistakes on
the ground can prove fatal when made in the air (Zeller, 1979).

Continuing studies demonstrated that the psychological stresses
fell into two distinct classes: those followed by an emotional response were called affective stress, while those which were primarily
of a non-emotional or intellectual kind were called cognitive stress.

Affective stress was induced experimentally by a technique devised by the Chinese and used on prisoners in the Korean War. A
deeply probing biographical interview is accompanied by electrodermal
recording which reveals those emotionally sensitive areas where the
subject is psychologically vulnerable. These 'touchy' topics are then
fed back later to the pilot through his earphones in a critical,
threatening and ego-destructive way (Haward, 1977). In its extreme
form, when the victim is kept in isolation, it is said to produce
madness. Even in the innocuous conditions of the laboratory it can
reduce tough airline captains to tears. Cognitive stress was induced
by overloading the channel capacity of the pilot (Haward, 1973). By
increasing the information flow within the cockpit for which action is
required, a state is reached in which the pilot can no longer process
the input efficiently. Mistakes and omissions in behaviour then occur
with increasing frequency and interfere with the ongoing task in the
form of positive feedback, thus augmenting the externally applied
stress (Haward, 1971).

Work overload is of course a common form of stress in aviation
(Nicholson, 1972) and the three types of stress employed experimentally therefore represented, albeit in exaggerated form, the stresses
encountered by pilots in their day to day activities.

The initial studies demonstrated that these 3 forms of stress
could be validly differentiated by virtue of their differing effects
in terms of strain. Physical stress produced predominantly physical
strain. Affective stress shows itself chiefly as a disturbance of the
autonomic nervous system, while cognitive stress leads to strain
mediated by the central nervous system.

Of special importance in this triad of stresses was the finding
that summation of stresses takes place within each class of stress,
but not necessarily across categories. The findings of Stopol (1954)
are ambiguous in this respect. As Fischer and Agnew (1954) point out,
the strain from one type of ongoing stress may be diminished by the
introduction of a different class of stress. Conversely, the person
under one stress can become hypo-reactive to an additional and different one. This phenomenon is well-known in psychiatry and psychopharmacology. The deranged patient may become lucid when beset with a
physical illness, well illustrated in the death bed scene of Don
Quixote. Some drugs such as methacholine chloride, produce profound
effects when injected into normal healthy people, but may behave as if
relatively inert when the person is already under stress. Other
anomalies exist, as when a symptom of strain, for example, insomnia,
produced by one class of stress, becomes a stress of a different class
in its own right. Moreover, the additive nature of stresses within
one category is not a linear function and cannot be accurately predicted in advance. Clearly any investigation of stress must take this
interplay of stresses into account.

In the first series of studies being overviewed, strain was
measured in the form it predominated, physical changes to physical
stress, psychophysiological changes to affective stress, and changes
in cognitive efficiency to cognitive stress.

In later studies, strain in the form of decrements in flying
proficiency became the central aspect of strain to be studied,
although concomitant monitoring of autonomic changes as well as the
use of psychometric measures were employed.

The relevant studies, reported elsewhere (Haward, 1965 - 1979)
demonstrated that all 3 forms of stress impaired flying efficiency,
although the nature of the impairment varied with the class of stress
(e.g., Haward, 1967), the personality of the pilot (Haward, 1976,
1977b) and the circumstances obtaining at the time (Haward, 1974). In
looking at ways in which the detrimental effects of stress could be
minimised, various forms of therapy were considered. Lifton (1953)

has shown traditional psychotherapy to be useful for pilots under stress, but this method does have some practical disadvantages. Ideally, a program was required which was not only therapeutic for strain but prophylactic for stress.

Autogenic Training was considered. This is a technique devised in Germany by Schultz (1930) between the wars, and consists of relaxation combined with self hypnotic somatic control. This method has since been successfully applied in the context of airline operations, by Hawkins, (1982, 1983) and a more detailed acount of his work is given later in these proceedings.

The value of autogenic training was first examined in relation to physical stress. A controlled trial undertaken in autogenic training was compared with two other forms of intervention, namely, behavior therapy and psychotherapy. Although all three methods had a beneficial effect upon the strain produced by physical stress, the greatest degree of improvement was associated with autogenic training. The study was not without flaws. For clinical reasons autogenic training was also associated with behavior therapy, and the effects of this had to be partialled out. Further, it was not possible to use a crossover design to enable all groups to have the same therapist, so some of the differences found betwen treatments could also be ascribed to the differential effect of the therapists. However, since the therapeutic role is objective in both autogenic training and behavior therapy, this factor was deemed to be relatively unimportant and it was concluded that autogenic training did have a significant effect upon strain-reduction (Haward, 1965).

A second trial was undertaken to examine the effect of autogenic training in affective stress. The sample consisted of 29 pilots with severe emotional difficulties. For clinical reasons, it was thought necessary to provide some form of stress control in addition to autogenic training, which was directed primarily towards alleviation of strain. Psychosynthesis (Tien, 1969 a,b) was used for the former purpose. Despite the combined treatment methods, it was possible to partial out the subjective treatment effects by psychometric methods. Among the latter, the use of the visual linear analogue scale, shown to be both valid and relaiable in this kind of research (Aitken, 1969, Bridges and Jones, 1973), was particularly useful. This analysis demonstrated that the significant improvements in the handling of strain could be ascribed to autogenic training (Haward, 1973a).

The third study involving autogenic training examined its effects upon flying performance. A flight simulator was used, its validity being established by putting pilots under stress in a real flying situation and correlating the results with those obtained from the same stress imposed in the simulator, which met all the criteria suggested by Pearson (1963).

Instrument flying, as required in flight simulators, imposes the highest level of naturally occurring stress encountered in flying, surpassing even that of combat stress (Paulson, 1975), so that the experimental situation already had a significant degree of cognitive stress built in before the experimental stress was introduced. The additional stress imposed was also cognitive, partly to avoid the inter-active effects of stress mentioned earlier, but also because it is the most common form of stress experienced by pilots and was therefore the most appropriate one to employ.

Strain was measured by errors in flying performance, automatically bled off from deviations in the instruments and recorded, together with reaction times and errors in a concomitant cockpit task, and details are given elsewhere (Haward, 1971). Because this type of stress also produces some degree of emotional response, this was monitored by recording autonomic variables (Haward, 1959, 1960, 1968).

For this trial, autogenic training alone was compared with progressive relaxation, using the same therapist. The possibility of therapist bias could not be eliminated. A weakness of all psychotherapy trials is that the double blind design is inappropriate. Using different therapists introduces even more interviewing variables, and a cross-over design is unsuitable because of the interactions between phases.

Twenty pilots entered the trial, as 2 groups of matched pairs, 3 being lost by postings during the trial, which extended over 12 weeks. This duration enabled both the standard and the meditative exercises to be introduced. At the 6 week mid-term assessment, no significant changes had occurred in flying skill, although physiological reactions had reduced in both groups. At 12 weeks, improvement in skill was demonstrated, the autogenic training group marginally better than the relaxation group.

The differences between the 2 groups showed itself mainly in the physiological measures, in which autogenic training proved significantly superior. These findings confirmed the results of previous studies that these methods of intervention are more successful in reducing the emotional components of strain than the cognitive components. Benson's (1974) own theory of the trophotropic response to relaxation provides a plausible explanation of why this should be so.

Finally, mention should be made of three ongoing studies which, although not yet completed, are sufficiently advanced to enable the data available to be analyzed. These trials examine differences in flying strain produced by the 3 different classes of stress. Healthy volunteer pilots are assigned to one of the 3 trials in strict rotation. Stress, one type for each trial, is administered by an automated programme, using the methods already mentioned, and strain is monitored as in the previous trial. Three sessions in the simulator have been employed, the first for base level recording, the second, after a 6 week interval, for acclimatization and learning effects to be evaluated, and the 3rd after a 6 week course of autogenic training. Thirty-seven pilots have entered these trials to date. Preliminary findings show a small improvement in all trials during the 1st control period, which is subtracted from gains in the second experimental period. During this period, strain from physical stress is reduced to a very significant degree, strain from affective stress, which possesses a much higher variance, is reduced by a significant degree, and strain from cognitive stress has not yet reached significant improvement. While it would be premature to draw any conclusions from these incomplete trials, the results so far are clearly in line with previous findings, and the overall indication is that autogenic training is particularly appropriate for somatic strain (expressed as a somatic disorder for example), is effective in reducing the emotional component of strain, but probably has little place in the reduction of cognitive strain.

One final caveat needs to be made. Novello and Yousseff (1974) have shown that pilots are psychologically homogenous. One pilot is more like another than anyone else. Even female pilots are psychologically more like male pilots than they are like members of their own sex. This means that any conclusion drawn from studies of pilots cannot be applied to other occupational groups without independant verification. The nature of stress and the effects of autogenic training briefly described here are therefore relative to the samples used. There is however considerable evidence to show the effectiveness of autogenic training in other conditions (Luthe, 1965). In particular, evidence is accumulating to demonstrate that autogenic training has an effective role in the management of stress and strain.

As aircraft increase in size and number, the crew suffer increasing stress and strain, and the potential to improve the health and safety of all who fly has never been greater. In some small way the studies described earlier support the view that autogenic training can make a useful contribution to this laudable endeavour.

REFERENCES

Aitken, R. C. B., 1969, Measurement of feelings using visual analogue scales, Proc. Royal Soc. Med., 62:989.
Basowitz, H., Persky, H., Korchin, S. J., and Grinker, R. R., "Anxiety and Stress," McGraw Hill, New York.
Benson, H., Beary, J. F., and Carol, M. P., 1974, The Relaxation Response, Psychiatry, 37:37-46.
Bridges, P. K., and Jones, M. T., 1973, Relationships between psychological assessment, body build, and physiological stress responses, J. Neurol. Neuro Surg. Psychiat., 36:839.

Bridges, P. K., 1974, Recent physiological studies of stress and anxiety in Man, Biological Psychiatry, 8:95–112.

Davis, D. R., 1948, "Pilot Error," Air Ministry 3139A, London.

Davis, D. R., 1949, Disorder of skill responsible for accidents, Q. J. Exp. Psycol., 1:136–142.

Fischer, R., and Agnew, N., 1954, Hierarchy of Stresses, J. Ment. Sci., 100:383–385.

Flack, M., 1918, "Flying Stress, Report No. 3 of the Air Medical Investigation Committee," Medical Research Committee, London.

Gillies, J. A., (Ed.), 1965, "Textbook of Aviation Physiology," Pergamon, Oxford.

Funkenstein, D. H., 1953, "Experimental Evocation of Stress, Proc. Symposium on Stress," Army M.S.G.C., Washington.

Hale, H. B., Duffy, J. C., Ellis, J. P., and Williams, E. W., 1965, Flying Stress in relation to flying proficiency, Aerospace Med., 36:122–116.

Haward, L. R. C., 1959, Speech and Respiration Recording, Brit. Med. J., ii 323.

Haward, L. R. C., 1960, Subjective meaning of Stress, Brit. J. Med. Psychol., 33:185–194.

Haward, L. R. C., 1965, Reduction of Stress activity by autogenic training, in: "Autogenes Training," W. Luthe, ed., Stuttgart, Thieme, pp.96–103.

Haward, L. R. C., 1967, Stress Tolerance Evaluation in Commercial Pilots, Log 27:233–234.

Haward, L. R. C., 1968a, Stress in Flying Training, Flight Safety, 2:19–23.

Haward, L. R. C., 1968b, Effects on Respiratory Disturbances of different types of Oxygen Face Mask, Flight Safety, 2:25–28.

Haward, L. R. C., 1971, Effects of DPH on Cognitive Efficiency in Pilots, in: "Aviation Psychology Research", J. D. Anderson, ed., W.E.A.A.P., Brussels.

Haward, L. R. C., 1973a, Application of psychosynthetic principles in supporting emotional stability in Pilots, Wld. J. Psychosynth, 5:18–20.

Haward, L. R. C., 1973b, Ideographic Study of Pilot Stress, Proc. 21, Int. Congress Aviat. Space Med., p.239.

Haward, L. R. C., 1973c, Effects of DPH upon Concentration in Pilots, Rev. Med. Aeronaut. Spat., 12:372–374.

Haward, L. R. C., 1974, Effects of Domestic Stress upon Flying Proficiency, Rev. Med. Aeronaut. Spat., 13:29–31.

Haward, L. R. C., 1975, Emotional Stress and Flying Efficiency, in: "Higher Mental Functioning in Operational Environments," AGARD CPP. 181 C8.

Haward, L. R. C., 1976, Rumination and Flying Impairment, Proc. Aerospace Med. Assoc., p. 170–172.

Haward, L. R. C., 1977a, Psychopathology and Flying Accidents, in: "Aviation Psychology Research," L. Carstedt, ed., W.E.A.A.P. Brussels.

Haward, L. R. C., 1977b, Impairment of Flying Efficiency in Anacastic Pilots, Aviat. Space Environment Med., 48:156–161.

Haward, L. R. C., 1978, Stress and Rumination in Pilots, in: "Wehrpsychologie Untersuchen," F. Fehler, ed., Min. of Defence, Bonn, German Fed. Republic. pp. 93-100.

Haward, L. R. C., 1979a, Stress Performance in Anacastic Pilots, Proc. Ann. Conf, CAMA, Orlando.

Haward, L. R. C., 1979b, Stress in the Air, Stress Today 4:5-7.

Hawkins, F. H., 1982, "Autogenic Training programme for aviation staff: an analysis of results," Hawkins, P. O. Box 75577, Schiphol Airport (c), Amsterdam.

Hawkins, F. H., 1983, Autogenic Training as a stress management tool in air transport operations, presented at: Second International Conference on Stress and Tension Control, Brighton.

Lacey, J. I., 1952, Autonomic Response Specificity, Psychosom. Med., 15-8.

Lifton, R. J., 1953, Psychotherapy with combat flyers, U.S. Armed Forces Med. J., 4:525-532.

Luthe, W., 1965, "Autogenes Training," Stuttgart, Thieme.

Mefford, R. B., and Wieland, B. A., 1966, Comparison of responses to anticipated stress and stress, Psychosomatic Med., 28:795-807.

Nicholson, A. N., 1972, Aircrew Workload, R.A.F. Quarterly, 12:273-280.

Novello, J. R., and Youssef, Z. I., 1974, Psychosocial Studies in General Aviation, I: Aerospace Med., 45:185-188, II: Aerospace Med., 45:630-633.

Paulson, N. W., 1975, "AGARD Advisory Report No. 69," U.S. Army Air Safety Centre, Fort Rucker.

Pearson, R. G., 1963, Use of Simulators for Biomedical Research, Navigation, 9:329-333.

Persley, H., Hamburg, D. A., Basowitz, H., Grinker, R. R., Sabshim, M., Korchin, S. J., Herz, M., Board, F. A., and Heath, H. A., 1958, Relation of Emotional Responses and changes in Plasma Hydrocortisone level after stressful interview, Arch. Neurol. Psychiat., 79:434-447.

Pronto, N. H., and Leith, W. R., 1956, Behaviour under Stress, Psychol. Rep. Monog. Suppl., No. 5.

Schultz, J. H., 1930, Uberdas Autogene Training, Wurzburger Abhdlg., 26:319.

Seley, H., 1951, "Stress of Life," Longmans Green, London.

Seley, H., (no date available) "Stress Without Distress," World Health, Geneva.

Stopol, M. S., 1954, Consistency of Stress Tolerance, J. Pers., 23:13 29.

Tien, H. C., 1969a, Pattern Recognition and Psychosynthesis, Amer. J. Psychotherapy, 23:1.

Tien, H. C., 1969b, Studies on Psychosynthesis, Wld. J. Psychosynthesis, 1:30-35.

Wiener, M., 1955, Effects of two experimental counselling techniques on performances impaired by induced stress, J. Abnorm. Soc. Psychol., 51:565-572.

Zeller, A. F., 1979, Stress, Flightfax 7, 2-5, U.S. Army Safety Centre, Fort Rucker.

REDUCTION OF STRESS BY PERSONNEL AT INSTITUTIONS

FOR CHILD CARE AND FOR THE MENTALLY HANDICAPPED

Anneli Leppänen

Department of Psychology
Institute of Occupational Health
Helsinki, Finland

INTRODUCTION

Professions that are characterized by intense social contacts
with clients, patients, or customers involve especially stressful
tasks that include contacts with those from exceptional groups
(Kalimo, 1980). Even routine tasks can cause stress among personnel
working with those whose behaviour strongly deviates from the generally accepted normative patterns in one way or another. Working with
people from exceptional groups requires emotional involvement. The
worker's personality becomes an essential work tool. The psychological consequences of work with exceptional personnel have not been
sufficiently clarified.

However, work which involves intensive social contacts causes
certain stressors characteristic to these professions. Typical of the
stressors are uncertainty about the objectives and outcomes of the
work, continuous emotionally negative contacts, and the constant fear
of failure, ineffectiveness, and helplessness (Pines et al, 1981;
Perlman and Hartman, 1982).

The aim of his study was to investigate how the stress experienced by professional personnel at reform schools and institutions
for the mentally handicapped could be reduced.

SUBJECTS AND METHODS

The study population comprised the professional personnel of
institutions for custodial child care (youth homes, nurseries, reception homes and children's homes) and 32 institutions for the
mentally handicapped in Helsinki.

The number of professional personnel (nurses, nursemaids, nurses
for the mentally retarded, instructors etc.) was 421 in the year 1982.
Every other member of the professional staff was selected for study
from an alphabetical list of the names of the personnel. One hundred
ninety-five persons (91%) of the 214 participated. Ninety-two of the
subjects worked at institutions for the mentally handicapped and 103
persons at institutions for child care. 82% were women, 18% were men.
Their mean age at the institutions for the mentally handicapped was
33.7 years (SD = 9.2) and 35.4 years (SD = 9.7) for those at the
institutions for custodial child care.

The design and methods of the study are described in Table 1. A
questionnaire was used in Phase 1 to study the stressors and stress
reactions. Assessment of suggestions for improvements to be carried
out in the work environment were also requested from the subjects in
the questionnaire. The questionnaire was completed at the work place
in the presence of the researcher. In Phase 2, small group sessions
were organized to discuss improvements. Forty groups with three to
six persons worked for half an hour during the five-day research
period. In Phase 3, a group of supervisors in the same institutions
was asked to comment on the actions proposed by the groups. A preliminary follow-up study was done a year after the first phase of the
study in order to find out if the changes proposed by the subjects had
been realized. Fourteen nurses from one youth home participated in
the follow-up study (Phase 4).

Table 1. The design and methods used to study work stress among
personnel in institutions for child care and the mentally handicapped.

Phase	Information collected	Method	N
1	Perceived work stress, subjective stress symptoms, and assessment of the given suggestions for improvement of work	questionnaire	195 subjects
2	Suggestions for improvements to be carried out in the work environment	small group sessions	195 subjects 40 groups
3	Supervisors' comments on the suggested improvements	small group sessions	27 supervisors 4 groups
4	A preliminary follow-up study	questionnaire	14 subjects

RESULTS

Perceived Work Characteristics

The challenge of their work load was considered as relatively great: 52% of the respondents were of the opinion that their work demanded ability to make decisions on various subjects simultaneously; 44% reported that their work included a lot of mental strain. One third of the personnel had the experience that their work often included too difficult tasks. Challenge of work is often connected with the variability of work. A majority (75%) of the respondents considered the work as suitably varying.

The quantitative work load was found somewhat problematic. About one half (46%) of the personnel considered the amount of work as suitable. The other half perceived the amount of work as too great. A majority (65%) of the respondents perceived the distribution of the work load as irregular. The peaks at work load were considered to be caused by the small number and great turnover rate of the personnel and a lack of substitutes. The fact that several duties had to be done during a short period in the mornings and evenings was perceived to cause periodic quantitative overload.

One third of the respondents perceived their role ambiguity as great. The causes of role ambiguity were different at different institutions. At the institutions for custodial child care, perceived role ambiguity was considered to be connected with educational problems and other difficulties at work. Uncertainty about the distribution of tasks among different professional groups was considered as the main cause of role ambiguity among personnel at the institutions for the mentally handicapped.

The quality of interpersonal relations among staff is very important in an organization where the possibility to reach the aims of the work essentially depends on the co-operation of the staff. 45% of the respondents considered the discussions with co-workers as easy. One half of the respondents sometimes experienced difficulties in discussions with co-workers. A majority (57%) of the respondents considered that collaboration among staff members ought to be increased.

Besides the relations among staff members the relations between supervisors and staff members affect the well-being of the employees and their clients. The relations between the supervisors and the staff of the institutions studied were considered as good: 60% of the personnel experienced their supervisors as supportive.

Table 2. Subjective stress reactions of professional personnel
  – percentage of 195 respondents mentioning each symptom.

Symptom	always or very often	sometimes
*** Burnout Symptoms ***		
prefers to stay at home instead of leaving for work	14	35
wants to 'forget' the future	6	23
feelings of passivity	13	32
perceives the future as hopeless	23	34
feelings of failure	9	40
exceptional fatigue	22	39
feelings of having discharged the duties	13	34
*** Psychological Symptoms ***		
tension	20	55
nervousness	15	56
depression	9	47
irritation	12	51
difficulties in decision-making	14	49
*** Somatic Symptoms ***		
stomach pain	11	25
irregular heart beat	6	23
headache	15	27
nausea	5	20
chest pain	4	17

Subjective Stress Symptoms

The professional personnel of the institutions had relatively many burnout symptoms and psychological stress reactions (Table 2). The amount of psychological stress symptoms was comparable to that of Finnish prison personnel (Kalimo, 1980) and greater than that of the Finnish teachers (Ojanen, 1982) and personnel at mental hospitals (Pöyhönen, 1982).

The amount of somatic symptoms was not exceptionally high. (See Table 2). The respondents experienced less somatic symptoms than the prison personnel (Kalimo, 1980), teachers (Ojanen, 1982) and personnel at mental hospitals (Pöyhönen, 1982). However, the amount of somatic complaints strongly depends on the age and sex of the respondents. The mean age of the subjects was lower in this study than in the other studies mentioned. Most of the subjects were, however, women as in the studies of teachers and personnel at mental hospitals.

Methods Of Stress Prevention At Work

The questionnaire included a number of optional proposals for the improvement of work. The subjects were asked to assess the actions proposed on a scale ranging from very necessary to harmful.

More than 90% of the respondents considered it necessary to hire substitutes for personnel in supplementary training or otherwise off duty. More than 90% of the respondents needed continuous professional training.

More than 80% of the respondents considered the following actions necessary.

1) the definition of an educational plan for each individual child;

2) reduction of the number of children at the departments of the institutions;

3) each employee should carry out his/her duties without additional requests.

More than 70% of the respondents considered the following actions necessary:

1) the opportunity for occupational counseling;

2) increased specialist consultation;

3) regular consultation with psychologists and social workers at the institutions;

4) more economic support for the guidance of children's
   activities outside the institutions;

5) hiring extra personnel for periods of overload;

6) personnel training in the group work skills;

7) management training for supervisory personnel;

8) regular meetings of the personnel.

Desired targets of improvement were further specified in the
group work sessions. The groups proposed 213 ideas for occupational
improvement. Only the main trends in the proposals can be presented
here.

Clarification and development of the aims, the contents, and the
methods of the work was considered a very important action. Clarification of the aims implies legislative work as well as discussion and
co-operation between the professional groups at the institutions.

The main obstacle to the development of the aims of the work at
institutions for child care is the insufficient professional training
of the personnel. The training of the reform school personnel ought
to be arranged according to the demands of the tasks. In order to
achieve the aims of the work, co-operation between the personnel, and
medical, psychological, and social advisers should also be increased.
Advisory help was needed both for educational problems and to support
and develop the work groups.

The relevancy of the content and the sufficiency of the amount of
feedback given at work were also considered as necessary elements for
the development of the work situation. It was also found necessary to
improve the ability to give and to receive feedback by training.

A great number of clients was an essential problem both at institutions for child care and institutions for mentally handicapped. The
mental and behavioral disturbances of the reform school children have
become more serious during the last two decades according to the reports of the personnel. It was considered as necessary to reduce the
combination of quantitative and qualitative work load. The amount of
children at a department should be diminished or the amount of personnel increased. A special care unit should be established for
children suffering from severe behavioral and mental disturbances.
One severely disturbed child should be considered to occupy the
places usually occupied by two normal children.

The need to improve professional skills was great. Continuous
professonal training was considered as one method to improve professional skills. The consulting specialists of the institutions should

give lessons on the topics of their specialty. Specialists from
outside the institution should also be requested to give consultation
and lectures.

Counselling was seen as a necessary way to reduce the personnel's
stress and stress reactions. A specialist's counselling is needed
both for solving the problems at work and for treatment of the emotional problems and cognitive conflicts the work evokes in the individual professionals.

THE COMMENTS OF THE SUPERVISORY PERSONNEL ON THE PROPOSED CHANGES

Occupational development implies the involvement of all personnel
groups. Therefore, the supervisors of the institutions were asked to
express their attitudes towards the proposed changes and to evaluate
the probability of their realization. They were also asked to assess
which administrative level ought to make decisions concerning the
realization of the different proposals.

The general attitude of the supervisors was very positive. They
supported most of the group proposals. They also believed that they
could themselves decide about the realization of some proposed
actions. Many of the proposals, however, were such that their realization required a decision of the City government, the City Council,
or the Ministry for Social Affairs and Health.

RECOMMENDED PLAN OF ACTION

The subjects made many proposals for actions on occupational
development, but they were considered to need guidance to be able to
continue the development process that had begun during the study.

As the needs for improvement vary according to the clientele and
the aims and the methods of work, it was not possible to make a
detailed plan of development to be followed at every institute. But a
plan of action was designed to help the personnel of the institutions
to start the utilization of the results of this study and to handle
occupational stress problems. The plan is presented in Table 3.

RESULTS OF THE PRELIMINARY FOLLOW-UP

Less than one year has passed since the dissemination of the
results of the study to the institutions studied. Therefore, it has
not been possible to systematically collect data about the trends of
development of work and working conditions at the institutions after
the first phase of the study. But it was possible to make a preliminary follow-up among the personnel of an institute for custodial

Table 3. Phases of occupational stress reduction at institutions for
custodial child care and institutions for mentally handicapped.

1) Problem recognition

The actual stressors of an institute should be identified,
as the stressors at one institute can deviate somewhat from
those found in the study on several institutions.

2) Definition of the order of priority for the solution of the
problems.

3) Proposal for actions for the reduction of the problems.

The practical implementation of the recommendations made in
the study must be adapted to the actual situation in any
single institution. The proposed changes must be made
concrete on the action level.

4) Identification of the administrative levels able to make
decisions about the realization of the proposed actions.

5) Plan of the actions of affect the administrative levels that
make decisions concerning the work at the institutions.

6) Planning of the timetable for the realization of the
proposed actions.

7) Realization of the improvements.

8) Organization of a program for continuous follow-up of the
realization of the proposed actions.

child care. The personnel of that institution had a course on occupational stress and mental health. After the course the participants
answered some questions about the realization of the proposed occupational changes.

The subjects reported whether the actions proposed had been discussed further or realized. The alternatives for responses were: the
realization of the action has started; nothing had been done to realize the action; the respondent did not know anything about the matter.
The results showed that seven important changes had been realized at
the youth home in one year. Regular counselling and consultation with
a psychologist, suplementary training during workdays, regular personnel meetings, and management training for the supervisors had been
arranged, and everyone carried out his/her duties without additional
requests. The number of the staff had also increased.

DISCUSSION

The personnel of the institutions for child care and institutions for mentally handicapped proved to be a risk group from the point of view of occupational health. The personnel studied had more burnout symptoms and psychological stress reaction than other Finnish groups in professions containing intensive social contacts. The personnel of the institutions experienced several characteristics of work as harmful. Work has characterized by quantitative and qualitative overload. Problems were found also at the organization of work as well as in the social contacts among the staff of the institutions. In regard to the amount of psychological stress reactions and the subjects' opinions about the work load, it would be very important to pay more attention to the development of the work and work conditions of the institutions of child care and the institutions for the mentally handicapped.

The study revealed that the members of the personnel of the institutions are well aware of their occupational problems. They are also active in trying to find solutions for the problems as the 213 proposals for the improvement of work indicated. Activity and interest towards one's work are known to be necessary preconditions for actions in prevention of harmful stress reactions. The proposals for the improvements of the work and working conditions at the institutions are relevant but not always easy or quick to realize.

The reduction of stress among personnel studied implied reduction of both quantitative and qualitative workload. The amount of children in departments should be decreased and new vacancies established to the departments. The reduction of qualitative work load is more difficult. Special attention ought to be paid to the basic professional training of the personnel of institutions for custodial child care. At the moment it is very heterogeneous. Some of the employees have studied three years at the university to become social workers, some have passed schools for social work, but some of the employees do not have any professional training. The majority of the personnel of the institutions for custodial child care has not studied education, psychology or social psychology, although they continuously work at situations that resemble group or individual therapy, teaching or nursing.

Continuous supplementary training should be arranged for the personnel studied. Systematic organizational development and increased co-operation between different groups of personnel ought to take place as well. The possibility for specialist counselling can also be considered as essential. The treatment of the emotional problems and cognitive conflicts evoked at work can otherwise be repressed.

The preliminary follow-up study proved that many important actions to prevent occupational stress had been carried out at one

institution within a year after the basic study. Of course, this
result cannot be generalized. A comprehensive investigation is needed
of the occupational changes and their effects on the characteristics
of the work and the stress reactions of the personnel at institutions
for custodial child care and institutions for the mentally
handicapped.

Small group activity for generation of new ideas in this study
proved to be effective. It could also be applied easily in problem
solving situations at other organizations.

REFERENCES

Kalimo, R., 1980, Conceptual analysis and a study on prison personnel,
       Scand. Journal Environ. Health, 6:suppl.3.
Ojanen, S., 1982, Stress of the teachers, in: "Reports 23 Serie A,"
       Institute of Education, University of Tampere, Tampere (in
       Finnish).
Perlman, B., and Hartman, E. A., 1982, Burnout: summary and future
       research, Hum. Rel., 34:4.
Pines, A. M., Aronson, E., and Kafry, D., 1981, "Burnout, from tedium
       to personal growth," The Free Press, New York.
Pöyhönen, T., 1981," The effects of the characteristics of the work and
       the worker on mental health among the personnel of mental
       hospitals," Institute of Occupational Health, Helsinki
       (unpublished, in Finnish).

STRESS AND TENSION CONTROL IN EDUCATION

AN EXPERIMENTAL STUDY OF RELAXATION TRAINING IN SWEDISH SCHOOLS:

PSYCHOLOGICAL AND PHYSIOLOGICAL RESULTS

Sven Setterlind and Goran Patriksson

Department of Educational Research
University of Gothenburg
Sweden

BACKGROUND

In recent years there has been a growing interest in different techniques of tension control in many fields, such as sport, education and medical treatment. In spite of this increased interest, it is remarkable that the amount of systematic research is small in relation to the frequency with which these techniques are used. This is particularly true with regard to research on children and adolescents, where few studies have been carried out. The studies that have been conducted have generally had a clinical approach and have usually been based on very few subjects. The situation is similar with regard to studies in school, where systematic large-scale investigations (Jacobson and Lufkin, 1968; Angers, Bilodeau, Bouchard, Tuthe, Mailhot, Trudeau and Vallieres, 1975; Engelhardt, 1976) are virtually non-existent. The lack of serious and well-documented research in the tension control area was one of the reasons why this project was started. Another, but quite different, reason was the belief that learning different kinds of tension control techniques could be of value for school pupils in our complex and often stressful modern society. A pilot study (Setterlind and Unestahl, 1978) showed some promising results, which strengthened the belief that such methods could be helpful in a variety of situations both in school and in leisure time. An important task in today's schools ought to be to increase the pupils' health consciousness by means of sport and other kinds of physical activity, to teach them the value of a well-balanced diet, the danger of drugs etc. These issues are usually emphasized in school curricula and other documents relating to the goals of education, but in Sweden, as in most other countries, the role of relaxation and other tension control techniques have been neglected (Setterlind and Patriksson, 1981).

THE AIM OF THE PROJECT

The project, entitled "Techniques of Tension Control and Their
Application in Educational Contexts", was started in 1979 and should
be completed in 1983.* Part of it has been documented in a doctoral
dissertation (Setterlind, 1983). The aim was to work out and evaluate
simple techniques of tension control for pupils in elementary school
and high school. Apart from this major aim, there were several other
questions to which, it was hoped, this project would provide answers:

- What short-term and/or long-term effects might a simple
  training program in relaxation achieve?

- Does age and/or sex affect the way a pupil experiences and is
  affected by relaxation?

- Are there any differences in the way a pupil evincing a high
  degree of anxiety experiences and is affected by relaxation
  compared to one evincing a low degree of anxiety?

- What stress reduction effects might be experienced in connection with relaxation?

- What general effects might a period of basic training in
  relaxation have on schoolwork and the situation outside
  school?

- Can training in relaxation affect body awareness and self-
  confidence?

- Is it possible to measure physiologically what happens to a
  child or adolescent during relaxation?

- Has relaxation any importance for recovery after physically
  strenuous activity?

- Why don't all pupils experience tension control as something
  positive?

- What form should an effective relaxation program take and how
  should relaxation in school be planned?

- What training is necessary for teachers and others concerned
  with pupil welfare, with regard to the application of relaxation in school?

- What should be included in training programs for teachers
  learning relaxation methodology?

---

* The project is financed by the Bank of Sweden's Tercentenary
Foundation.

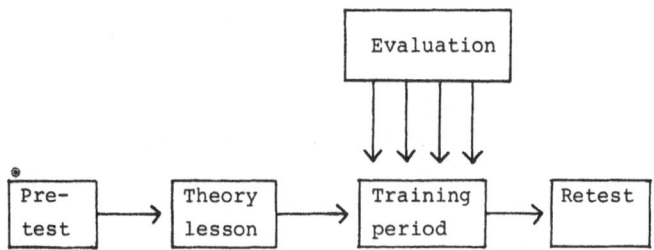

Figure 1. Schematic presentation of the main study.

A further aim of the project was to attempt to survey and evaluate the more common tension control techniques, in order to create some order in the mass of techniques and concepts at present used in connection with relaxation and self-control. The methods or techniques which were used in their entirety or in part in the investigation into tension control techniques were progressive relaxation, autogenic training and meditation.

THE DESIGN OF THE PROJECT

In the main investigation approximately 300 pupils between the ages of 12 and 17 received relaxation training three times a week for six weeks during the autumn term of 1979. The relaxation program, which was recorded on tape, was practised at the end of each physical education lesson. A schematic presentation of the main study is given in Figure 1.

In the first place, questionnaires were used to measure experiences, somatics and behaviour. The physiological effects, both as regards relaxation and recovery after physical effort, were registered in a special study in autumn of 1980. Further, a survey of the requirements for training and further training of teachers and other staff concerned with pupil welfare was made.

MAIN STUDY

The overall plan of the main study is shown in Figure 2.

Questionnaire AI consisted of 32 questions and was filled in during the week prior to the commencement of tension control training. At that time, pupils did not know that a study would be carried out at their school. Both the control group and the experimental group answered the questions, which were intended to measure the following:

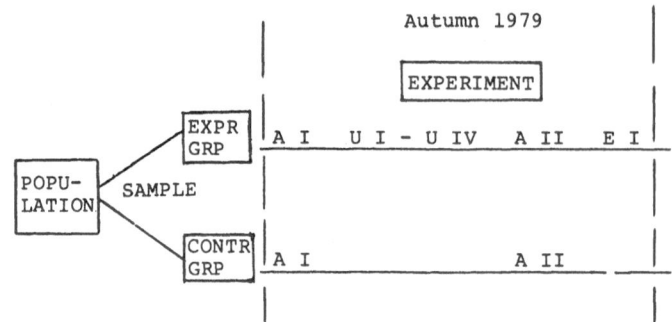

Figure 2. Plan of the main study.
(A = general questionnaire)
(B = questionnaire on experiences during relaxation)
(C = questionnaire on more long term results)

1. Previous experience of and motivation for relaxation.

2. Attitude to and experience of school.

3. Sport, exercise and smoking.

4. Worry, anxiety and psychosomatic problems.

5. Mood test.

Questionnaire AII was identical to AI except for the fact that the questions on previous experience of and motivation for relaxation were deleted. This questionnaire was answered during the week after the completion of relaxation training. Together with questionnaire AI, it was the only one that was filled in by both the experimental and control groups. Questionnaire B was intended to provide a picture of pupils' experiences while they were learning relaxation. The questions were answered immediately after the pupils had concluded relaxation sessions on four different occasions during the six-week long training period. Each questionnaire U included four types of questions:

1. General experiences of relaxation training.

2. Special experiences during relaxation.

3. Degree of physical and mental relaxation.

4. Mood test.

The final questionnnaire (E) in the main study was handed out just before the Christmas holiday of 1979. The questions, a total of 26, were constructed to measure what the pupils got from relaxation training, i.e.:

1. The experience of learning relaxation.

2. Views on the relaxation program.

3. Views on the design of relaxation training.

4. The effects they received from relaxation training.

SELECTION OF SCHOOLS AND SUBJECTS

One middle school (6th grade), two upper level schools (7th and 8th grades) and two high schools (1st and 2nd years) with a total of 294 pupils took part in the experiment; there were also 287 pupils in the control group. The distribution of the sexes was as follows: in the experimental group 48% were boys and 52% girls, in the control group 52% boys and 48% girls, as in Table 1.

Table 1. Numbers and distribution of pupils in the main study.

Level	Experimental Group	Control Group
6	51	53
7	55	54
8	63	60
I	56	58
II	69	62
	294	287

APPLICATION OF THE TECHNIQUES IN SCHOOL

In order to discover the most suitable method of teaching relaxation in school, a detailed study of various tension control techniques was made with regard to learning and effects. Since we did not know of any comparative studies which provided evidence for the advantages of any one technique over another, various methods of learning relaxation were combined. We thus commenced with an abridged version of progressive relaxation, followed by a tension control

progam with some autogenic exercises such as heaviness and warmth.
The pupils also learned to start tension control rapidly, a technique
that can be used in situations of particular stress. They also
learned to deepen relaxation and, further, to use simple meditation
techniques of the type in Benson's (1975) "The Relaxation Response".
The aim of the relaxation training in school was to allow the pupils
to test various techniques in order to choose, at a later stage, the
technique or combination of techniques that would suit them best. It
was hoped that they would then be able to continue training on their
own without the aid of a cassette or a teacher. The order, contents
and length of the programs are described below:

		Number of Minutes	Week Number
1.	Muscular relaxation	12	1
2.	Muscular relaxation	8	2
3.	Mental relaxation	10	3
4.	Mental relaxation	8	
5.	Conditioning by a signal	8	4
6.	Testing of methods	8	
7.	Choice of individual method	8	5
8.	Self-instruction	7	6

A MODEL FOR LEARNING RELAXATION

As with all complex learning, it may be presumed that components
from various theories of learning are important when learning relaxation. A comparison may be made with Gagnes (1970) hierarchic construction of the learning process. With reference to the above,
learning relaxation can, therefore, be explained by means of a number
of factors. The final theoretical model for learning consists of
eight different phases, which include parts of different theories of
learning. The phases are presented below:

Phase:

1. Inform and motivate the pupils in a lesson before starting
   relaxation training in the physical education lessons.

2. Give experience of what relaxation is. (Programs 1 and 2).

3. Learning relaxation by conditioning processes. (Programs 1 and 2).

4. Passive concentration on certain muscles or groups of muscles gradually produces, after a period of training, a relaxation effect by means of the "reversed ideomotory tendency". (Programs 3 and 4).

5. Conditioning by a "trigger", a special signal, which rapidly stimulates the relaxation reaction. (Program 5).

6. Consciously test various techniques of relaxation and of deepened relaxation. (Program 6).

7. Individual choice of techniques. (Program 7).

8. Further independent training. (Program 8).

It is also possible to identify another phase, namely, generalization to other situations, but this is considered as lying outside the learning situation.

SOME RESULTS FROM THE MAIN STUDY

The major aim of the study was to measure the subjective experiences and effects of relaxation training during a six-week period. A large battery of questions, which were intended to complement each other, was used to provide a complete picture. Significance testing was not carried out. Instead the trends were described by calculating the percentage of change. Many researchers are sceptical of significance testing, particularly with field investigations of the survey type, since they cannot, then, test the data from an overview perspective. Furthermore, there is an obvious risk of type-2 errors (Galtung, 1967). The material from the various areas was instead, collected and totalled for analysis, even on the subgroup level, sex, school levels and degree of anxiety. The analysis of individual questions was not considered as important as the mapping of tendencies during the learning of relaxation.

Since the data are so extensive, only a survey of the results are given here, together with certain comments. The analysis shows that the dropout, both totally and on the subgroup level, was very little (5–10%) and should, thus, not bias the results.

Experience of relaxation training was measured on four different occasions in the experimental group, using questionnnaire U. The results of two questions are shown in Table 2.

Table 2. Two questions from questionnaires of experiences UI - UIV.

Q & A	UI	UII	UIII	UIV
Q1 "What is your experience of the relaxation training?"				
Very pleasant	50	53	51	48
Quite pleasant	44	39	40	42
No opinion	5	7	8	7
Rather unpleasant	1	0	0	1
Very unpleasant	0	1	1	2
Q2 "I feel that it is:"				
Very easy for me to relax	25	27	28	31
Quite easy for me to relax	54	57	56	52
No opinion	18	13	12	13
Rather difficult	3	2	2	3
Very difficult	0	1	2	1

About 90% of the pupils found that, on each occasion, relaxation was pleasant and positive. Over 80% thought is was very easy or fairly easy to relax. Only about 3% found it was difficult for them to relax. About 70% felt they were completely or almost completely relaxed on the first two occasions whereas there was a slight reduction on the next two occasions. The girls tended to be more positive than the boys on the first two occasions.

The experience of relaxation was measured by means of carefully tested questions immediately after the pupils had finished listening to the relaxation program and had sat up. Most of the pupils (80-90%) generally experienced calmness and rest on the first three occasions; on the final occasion there was a slight reduction.

The pattern was more or less the same for all the experiences detailed in questionnaire U. Most of the experiences were concerned with bodily sensations, such as a feeling of heaviness, which an average of about 70% experienced to a high degree. Other bodily experiences and average values were: the release of various tensions (63%), a feeling of sinking (59%), warmth in various parts of the body (43%), various parts of the body became numb (41%), and lightness in the body (30%). About half the pupils experienced mental sensations, to a high degree; their surroundings "disappeared", their mind became blank and they had difficulty in estimating the time they had been in a state of relaxation. About 50% felt tired when the session ceased and as many felt very rested. For further details see Setterlind (1983, p. 283).

An analysis of the results for various subgroups reveals a slight tendency for boys, pupils in the upper level schools, and pupils evincing a high degree of anxiety to show positive development throughout.

Various methods of measuring the degree or depth of relaxation were tested. By noting all those who indicated throughout that they experienced the different variables to a high degree, a categorization of the pupils as regards depth of relaxation gave the following results: about 25% experienced light relaxation, 50% medium and 25% deep (Setterlind, 1983).

The final measuring device on the questionnaire was a mood test, (Svensson, 1978) covering eleven positive emotional states and their opposites. The emotional states that were most relevant for the study received consistently high values. Thus, about 90% of the pupils felt assured, satisfied, happy, friendly, relaxed and calm immediately after relaxation exercises during the physical education lessons. These emotional states were more or less stable but, on the final occasion there was a slight reduction in most of them.

Questionnaires AI and AII were used to measure the possible effects of relaxation in the experimental group as compared with the control group. The first questions concerned the pupils' attitude to and experience of school. Almost 60% thought that it was fun to go to school every day or at least 3-4 days per week. A little more than 30% thought that it was fun 1-2 days per week. Just under 25% wished they didn't have to go to school every morning or at least 3-4 days per week. About 15% experienced stress at school every day or at least 3-4 days per week. About 35% experienced this 1-2 days per week. The retest revealed some changes to the advantage of the experimental group. These changes are strengthened if an analysis is made on the subgroup level. It was particularly the boys in the experimental group that changed; they became more positive to school. Pupils at upper level schools and pupils evincing a high degree of anxiety were most negative to school but those in the experimental group revealed some positive change. The girls, pupils at the upper level schools and high schools, and those evincing a high degree of anxiety found school a greater stress situation than other groups did. The retest revealed more and greater changes in a positive direction for the experimental group than for the control group.

A number of questions in the tests were collected under the heading: worry, anxiety and psychosomatic problems. There seemed to be some difference between the groups in some questions right from the start. Thus, more pupils in the control group were calm, assured and

not particularly nervous before examinations than in the experimental
group. The retest revealed however, that the pupils in the experimental group had more often changed in a positive direction whilst
those in the control group hadn't experienced any change at all. On
the subgroup level, the greatest changes were found in the girls,
pupils in the middle school, and upper level schools, and those evincing a high degree of anxiety in the experimental group, as regards
the questions on calmness, security, nervousness before examinations,
uneasiness about being alone and about going to the dentist, and
satisfaction or dissatisfaction with themselves.

As regards the questions on psychosomatic problems, however, the
changes were very small and went in both directions. This was the
only point in this study where the trend was not quite clear. It is
difficult to give any definite reasons for this. Probably the instruments for testing were not sensitive enough.

The mood test (Svensson, 1978) seemed to be a sensitive measuring
device for detecting changes in emotional states before and after the
test period in school. In general, it may be stated that there were
many positive changes in the experimental group and many of them were
relatively large in comparison with the control group. The total
"points of change" in the 22 variables was almost three times as great
for the experimental group as for the control group (159 to 58). The
eight most relevant emotional states for the study were combined in
order to calculate the points of change for the subgroups. These
emotional states were: relaxed, tense, unstressed, stressed, alert,
tired, self-assured and shy (Setterlind, 1983).

Table 3. Points of change for the high and low anxiety pupils
in the experimental and control group.

	Experimental group		Control group	
	High anxiety	Low anxiety	High anxiety	Low anxiety
Relaxed	19	1	9	4
Tense	27	5	-11	-4
Unstressed	29	5	-10	12
Stressed	22	3	-10	-1
Alert	22	15	1	10
Tired	11	26	5	21
Self-assured	3	11	6	10
Shy	4	9	1	-2
Total	115	75	22	57

A short summary of the results of the mood test shows that the girls, high school pupils and those evincing a high degree of anxiety in the experimental group changed most from the pretest to the retest. The total points of positive change for the three subgroups in the experimental group was 645 and for the control group 184. To this must be added 106 points of negative change for the control group as against 0 for the experimental group. As an example the results from the high and low anxiety pupils are shown in Table 3.

When the experimental period was over the pupils were asked to answer questionnaire E, which dealt with their experiences of the relaxation programs and of the way the training was carried out. There proved to be a clear majority who thought that relaxation had worked and that they had learnt to relax. 56% were absolutely convinced of this whilst 1% didn't at all feel that they had learnt to relax. Most of them were now able to relax themselves without assistance. Only 3% considered that they couldn't manage this. Almost half

Table 4. Subjective experiences of the relaxation training (%).

I feel less irritable	44
I feel more irritable.	2
No change	54
I feel more rested and alert	46
I feel less rested and alert	8
No change	46
I find learning easier	25
I find learning more difficult	4
No change	71
I manage school work better	52
I manage schoool work worse	2
No change	46
I sleep better	33
I sleep worse	3
No change	64
I feel less stressed	60
I feel more stressed	2
No change	38
I feel more at ease	40
I feel less at ease	4
No change	56

stated they had definitely found it useful to be able to relax. About 25 pupils or 10% didn't want to continue with relaxation training during the spring term.

One third of the pupils in the experimental group thought that relaxation training had been of great value and 8% considered it worthless. Over 90% felt that it had been easy to learn to relax using the model they had tried. Just over half of the pupils had been positive to relaxation exercises from the beginning and still were, whilst over 40% had become more positive. 7% were negative to tension control exercises.

Finally the pupils in the experimental group were asked to answer questions about what they thought had been improved or worsened by the six-week relaxation training. Some of the noteworthy results were that over half said that they managed their schoolwork better and 1/3, that they slept better. 60% felt less stressed, 44% less irritated and 46% more rested and alert than previously. Many pupils (24%) felt it was easier for them to learn things in school (See Table 4).

Complementary interviews with pupils and teachers reinforced the picture presented by the questionnaires (Setterlind, 1983, pp. 336-347).

DISCUSSION

The experiment with relaxation training in school seems to have been very successful if one takes the results of the pretest and retest into account. However, certain reservations concerning the measuring techniques should be made. As the sample of subjects in the experimental group and control group was not entirely random, but was regulated by organizational and timetable considerations, regression effects pose a threat to unambiguous results (Stukat, 1970). These effects are, however, most noticeable when the individuals are taken from extreme groups and when it is a question of one-group studies. In this study it should be the group of pupils evincing a high or a low degree of anxiety that are most exposed to regression effects, as these groups were made up of the approximately 15% who evinced the highest or the lowest degree of anxiety according to the criteria used in this study. There is also a risk in studies like this that the individuals distributed non-randomly differ with regard to spontaneous individual changes.

In all experimental field studies where new methods are tested, irrelevant environmental influence is always present because of, for example, the special attention the experimental group receives. This so-called "Hawthorne-effect" is expressed in the form of positive change. Another source of error is the effect of expectation and suggestion unconsciously conveyed by the teachers involved. To avoid these sources of error the individuals in the control group should receive the same attention as the experimental group or a similar treatment with a neutral content, the so-called placebo. In a pilot study, (Setterlind and Unestahl, 1978), such a method study was carried out; the pupils in the control group were asked to lie down on the floor and chose their eyes, but they received no stimulus. They soon got bored and were not interested in participating in the program for the full length of the experiment. Just over 40% were satisfied with this training, as compared to 90% in the experimental group, who listened to the relaxation program on tape. Therefore, it was not judged meaningful to use a placebo control group in this extensive study because of the risk that the pupils would not want to complete the experiment. The control group did in this study, however, receive some "treatment" since they took part in the same physical education program with the exception of the relaxation training. It may also be said that many in the experimental group thought that this training took too much time.

Disturbances in the measuring process may also have influenced the result. It appears that there was a tendency for many pupils to get tired with the many tests while they were learning relaxation. They were, in particular, negative to the last test, something that was confirmed in the interviews with pupils and teachers. It seems that the pupils in the middle school were especially sensitive to this. The results also suggest that the various age groups experienced relaxation differently. Thus, the middle school pupils were most positive to muscular relaxation but they were less interested in the programs where the pupils started tension control and trained on their own. Upper level pupils and, in particular, pupils in the high schools were positive throughout the period of learning and were especially interested in the mental and meditative programs.

It is not likely that these sources of error are the only explanation of all the major changes to the advantage of the experimental group. Some of the changes that took place can possibly be explained by an artificial flexibility in the material, but such great and systematic changes as those found in this study must suggest real changes as a result of relaxation training. Furthermore, if fewer measurements had been taken during the period of learning, the results would probably have been more positive.

The studies show that information about stress and instruction in stress reducing methods, such as relaxation, are comparatively easy to administrate in the course of the regular Physical Education classes at school, and should therefore constitute a fundamental part of Health Education.

PHYSIOLOGICAL STUDIES

    The aim of the special studies, which were carried out in the
middle of the autumn term of 1980, was to register the recovery phase
in a group of 14-15 year old pupils after strenuous activity on a
bicycle ergometer. A comparison was made between an experimental
group which had trained-relaxation over a five-week period, and a
control group, which had not received any relaxation training. One
study covered 40 pupils (recovery Study I) and the other 20 (recovery
Study II). Some of the questions raised in connection with these
studies were:

- Is it possible to measure what happens to children and
  adolescents during relaxation and rest, by physiological
  methods?

- Is it possible to establish any differences in effects and
  experiences between the experimental group's relaxation and
  the control group's rest?

- Has relaxation any significance for the feeling of recovery
  after physically strenuous activity in children and
  adolescents?

    In an attempt to find answers to these questions, a test battery
consisting basically of physiological, but also of psychological measuring devices was used.

DESIGN

    The studies were carried out with 8th grade pupils. They were
from the same group that had taken part in the main study during the
year 1979-80, when they were in the 7th grade. In the middle of
October, all the pupils in the 8th grade (98 altogether) underwent a
sub-maximum work test on a bicycle ergometer to calculate their ability to absorb oxygen (Astrand and Rodahl, 1970). With the aid of the
test values and the weight of the individuals twenty pairs of pupils
were matched, ten pairs of girls and ten pairs of boys, thus forming
an experimental group and a control group.

    In the next week the matched pairs were submitted to two pre-
tests, consisting of 4 minutes warm up on a normally loaded bicycle
ergometer and then 10 minutes work at 80% of the calculated maximum
ability to absorb oxygen. Before the testing began, the pupils had
lain down and rested for 4 minutes while their pulse rate at rest was
noted. While they worked on the bicycle ergometer, the pupils' heart
rate was continuously monitored by a cardiometer. After five and ten
minutes the pupils made a subjective assessment of their exertion
(Borg, 1973). The load was on average about 790 kpm in both the

pretest and retest for both the experimental group and control group. In the pretest the load varied between 600 and 1050 kpm and in the retest between 600 and 900 kpm.

After 10 minutes' work on the bicycle ergometer, the pupils lay down on mats to rest for 10 minutes. Their heart rate was measured after 1, 3, 5, 7, 8, 9 and 10 minutes. The rate of breathing was measured with the help of a piece of paper that was folded and placed on the chest of the subject, and it was counted over a period of 60 seconds. Immediately after their 10 minutes' rest, the pupils were asked to complete a "rest test" where they were asked about their state of rest and recovery.

During the following five weeks the ten boys and ten girls in the experimental group were urged to train relaxation at home five times per week. They were instructed by the project leader how they should train with the structured programs used during the autumn term of 1979. These were recorded on a cassette so that they could use them at home. They also received a simple record card to be filled in after each training session. This consisted of three simple questions about the frequency and the experience of physical and mental relaxation.

The following week two retests were carried out and during these the pupils in the experimental group consciously applied their relaxation techniques while resting. Otherwise, the test took place under the same controlled conditions as in the pretest.

A further special study was carried out in the following week, recovery Study II, on five boys and five girls from both the experimental group and the control group. The criteria for the selection of these matched pairs were that the pupils had to have taken part in all the tests during both the pretest and retest so that there would be no dropout in the matched pairs.

The reason why two pretests and retests were made in the same week with a day between them was to provide as reliable results as possible and to remove all potential sources of error, such as mistakes in measurement and variations in form. In the presentation of the results the mean values for the two test occasions will be used throughout.

Recovery Study II was carried out at the pupils' own school, in the same way as the pretests and retests with the larger group, but this time blood pressure and blood tests were also taken. Before the pupils started work on the bicycle ergometer, a blood test at rest was taken; then a test was taken during the last minute's work and again on four occasions during the 10-minute rest period. The blood tests were taken from a vein catheter placed in the bend of the arm 10-15 minutes prior to the start of the experiment. The blood pressure was registered by an indirect method (sleeve and auscultation) at rest, just before the work was completed and again on four occasions during the rest period. As before, the pupils were asked to give a subjective assessment of their exertion after five and ten minutes' work on the ergometer, and after the rest period, to give an assessment of their degree of recovery.

The blood tests were taken for later analysis of lactic acid, (Lowry and Passonneau, 1973), adrenaline and noradrenaline (Christensen, 1973). The blood samples were centrifuged immediately after the experiments and the plasma kept at −80° C until analysed. Catecholamine analyses were made on a limited number of pupils (five boys and five girls) from the experimental group and the control group. As in the previous study, the heart rate was determined with a cardiometer and the rate of respiration counted over a period of 60 seconds.

RESULTS OF THE PHYSIOLOGICAL STUDIES

The results from both of the special studies give a clear picture of recovery from strenuous physical activity on the bicycle ergometer. In recovery Study I the differences between the experimental group and the control group as regards rate of respiration were significant at 0.01 level (See Figure 3).

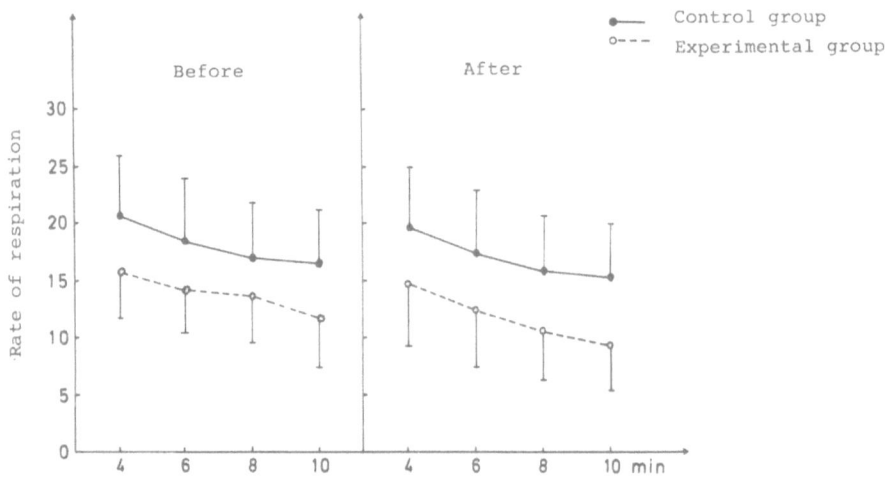

Figure 3. Rate of respiration before and after the test period in the experimental and control groups. (Recovery study I)

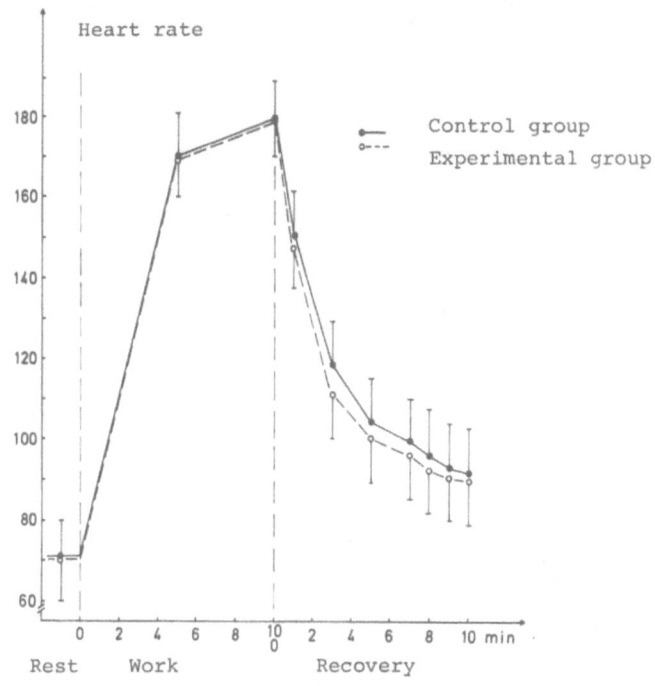

Figure 4. Heart rate in the experimental and control groups
(Recovery study I)

Even at the pretest stage, the test group's rate of breathing was considerably lower than that of the control group. A probable explanation for this is that the pupils in the experimental group took part in the experiment in relaxation training the previous year. The results of the pretest may be, therefore, be an indication that learning relaxation was successful, even in a long-term perspective. In spite of this, the experimental group lowered their rate of breathing further and even more than the control group. With regard to the heart rate, the tendencies were clear, but, in contrast to the rate of breathing, no significant differences were noticeable between the two groups (See Figure 4).

The subjective experiences of rest and recovery were much stronger in the experimental group than in the control group (Table 5).

The picture was relatively uniform in recovery Study II. During recovery, major significant differences between the groups were only noted for the rate of respiration, whereas for variables such as blood pressure, heart rate and catecholamines in the plasma there was a slight tendency towards a more rapid return to normal in the experimental group. The results for blood pressure and catecholamines are presented in Figures 5 and 6.

Table 5. Experiences of rest for the experimental and control group.
(In percent) – (Recovery study I)

Q & A	Pretest		Retest	
	Experimental Group	Control Group	Experimental Group	Control Group
Q1 "How have you managed to rest during these 10 minutes?"				
Very well	26	50	85	36
Quite well	74	40	15	32
Not so well	0	5	0	16
Rather badly	0	5	0	11
Very badly	0	0	0	5
Q2 "How recovered were you after the rest?"				
Completely	37	20	65	21
To a high degree	58	75	35	68
To a certain degree	5	5	0	11
Not at all	0	0	0	0

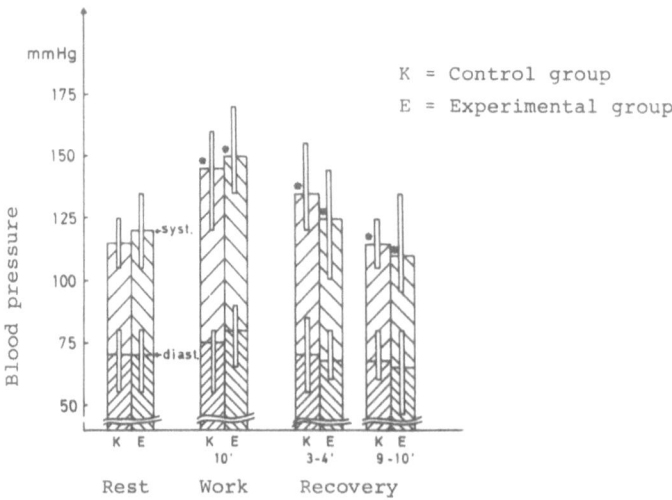

Figure 5. Blood pressure in the experimental and control groups.
(Recovery study II)

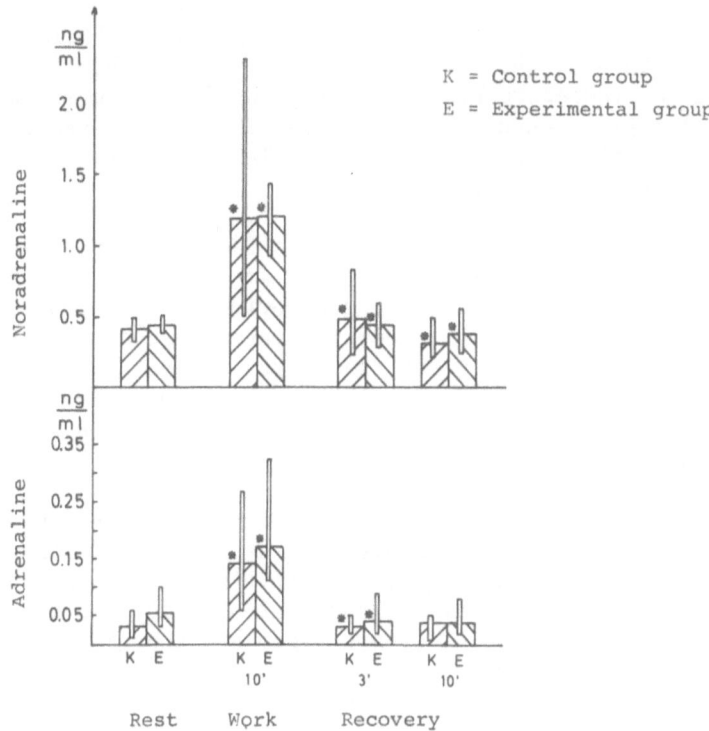

Figure 6. Catecholamines in the experimental and control group.
(Recovery study II)

Even if the group mean values only showed small differences,
there were some individual cases in the experimental group where the
return to normal for all the relevant variables was more pronounced,
as can be seen from the greater deviations round the mean values. No
differences were found concerning lactic acid.

DISCUSSION

The clear difference in rate of respiration with a more rapid
return to normal in the experimental group is linked to the fact that
breathing is more easily influenced by the will and that this form of
relaxation training focused on breathing. The subjects worked on
similarly loaded ergometers. Mechanical effects showed only small
individual variations, so it can be assumed that both oxygen consumption and ventilation were the same for both groups at the commencement
of the work. It is not likely that the rest level for ventilation was
achieved in either group during the 10-minute rest period. This
conclusion is also supported by the fact the lactic acid concentration

in the blood was heightened after the ten-minute-rest. This means
that the experimental group probably compensated for their low rate of
respiration with a slightly greater volume per breath. Alveolar
ventilation was therefore greater, which could lead to somewhat better
conditions for gas exchange. It is more difficult to assess to what
degree the deeper breathing helped to explain the differences in the
reduction of the heart rate.

The identical reduction of catecholomines suggests that the dampening of sympathetic activity after work was the same in both groups.
The somewhat lower heart rate in the experimental group was, however,
connected with a somewhat more powerful parasympathetic activity. In
this connection, it can be noted that vagus activity is primarily
significant for frequencies under 120-130 beats per minute. It was
also here that the small differences between the groups was most
obvious.

In spite of a more extensive and disturbing measuring procedure
in recovery Study II than in recovery Study I, many of the pupils in
the experimental group found no difficulty in resting and they felt
rested afterwards. The pupils in the control group did not feel
anything like as recovered afterwards and thus thought it was more
difficult to rest.

The conclusion is that most of the results show that relaxation
training with adolescents seems to have a certain physiological effect
but, in particular, a psychological effect with regard to the feeling
of rest and recovery.

GENERAL CONCLUSION

An important task in today's school is to make the pupils more
aware of their health. This can be done by increasing their awareness
of their body and its possibilities. This includes teaching them to
listen to their own body and to be sensitive to its signals. The
delicate interplay between mental and physical factors is easily
disturbed and disturbances in the form of acute or chronic stress can
lead to ill health. However, not enough information is given about
the various methods of preventing and dealing with the stress and
overstimulation our complicated and easily-disturbed nervous system is
exposed to in today's society. Relaxation could be included under the
heading "mental prophylaxis" and could play an important part in a
holistic health care perspective.

REFERENCES

Angers, P., Bilodeau, F., Bouchard, C., Luthe, W., Mailhot, D.,
    Trudeau, M., and Vallieres, G., 1975, Application of autogenic
    training in an elementary school. Part I and II. In Luthe and
    Antonelli, red., Therapy in Psychosomatic Medicine, Proceedings
    of the 3rd Congress of the International College of
    Psychosomatic Medicine, September 16-20, 1975, Rome, Italy.
Astrand, P-O., and Rodahl, K., 1970, Textbook of Work Physiology.
    McGraw-Hill Book Company, New York.
Benson, H., 1975, The Relaxation Response. William Morrow and
    Company, Inc., New York.
Borg, G., 1973, Perceived exertion: A note on "history and methods".
    Medicine and Science in Sports, 5, 90-93.
Christensen, N. J., 1973, Plasma noradrenaline and adrenaline in
    patients with thyrotoxicosis and myxoederna. Clin. Sci. Mol.
    Med., 45, 163-171.
Engelhardt, L. J., 1976, The application of biofeedback techniques
    within a public school setting, presented at: Biofeedback
    Society, Seventh Annual Meeting, Colorado Springs, Colorado.
Gagne, R. M., 1970, The Conditions of Leaning. Holt, Rinehart and
    Winston, Inc., London.
Galtung, J., 1967, Theory and Methods of Social Research.
    Universitetsforlaget, Oslo.
Jacobson, E., and Lufkin, B., 1968, Tension control in public schools.
    A research project in physical education. Part 4. Foundation
    of Scientific Foundation, Chicago.
Lowry, O. H., and Passonneau, J. V., 1973, A Flexible System of
    Enzymatic Analysis. Academic Press, New York.
Setterlind, S., 1983, Teaching Relaxation in School. A survey of
    research and empirical studies. Department of Educational
    Research, University of Goteborg, Sweden (in Swedish).
Setterlind, S., and Unestahl, L-E., 1978, Introducing relaxation
    training in Swedish schools, presented at: First European
    Congress of Hypnosis in Psychotherapy and Psychosomatic
    Medicine in Malmo, Sweden.
Setterlind, S., and Patriksson, G., 1981, Teaching children to relax.
    Journal of the Society for Accelerative Learning and Teaching,
    6(3), 206-211.
Stukat, K-G., 1970, Research Methods in Education. Almqvist and
    Wiksell, Stockholm (in Swedish).
Svensson, E., 1978, Mood: Its structure and measurement. Psychological
    Reports, Number 6, Volume 8, University of Goteborg, Sweden.

SOME USES AND LIMITATIONS OF STRESS CONTROL METHODS

IN A LOCAL AUTHORITY EDUCATIONAL PSYCHOLOGY SERVICE

David Cowell[*]

Senior Educational Psychologist
Wiltshire County Council
Trowbridge, England

ABSTRACT

This paper outlines the ways in which methods of stress control
have been used by educational psychologists and allied professional
groups over a period of six years. Reference is made to individual
case studies. Limitations are considered, and alternate strategies
suggested for some difficult-to-help pupils.

CONTEXT

The relationship of physiological arousal to psychological processes has been the concern of applied and academic psychologists for
many years. Geen, Rakosky, and Pigg (1972) discuss a viewpoint which
has received widespread support. A number of subjects are placed in a
situation where their arousal is increased by others. Their perception of the motivation involved influences the nature of their response, i.e., where the perceived intention is to annoy, this does
tend to occur. Where arousal is perceived as being accidental, there
is much less feeling of annoyance and aggression. As educators and
psychologists, we can therefore derive a practical plan for attempting
to deal with a continuing problem in schools, that of the 'disruptive'
pupil. Teachers would normally be advised to "avoid confrontation",
thereby not physiologically arousing potentially difficult pupils.
Teachers should attempt to explain the reasons for points of conflict.

[*] Acknowledgement: I should like to express my appreciation to
the following colleagues who have helped in various ways:
        J Franklin, L Harlatt, J Nicholas, J Roberts,
        R Springall, J Walker, R West and P Wingfield.

This model has recently been developed by Apter (1982) in his theory
of telic/paratelic dominance. The above is commonly accepted by
educational psychologists, although ways of seeking to control stress
have received little attention. The author was responsible for Day
Adjustment Classes for Junior and Secondary age pupils. It was important to give appropriate advice to the experienced teaching and welfare staff, and this initially included discussion of the nature of
arousal, and the role which could be played by counselling which aimed
to bring about a degree of cognitive restructuring. Subsequently, the
team used galvanic skin-recording machines, or 'relaxometers' on a
fairly regular basis (Cowell, 1982; Cowell and Franklin, 1983).

We were encouraged by reports that relaxation-training had been
shown to have some degree of effectiveness in modifying hyperactive
behavior in boys (Dunn and Howell, 1982; Omizo and Micheal, 1982), in
promoting more desirable types of classroom behaviour in learning-
disabled children (Amerikaner and Summerling, 1982) and in temper-
control (Collins and Hodgkinson, 1982).

Cobb and Evans (1981) review forty-four studies investigating the
efficacy of biofeedback procedures in treating behavioural and learning disorders. Various methodological weaknesses were identified and
the data did not suggest that biofeedback techniques are superior to
more conventional treatments in remediating learning or behavioural
disorders. However, many studies reported eclectic procedures, making
it difficult to isolate experimental variables. This need not be due
to lack of respect for research purity, but rather because of an
interest in developing effective treatments, where concerned practitioners may wish to individualise therapeutic programmes. An interesting example is provided by Smith and Denney (1983) who effectively
combined relaxation-training with assertiveness training in helping a
woman with a history of chronic headaches following a traumatic injury
to her head and neck.

Winfield (1983) notes that biofeedback appears to be used as an
aid for specified counselling goals wherein the counsellor gains a
better understanding of the client because he/she can be in direct
contact with the client's physiological reactions. We similarly found
that biofeedback was efficacious for both the clients and the staff.

POPULATION AND METHODS

Stress control procedures were conducted on three groups
including:

(1) adjustment class pupils,
(2) counselling students and
(3) adult education.

Adjustment Class Pupils

We were particularly interested in using stress-control procedures with pupils who agreed that they were bothered by tension or who
appeared to show high or free-floating anxiety on a counselling questionnaire. Initially, the author (an educational psychologist) ran
the sessions, but they were also quite effective when supervised by a
welfare assistant who used a written script. Some pupils preferred to
use a tape-recorded script, and a few seeemd to respond quite well to
progressive relaxation practice paying attention only to the buzzer on
the machine. Initially, we used the sessions as a form of 'time-out'
to avoid conflicts, but our more general aim was to teach the pupil to
apply tension-control whenever it was needed.

A number of quite experienced teacher and other staff at first
found great difficulty in practising as relaxation therapists. They
needed continuing support, as well as more experience as students in
the relaxation-procedures. A number of scripts were devised, some of
them based on Hartland (1971), and these are included in the appen-
dices. Appendix I "Quick Relaxation" was found to be most popular,
although Appendix II "Deep Relaxation" has also been used quite extensively and effectively.

The above procedures refer mainly to Secondary School age pupils.
Individual methods have been used with younger pupils, particularly
those over nine years old who have sufficient understanding of the
method, and who can see that they experience feelings of tension.
With those of Junior School age, a group relaxation procedure was
additionally used by the teacher in charge of the class.

One ten year old currently attending an Adjustment Class was
first referred to the educational psychologist in 1981 for restless
and difficult, but not delinquent, behaviour in school. There were
various family problems including the serious illness of a close
relative. He was seen informally by the educational psychologist in
the Centre during the course of which the following responses were
given as part of a counselling inventory.

"I would like to be better behaved. I need to relax more. I
sometimes find it hard to sleep at night. I sometimes have a stomach
ache, feel stressed ('tight') and feel bad tempered for no reason. I
have lots of arguments with my friends and my teachers at school. I
frequently forget things and I find it hard to concentrate on school
work, I frequently feel tired, usually for no reason and I am sometimes moody."

Clearly, this particular pupil was suffering from tension, but he
was able to gain some insight into his situation, and is thus one of
the group of pupils for whom there is a good prognosis for teaching
stress control. Relaxation practice was suggested as part of a time-
out procedure. There appeared to be a reduction in problems but

following a resurgence of difficulties in school in 1983 he was admitted to the Adjustment Class where his behavior rapidly improved. At
the time of writing, July 1983, there has already been an encouraging
report of his progress.

Counselling Students

We have previously reported cases including social phobias with
poor school attendance, outbursts of weeping at school, insomnia with
associated school problems, a speech stoppage, and a wide range of
other difficulties (Cowell and Franklin, 1983). Winfield (1983) asserts that biofeedback may have "come of age", and says that in
America, at least, it is regarded by many practitioners as a well-
tried and respected adjunct to their counselling endeavours, however
the situation appears to be different in England. Daws (1976) recommends a greater emphasis on guidance and a more eclectic approach.

Lago (1981) illustrates the use of progressive-relaxation in
systematically desensitizing an adult client who experienced phobic
anxiety. The tension-control method here entailed the usual kind of
spoken technique, and subsequently the use of tape-recorded scripts.
A technique involving imagery was used to bring about unlearning of
the learned response by the repeated presentation of the conditioned
stimulus in a modified form. Once this theory and technique has been
grasped by the counsellor or the psychologist, it can be used in a
wide range of counselling interactions with not only adults but young
people. A controlled projection technique seems to be useful, where
for example the counsellor might say - "I want you to relax and I will
mention certain situations which we have already discussed. I want
you to indicate to me those which make you feel particularly tense."
Imagery can be used to overcome anxieties, and various behavioural
strategies may also be devised. These might include setting-up discussions or situations with the prior agreement of teachers and
others, perhaps to negotiate or apologise.

Part of the service of an Educational Psychology Department is
to schools and teachers. Some pupils may not be considered by their
teachers as either anxious or disruptive, but may be worried by feelings of tension. Wherever possible, such young people are considered
to be priorities, and often benefit from advice on a simple relaxation
routine, using breathing, muscular relaxation, particularly of the
muscles of the upper body and face, and appropriate focusing of attention. This is normally practised as part of a counselling session,
which could be held either in school or at some other more appropriate
location. In general, we have found that pupils attending normal
schools need to have tension-control programs taught by a member of
the department. Attempts have been made to prepare teachers as qualified instructors in the school, with very limited success so far. In
the absence of qualified teachers/instructors in tension control,
individually-made tape-recordings can be used in school to to make
economoical use of scarce resources.

Adults

Professional staff of educational institutions have quite frequently taken an interest in tension-control techniques for personal
reasons and Ross (1981) reports success with a number of individual
cases. It has also been suggested, albeit in a half-satirical fashion, that training for senior posts in the Education Department should
involve a study of how to relax under pressure (Cowell, 1982). It may
be that a number of books on tension-control have made people aware of
the possibilities of such techniques (Rosa, 1976; Norfolk, 1979;
Parrino, 1979; Mason, 1980).

The situation is a little different for the parents of handicapped children. In spite of emotional stress which many of them
experience (Wilks, 1974; Gath, 1978; Van der Hoever, 1978) the use of
relaxation procedures has been referred to only very rarely. There
are signs, however, that relaxation may be increasingly regarded as an
important part of a professional service (Kozloff, 1979). As a local
authority psychological service we have an ongoing involvement with
parents of handicapped children through the special schools. In
'workshop' situations similar to those of Clayton (1982), we feel that
the following methods and aims are important (Cowell 1983). These
include:

(1) to outline some basic facts about the physiology
    and phenomenology of 'anxiety',

(2) to describe and practise tension-control
    techniques and

(3) to consider aspects of daily life which could be
    modified to assist in tension-control.

The latter aspect included the appropriate setting of realistic objectives, the planning of time, and deciding upon priorities. These
are described more extensively in behaviour modification manuals
(Westmacott and Cameron 1981) and management texts (Drucker 1967)
which do not concern themselves with more clinically-oriented tension-
control methods.

DISCUSSION

It should be noted that some pupils are difficult to help, often
those with behaviour disorders or delinquent tendencies. This has
recently been described by Hare and Schalling (1978) and West (1982).
One fifteen year old girl was admitted to the Day Adjustment Class
following outbursts of aggressive behaviour in school. She seemed
reasonably well controlled in the Adjustment Class but remained sullen
and aloof. Individual counselling with the educational psychologist
was arranged, and after a few sessions, progessive-relaxation training

suggested. Although she was superficially co-operative, she made
little real progress and was often seen to be in a bad mood after
sessions, asserting that she had been given "treatment" which implied
that she was "mental". Further work was discontinued and this pupil
continued to attend the class but without making any real progress.

Our experience is that many pupils with conduct disorders can
respond well, and it is not possible to predict success from school or
other reports. Normally, advice on tension-control is preceeded by a
more general counselling approach, appropriately in view of the wide
range of problems which pupils may be noted to have (Mitchell and
Rosa, 1981; West, 1982; Stewart et al, 1980). In general, those who
respond are those who admit to being bothered by feelings of tension
and wish to reduce them.

Lehrer (1982) considers the possibility of negative effects of
progressive relaxation. He asserts that Jacobson and his followers do
not mention these, although some individuals reportedly are afraid to
relax (Lehrer, 1979; Borkovec and Heidi, 1980), and a report is cited
in which progressive relaxation-training increased levels of "physiological activation" (Delman 1975). Lehrer suggests that poor training
techniques may be a cause, but adds that some individuals may attribute unpleasant properties to the bodily feeling of relaxation.* They
may feel that it signifies vulnerability, lack of control over anger
or sexual desire, overpassivity etc. In support of this interpretation, Abromowitz and Wieselberg (1978) report five cases of treatment
failure with desensitization, in which the patients experienced angry
feelings when they were asked to relax. It is possible that they
would not have felt these unpleasant feelings if they had learned the
technique sufficiently. The intense feelings aroused by initial instruction may have made them unwilling to continue their training,
however. This may have been the case with the Adjustment Class pupil
mentioned above.

A number of pupils referred to the Educational Psychology Service
for aggressive behaviour, restlessness, difficulties in concentrating
on school work and other behaviour problems appear to be remarkably
under-involved in their problems. This often manifests itself in a
dismissive attitude to counselling, or taking a simplistic view of the
situation in which they find themselves. Sometimes it seems as though
the individual is abnormally relaxed when confronting difficulties
which have potentially serious effects on their future place in society. Mitchell and Rosa (1981) in their fifteen year study of boys
with behaviour problems report that those most likely to appear in
court are the "apparently problem free", those described by their
parents as "carefree" and "hearty eaters".

---

* Editor's Note: When one is totally relaxed, there _is_ no feeling.
Hence this statement is a self-contradiction and indicates that the
individuals were probably detecting tension states when they should
relax them away (Jacobson, 1929, cited elsewhere in this book).

It is clearly a difficult treatment problem for the applied psychologist when confronted by a young person reported to be causing severe problems to himself and others by restless, over-reactive behaviour and learning difficulties, when the referred individual himself is inclined to dismiss all reports and attempts to help him as some kind of meaningless conspiracy.

A number of writers have approached such difficult-to-help clients by following a treatment plan which systematically asserts reality demands. It has already been suggested that teachers should formulate a school policy on stealing (Cowell 1981) deciding on 'consequences' and using them systematically. This would involve reminding pupils of the possible consequences of their actions, in this case further involvement in stealing.

Levis (1980) notes that many phobic or other states can, from a theoretical viewpoint, be associated with anxiety, and yet to attempt to help such individuals will sometimes initially increase their fears, undermining the treatment. His outline of implosive theory (IT) can be regarded as a method of tension control, in that the aim is to intensify and direct emotional responding as a motive force for change. Levis states that "the implosive procedure involves the subject being exposed to high degrees of anxiety." As treatment proceeds, the individual is enabled to overcome short-term fears in order to achieve longer-term goals. It is an advanced technique which needs to be applied with care, and yet has demonstrated efficacy when used with a wide range of resistant client groups. Cowell (1983) has also described a technique based on IT to overcome certain treatment difficulties in dealing with a Secondary School pupil who set fires and was difficult to help. The notion of 'realistic anxiety' was suggested as a useful concept which unified various therapeutic approaches.

Cade (1980) describes a way in which a family therapy team broke through a "therapeutic deadlock" with "highly resistant families". He presents the problem of the continuing "game without end" typical of a maintained "chronic symtomatology". This is broken by the presentation to the family of a simulated conflict within the team regarding the methods and efficacy of the treatment. That this method evokes a realistic anxiety is clear: for example a family became tense and one member stayed away, but there was an increase in desirable behaviour.

Miller (1983) outlines "motivational interviewing" with problem drinkers. He criticises the concept of denial of the problem, often said to be typical of the alcoholic, and the associated view that no improvement can be expected until they 'bottom-out' i.e., become much worse. He recommends an approach based on cognitive restructuring, including the process of "internal attribution". That is: "If the individual sees himself as being responsible for having accomplished a change, then it is more likely that the change will maintain." To

bring this about, responsibility is ascribed to the individual rather
than the therapist or others. But Miller also refers to the use of
"objective assessment", in which the individual has medical and other
objective information presented to him, agreeing that this is similar
to traditional and confrontational methods. The counselling plan
further considers the possible outcome should the alcoholic continue
to drink as before, and this should be discussed openly, even introduced by the counsellor. The client may be asked what he or she
anticipates would occur if drinking continued unchanged.

CONCLUSION

Some uses of stress-control methods to help children and adults
have been outlined. A wide range of clients have been involved in an
eclectic approach to treatment, based upon a consideration of published research. Clients most likely to succeed with the methods are
usually those who agreed that they are bothered by feelings of tension
which they would like to control or lose. Some individuals who show
uncontrolled emotional reactions are resistant to help, and present
some well-observed treatment difficulties. It may be appropriate to
use 'paradoxical' procedures, aiming to break through defensive or
habit barriers by emphasising reality demands. These may initially
increase stress, and should therefore be approached with care.

REFERENCES

Abromowitz, S. I., and Wieselberg, N., 1978, Reaction to relaxation
        and desensitization outcome: five angry treatment failures,
        Am. J. Psychiat., 135:1418-1419.
Amerikaner, M., and Summerling, M. L., 1982, Group counseling with
        learning disabled children: effects of social skills and
        relaxation training on self-concept and classroom behaviour,
        J. Learn. Disab., 15:6.
Apter, M. J., 1982, "The Experience of Motivation: The Theory of
        Psychological Reversals," Academic Press, London.
Borkovec, T. D., and Heidi, F., 1980, Relaxation-induced anxiety:
        psychophysiological evidence of anxiety-enhancement in tense
        subjects practicing relaxation, presented at: the Annual
        Meeting of the Association for the Advancement of Behavior
        Therapy, New York.
Cade, B. W., 1980, Resolving therapeutic deadlocks using a contrived
        team conflict, Int. J. Fam. Th., 2:4.
Clayton, T., 1982, The behaviour modification workshop: an antidote to
        mindless technology, Behav. Appr. with Children, 6:2.
Cobb, D. E., and Evans, J. R., 1981, The use of biofeedback techniques
        with school-aged children exhibiting behavioral and/or learning
        problems, J. Abn. Chld. Psychol., 9:2.

Collins, T., and Hodgkinson, P., 1982, Biofeedback, behavioural
        counselling and temper control, New Growth, 2:1.
Cowell, D., 1981, Stealing – A brief guide for teachers, Links, 7:1.
Cowell, D., 1982, The use of hypnosis and allied procedures in
        counselling secondary school pupils, presented at: Fortieth
        Annual Convention – International Council of Psychologists,
        University of Southampton, England.
Cowell, D., 1982, A model screening procedure, AEP Journ., 5:8
Cowell, D., 1983, Relaxation-training for parents of handicapped
        children, AEP Journ., (In press).
Cowell, D., 1983, The use of an implosive therapy method to create
        'realistic anxiety' in the counselling of a secondary school
        pupil who set fires, (unpublished).
Cowell, D., and Franklin, J., 1983, The use of progressive relaxation
        and hypnosis in counselling secondary school pupils,
        Brit. J. Guid. and Counselling, 11:2.
Daws, P. P., 1976, "Early Days," CRAC Hobson, Cambridge, pp. 15,44,46.
Delman, R. P., 1975, Biofeedback and progressive relaxation: a
        comparison of psychophysiological effects, (Abstract of
        unpublished Ph. D. thesis), Dissert. Abstr., 1976, 36B, 4150.
Drucker, P. F., 1967, "The Effective Executive," Heinemann, London.
Dunn, D. M., and Howell, R. J., 1982, Relaxation training and its
        relation to hyperactivity in boys, J. Clin. Psychol., 38:1.
Gath, A., 1978, "Down's Syndrome and the Family, the Early Years,"
        Academic Press, London.
Geen, R. G., Rakosky, J. J., and Pigg, R., 1972, Awareness of arousal
        and its relation to aggression, Br. J. Soc. Clin. Psychol.,
        11:115-121.
Hare, R. D., and Schalling, D., 1978, "Psychopathic Behaviour," Wiley,
        Chichester.
Hartland, J., 1971, "Medical and Dental Hypnosis and its Clinical
        Applications," Balliere Tindall, London.
Kozloff, A., 1979, "A Program for Families of Children with Learning
        and Behaviour Problems," Wiley, Chichester.
Lago, C. O., 1981, Systematic desensitization: a case history
        including some developments in the use of fantasy,
        Brit J. Guid. and Counselling, 9:1.
Lehrer, P. M., 1979, Anxiety and cultivated relaxation: reflections on
        clinical experiences and pyschophysiological research, in:
        "Tension Control, Proceedings of the Fifth Annual Meeting of
        the American Association for the Advancement of Tension
        Control," F. J. McGuigan, ed., AAATC, Chicago.
Lehrer, P. M., 1982, How to relax and not to relax: A re-evaluation of
        the work of Edmund Jacobson – I, Behav. Res. Ther., 20:417-428.
Levis, D. J., 1908, Implementing the technique of implosive therapy,
        in: "Handbook of Behavioral Interventions,"
        A. Goldstein and E. B. Foa, eds., Wiley, Chichester.
Mason, J. L., 1980, "Guide to Stress Reduction," Peace Press,
        Culver City.

Miller, W. R., 1983, Motivational interviewing with problem drinkers,
        Behav. Psycho. Th., 11:147-172.
Mitchell, S., and Rosa, P., 1981, Boyhood behaviour problems as
        precursors of criminality: a fifteen year follow-up study,
        J. Child. Psychol. Psychiat., 22:19-33.
Norfolk, N., 1979, "The Stress Factor," Hamlyn, London.
Omizo, M. M., and Michael, W. B., 1982, Biofeedback induced relaxation
        training and impulsivity, attention to task and locus of
        control among hyperactive boys, J. Learn. Disab., 15:7.
Parrino, J. J., 1979, "From Panic to Power," Wiley, Chichester.
Ross, P. J., 1981, private communication.
Rosa, K. R., 1976, "Autogenic Training," Gollancz, London.
Smith, T. W., and Denney, D. R., 1983, Relaxation training in the
        reduction of traumatic headaches: a case study,
        Behav. Psychotherapy, 11:109-115.
Stewart, M. A., DeBlois, C.S., and Cummings, C., 1980, Psychiatric
        disorder in the parents of hyperactive boys and those with
        conduct disorder, J. Child. Psychol. Psychiat., 21:283-292.
Van der Hoever, J., 1978, "Slant-Eyed Angel," Smythe, Gerrards Cross.
West, D. J., 1982, "Delinquency, Its Roots, Careers and Prospects,"
        Heinemann, London.
Westmacott, E. V. S., and Cameron, R. J., 1981, "Behaviour Can
        Change," Globe, Basingstoke.
Wilks, J., and Eileen B., 1974, "Bringing up our Mongol Son,"
        Routledge and Kegan Paul, London.
Winfield, I., 1983, Counselling with biofeedback: a review,
        Brit J. Guid. and Counselling, 11:1.

APPENDIX I - QUICK RELAXATION

Relax as comfortably as you can wherever you are / close your eyes /
just let yourself relax and unwind / concentrate on the key word relax /
let the feeling of relaxation take over / enjoy the sensation of
letting go / relax your breathing / don't hold your breath / every
time you breathe out feel yourself becoming more and more relaxed /
just relax as deeply as you can / let all the tensions ease away from
your mind and body / comfortable and relaxed / no tension in your
hands / or your arms / let your hands and arms unwind completely /
no tension in your shoulders / let your shoulders drop and relax as
your breathing relaxes / no tension in your neck / let your head rest
gently back / let the muscles in your neck relax / and unwind completely / no tension in your forehead or face / smooth out all those
lines of tension in your face / let your eyebrows drop and relax /
let the muscles round your eyes relax / no tension / eyelids lightly
closed / eyes looking straight ahead / no tension in your jaw / teeth
slightly parted / and let your jaw unwind and relax / more and more /
no tension in your face / anywhere / just let your breathing work on
its own / every time you breathe out feel more and more relaxed / let
your whole body relax more and more / no tension in any part of your
mind and body / your body is relaxing / and letting go more and more /
your mind concentrating only on the key word relax / no tension /
relax / that is fine / carry on letting your body relax / and unwind
more and more deeply / relaxed / breathe in relaxed and at ease /
concentrate on the word relax / allow yourself to feel more and more
comfortable / more and more relaxed / just carry on letting go /
(a pause may be introduced here) / now I am going to count backwards
from seven to one / when I reach one you will wake up / and feel
perfectly alert / and very relaxed / seven / six / waking up /
perfectly alert / five / four / waking up more and more /
very relaxed / three / two / one.

APPENDIX II - DEEP RELAXATION

Make sure you are comfortable / sitting or lying with your eyes
closed / relax your arms by your sides with your legs uncrossed / it
is important to focus on the key word relax when you are relaxing each
muscle group / try to let the relaxation happen without forcing /
don't worry about how well you are doing / simply let the relaxation
flow over you and deepen at its own pace / keep your breathing regular
- shallow and relaxed / notice how as you breathe out you relax a
little more / breathe in and out through your nose / and each time
you breathe out relax more and more / don't hold your breath / keep
your breathing steady and regular / relax more and more / try to
concentrate on the key word relax / each time you breathe out relax
more and more / breathe out and relax more and more / notice the
difference between tension and relaxation / tense the muscles in your
hands and forearms / clenching your fists / clench them as tightly as
you can / feel the tension in your hands and forearms / clench them

tight / tighter / feel the tension in your hands and forearms / now
relax / relax your hands more and more / notice the difference between
tension and relaxation as you relax more and more / focus more and
more on the key word relax / as the muscles in your hands and forearms
relax more and more / concentrate on the feeling of letting go / now
tense the muscles in your biceps / bend your arms at the elbows / try
to touch your arms to your shoulders / feel the muscles become more
and more tense / more and more tense / now relax / feel your biceps
relax more and more deeply / concentrate more and more on the key word
relax / let your arms fall back quite loose by your sides / you will
feel them grow heavier and heavier / notice the difference between
tension and relaxation / carry on the feeling of letting go as the
muscles relax more and more / now the muscles in your shoulder / tense
these muscles by shrugging your shoulders and drawing them up into
your neck / now shrug your shoulders / tight / tighter / feel the
tension / hold it and relax / relax more and more / feel yourself
growing lighter* and more relaxed / now the muscles in your leg /
tense the muscles / make your legs become very hard and stiff / tense
the muscles more and more / then relax / now feel your legs relax /
and grow lighter and lighter / no tension / concentrate on the word
relax / notice the difference between tension and relaxation in your
legs / now the muscles in your stomach / concentrate them hard and
make your stomach hard and solid / tighten them more and more – more
and more / now relax / feel your stomach relax / feel your stomach
relax as you feel lighter and more deeply relaxed / now let's concentrate on the important muscles in the face / first the muscles in your
forehead / use these by raising your eyebrows / feel the tension in
your forehead / hold it / hold it / feel the tension / and now relax /
let your eyebrows drop and relax / no tension / concentrate on the key
word relax / now the muscles in your eyes / frown as hard as you can /
squeeze your eyes / shut them and frown as tightly as you can / feel
the tension in your face / feel the tension / hold it / and relax /
continue to notice the difference between tension and relaxation /
now the muscles in the lower part of the face / squeeze your teeth
together / bite tightly / tight / tighter / feel the tension in your
jaw / feel the tension / hold it / and relax / part your teeth
slightly and release all pressure / feel your face grow more and more
relaxed / feel your breathing become more and more relaxed / as you
breathe in and out / in and out / each time you breathe out / feel
yourself relaxing a little more /each time you breathe out /
and relax / focus on the key word relax / continue to let all the
muscles which you have tensed unwind / more and more / feel them
unwinding / and relaxing / more and more / feel as though you are
sinking more into relaxation / be comfortable / calm and relaxed /
no tension / enjoy the feeling of becoming light and relaxed / (Pause
may be introduced here) / and now I am going to count backwards slowly
from seven to one / and when I reach one you will sit up and feel
alert and refreshed / perfectly alert / seven / six / five / four /
I can see you waking up / three / two / you are awake / one.

---

* the word 'heavier' may be substituted if desired.

IMPROVED STRESS MANAGEMENT AND TENSION CONTROL:

A MODEL PROGRAM IN A COMMUNITY COLLEGE

Stephen R. Germeroth

Center for Improved SMTC
Catonsville Community College
Baltimore, Maryland

'Student Shock' . . . . a symptom of excess stress among
collegians is more severe today as counseling centers
around the country report more numerous psychological
problems among students.

Gottschalk, 1983

INTRODUCTION

The purpose of this paper is to describe the origin, content, and
results of a model program to teach improved stress management and
tension control at a community college. Much of this text will deal
with methodology, allowing for only a brief summary of the results of
several years' research using program participants. I hope to illustrate the form and function of an effective adult educational program
which is intended to enhance the participant's progress toward the
goal of high-level wellness.

RATIONALE

Given the level of American technology and affluence today, it is
shocking to learn that the general health of the average citizen is
not good. Evidence which is readily available abounds to support this
statement. For example, most people will experience two acute illnesses per year, usually requiring a physician's attention on four
occasions and tending to keep the individual feeling ill for the
equivalent of one week (Holmes and Holmes, 1974). Also, during any
given year, in excess of ten percent of the population, excluding
maternity cases, will spend some time in the hospital.

331

Death rates illustrate the point further. According to 1980 data, males in twenty-one other countries and women in nine live longer than American men and women respectively, even though this nation is proclaimed the richest on earth and spends much of that wealth, 286.6 billion dollars in 1980, on health care (Newspaper Associates Enterprise, 1983). It is generally accepted that the average life expectancy of American men and women at birth is 70 and 77.7 years respectively, far short of the much touted 100-year genetic potential of humankind. Fadiman (1977) suggests that significant progress in longevity has been unimpressive in the United States over the past 120 years.

Unimproved mortality rates are not the only indicator of a stagnating health picture. Western morbidity statistics are equally shocking, exposing an excessive incidence of disease. However, epidemiological data for western countries show that disease patterns have shifted over the past century and a half (Stoyva, 1975). The change has been from epidemic, communicable diseases caused by bacteria to epidemic, stress-related and degenerated diseases believed to be caused, in part, by a life-style awash with excessive stress and chronic tension.

Currently, the illness most responsible for reduced life-expectancy and serious disability is cardio-vascular disease, affecting the lives of one out of two Americans and ultimately responsible for over fifty percent of the annual death rate. Cancer, historically not considered stress related, affects 53 million Americans and is the second most frequent cause of death annually (American Cancer Society, 1974). Contemporary medical researchers are suspicious of the relationship betwen metastatic disease and excess stress. The list of other frequently reported medical complaints is lengthy.

As a health science educator, I am vitally concerned with facilitating the reduction of premature death and abating the incidence of serious, disabling disease. Beyond this customary and often stated goal, I am driven even more by the desire to help people acquire whole person excellence through intelligent self-directed behavior.

Accumulating evidence clearly indicates that the greater abundance of health which is the birthright of every individual cannot be achieved by more of the same thing more immunizations, more medicines, more surgery, more sanitation, and the like (Switzer, 1977, p.9).

We are past the time when it was enough to deal with the symptoms of, and therapies for, crisis situations. Our knowledge about the etiology of disease now makes a preventative approach on an individualized basis necessary and possible. The next major advances in health will come from the assumption of individual responsibility for one's own health and a necessary change in the dominant life-style of an individual.

Herein lies the rationale for offering <u>Improved Stress Management and Tension Control</u> to students of health. <u>The inability to deal effectively with stress</u>, i.e. to manage it, evokes all kinds of coping behavior that is recognized as self-destructive and illness-causing. One observes the uptight and nervous individual misusing and abusing substances — everything from aspirin for perceived pain to excessive caffeine, nicotine, and ethanol. He overeats, especially highly sugared junk foods, and is typically under-exercised, claiming, "I'm just too tired." The result is often an unhealthy weight gain. He proceeds, day by day, week by week, repeating behaviors learned long before that seemed to help him deal with the stressors of life. Most of these behaviors are in conflict with sound principles of well-being. Physicians meet this typical patient increasingly earlier in life, the patient complaining of angina, chronic fatigue, colitis, diarrhea, hypertension, nausea, ulcers, and a thousand aches and pains that will never be seen on an x-ray. Too often he dies "unexpectedly" early in life.

The link between stress and tension and disease has been recognized for many years. One of the earliest observations was recorded in 1838, when Dr. James Johnson described the "wear and tear complaint" experienced by many of his fellow Englishmen (Johnson, 1938). The history of stress research is replete with brilliant scientists: Walter Cannon experimented with cats and dogs to study the physiologic responses to stress, later coining the phrase "fight or flight" (Cannon, 1939); Hans Selye spent his lifetime researching the normal aspects of the stress response (Selye, 1975), coining the term "STRESS" and developing the concept of the General Adaptation Syndrome; Edmund Jacobson, having recognized the relationship between excessive sress and tension and illness, perfected a technique to control tension, a component of the stress response (Jacobson, 1976).

With the knowledge of prior research findings, the model program herein described was developed. The literature abounds with techniques designed to treat the excessively stressed, overly tense patient, usually <u>after</u> some dysfunction manifests itself. Examples include autogenic training, biofeedback, hypnosis, progressive muscle relaxation, relaxation response, simple meditation, systematic desensitization, transcendental meditation, and yoga. This project is unique in that it is intended to enhance one's ability to "live a more relaxed life style," thereby <u>reducing the likelihood of developing stress and tension-related disorders</u>. Almost everyone can benefit from the increased ability to manage stress and control tension (Bernstein and Borkovec, 1973). If this skill and attitude can be successfully taught to groups of 20 to 25 people in an educational setting, and if an attempt is made to get the program to the masses, it is possible that modern man might yet realize improved well-being and a longer, more joyful life.

THE PROGRAM

In 1975, I designed a 15-week course of study for an experimental section at Catonsville Community College in Baltimore County, Maryland; its focus was the improved management of stress and the reduction of human tension. The basic philosophy of the program has not changed since then, although many revisions have been made of the content and methods. The central objective remains, <u>to help students live a more relaxed life style</u>. The benefits derived from this one modification are innumerable.

The classic learning model was used to design the process. In this model it is assumed that learning will take place when: 1) the student prossesses certain understanding -- meaning that he or she has <u>knowledge</u> relevant to the task; 2) the student is <u>trained-in skills</u> necessary for the task; 3) the student <u>must practice</u> the skills to develop mastery; 4) the student must <u>apply the knowledge and skills</u> in situations that require their use. With this in mind, the program as illustrated in Figure 1 and described with specific cognitive and performance objectives in Figures 4 and 5 was evolved.

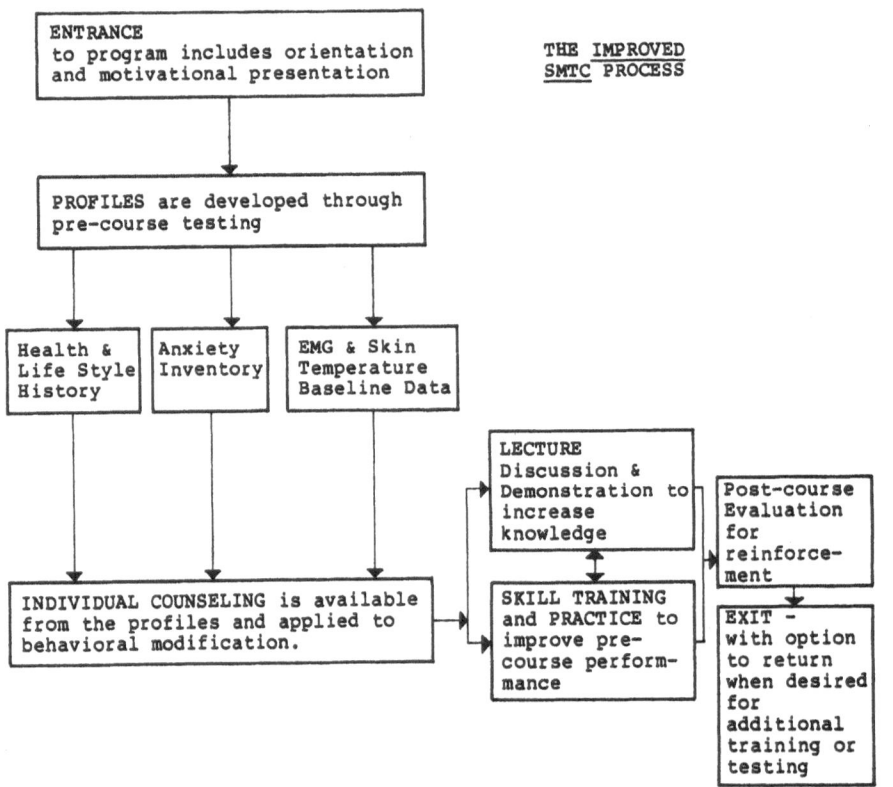

Figure 1. This graphic illustrates the process each participant undergoes within the fifteen-week <u>Improved SMTC</u> program.

Since the beginning, three meeting arrangements have been used. One section meets three times per week for 50 minutes each meeting, and another section meets two times per week for 75 minutes each meeting. Other sections meet once per week for 2-1/2 hours. I believe the best arrangement for learning is the 50-minute meeting every other day. Class meeting only 1 or 2 times weekly are designed and offered solely for scheduling and student convenience.

The semester's class time is devoted to 1/3 cognitive/affective tasks and 2/3 psychomotor training. Students are asked to commit themselves to the goals of the course at the outset. If they honestly feel they cannot make that pledge, they are invited to withdraw from the program. Anything less than sincere effort will produce little if any benefit -- in essence a waste of time. The commitments required are these:

    I am willing to practice - 1 hour per day

    I am willing to read - all assignments

    I am willing to attend - every class

My experience has been that students who remain in the course, and have therefore agreed to the commitments, do remarkably well as measured by their increased knowledge, their increased skill, and their increased ability and tendency to relax in the scientific sense. Also, they typically produce an improved (that is, reduced and controlled) frontal EMG value and a sustained, increased, combined hand temperature. Without fail, students who complete the course are impressed with their own success. And it is theirs, for they have learned in the true sense.

At this juncture, let me explain that even the program title has meaning which goes beyond a merely "catchy" label. The title, Improved Stress Management and Tension Control, is the first indication to the student that this is a course designed to help her or him do better with the problem of excess stress and tension. I see stress management as taking two forms. First, the student learns to identify the stressors in life that are manageable (The Negotiables) and begins to take action to reduce their frequency of occurrence. (Example: Avoid those things that can be avoided.) Second, the student learns to manage stress-reactivity to those stressors that are not negotiable (The Non-negotiables) -- those troubles or problems that are inevitable but which one can learn to view in a less injurious way. Next, I see tension control as a separate but connected issue, a motor skill to be learned like playing the piano or riding a bike. Figure 2 illustrates how the program title takes into account the major concepts of the program and helps to organize its thrust.

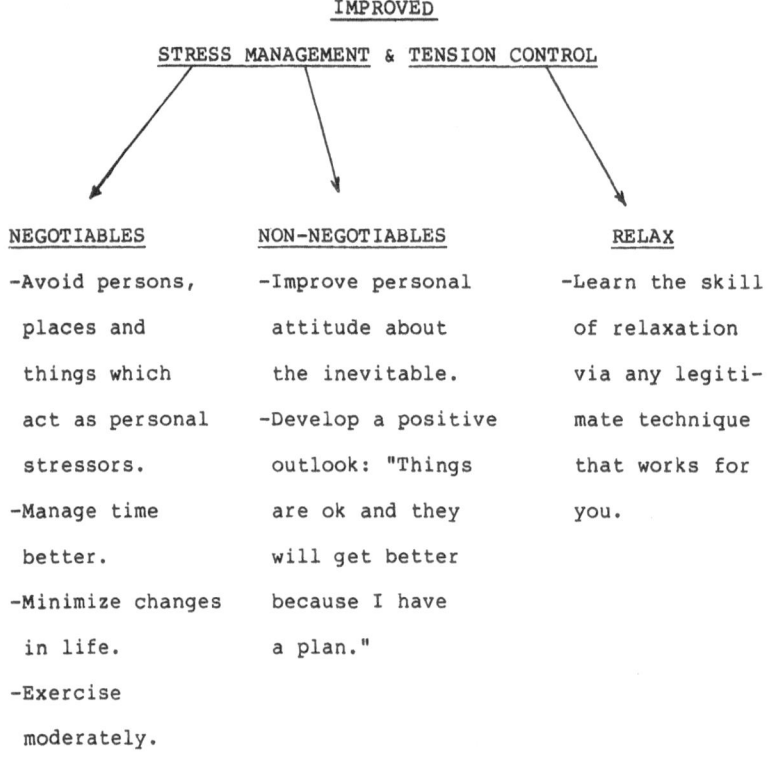

Figure 2. This graphic outlines some of the important points which
are emphasized under the concepts SM&TC.

There are four major components in this program, each of which is
designed to help students live a more relaxed life style. Although
this model may be useful in other populations, it was admittedly
designed for young adults and older students in a typical collegiate
setting.

The first component consists of a series of lectures intended to
inform and motivate. Topics of importance include:

- A review of philosophy and theory as it pertains to
  achieving high-level wellness.

- A brief history of stress and tension research over
  the years.

- A description of the general need to study stress
  management and practice relaxation.

- A description of stress-related illness.

- An examination of how substances in the environment, including food, may impact on one's stress level.

- An examination of how the lack of adequate exercise increases one's stress reactivity.

- A description of stress as psycho-physiologic interaction in humans.

- A comparison of currently proposed methodologies which claim to effectively ameliorate stress and tension.

- A presentation and discussion of "common sense" ways of dealing with stress.

The second component consists of values clarification tasks, usually directed by personnel from the college counselling staff. It is my opinion that physical relaxation without an understanding of what drives us, often at blurring speeds and with great confusion, does little to reduce one's stress and/or tension level. The values component is designed to help students put priorities in order, to resolve conflict, and to become the "captain of one's ship, the chairperson of one's own board." In this way one begins to manage stress more effectively. We would not want to live stress-free, even if we could, which we can't. Managed stress is fun and exciting -- the highs of the day. It's stress gone wrong, distress, that we wish to control. The component of values clarification has been well received with many students often finding their way into improved communication with specific counselors.

The third component of this model involves the training of students in the-technique (skills) of Progressive Muscle Relaxation which was developed by Dr. Edmund Jacobson, Director, Laboratory for Clinical Physiology, located in Chicago, Illinois. During his lifetime, Dr. Jacobson published numerous articles and books dealing with the topic. The most popular book is "You Must Relax" (Jacobson, 1976). The Jacobson technique is simple, clearly a teachable skill, and most importantly, it works. Our approach is to teach the technique in a slightly modified form, a necessity when faced with limited program time.

The purpose of the skill training sessions is to teach each participant how to practice each skill properly and to recognize the correct tension signal when it is appropriate. Each movement, clearly illustrates the difference in the feeling one has when a muscle group is tense or relaxed. The specific objective is to develop an auto-sensory perception of this phenomenon, often referred to as "The Muscle Sense of Bell." The student of PR practices each group as presented in class during the time he or she has set aside for the homework — an agreed-upon one hour per day. As the student becomes more skilled in recognizing when and where tension exists, he or she proceeds to control it by willfully turning it off. The practice and skill proceeds progressively as the student learns to:

1. Just relax at all (any level)

2. Relax in a no-stress situation

3. Relax in a mild-stress situation

4. Relax in a great-stress situation

5. Achieve relaxation as a customary behavior

Clearly we are tackling a difficult task. We are attempting to unlearn a lifetime of learned tensing behaviors (bracing) in favor of an unusual habit, to relax. We are dealing with the whole individual, focusing on a life style which exemplifies hurry, anxiety, confusion and complexity. Also, we are tampering with the basics of diet, exercise, priorities, and values.

The fourth component of this model is feedback in many forms, from teacher to student and student to teacher. The most dramatic feedback tools used in this program include two Cyborg physiometric devices; the P303 electromyograph (EMG) and the P642, a thermal instrument to measure hand temperature. For an initial assessment and later pre-post comparisons, the student's frontal tension and combined hand temperatures are obtained in the SMTC laboratory during a 15-minute procedure. During the assessment, the student is in a reclined position, told to relax as much as possible, with bilateral earphones attached and surface electrodes placed on the forehead, a thermistor attached to the forefinger of each hand. Continuous readings are averaged at one-minute intervals over the 15-minute time period and automatically recorded. The protocol for the assessment is: 1) five minutes of "resting" during which no sound or external intrusions occur, 2) five minutes of "stress" during which a recorded "stressor" or irregular and loud noise is heard by the participants through the earphones, and 3) five minutes of "recovery" during which the stressor is removed. Two pre-tests and one post-test of this sort are required of each student.

Additionally, students visit the lab during the middle weeks of training at least four times for biofeedback exposure. Using the same devices, each student has an opportunity to try out audio and visual biofeedback, and if it complements their results, the lab technician will direct a student through feedback training of his or her choice. It should be remembered that this is an educational program, not clinical therapy. The lab experience is designed to assess, inform, and reinforce students, not treat problems per se. In this capacity,

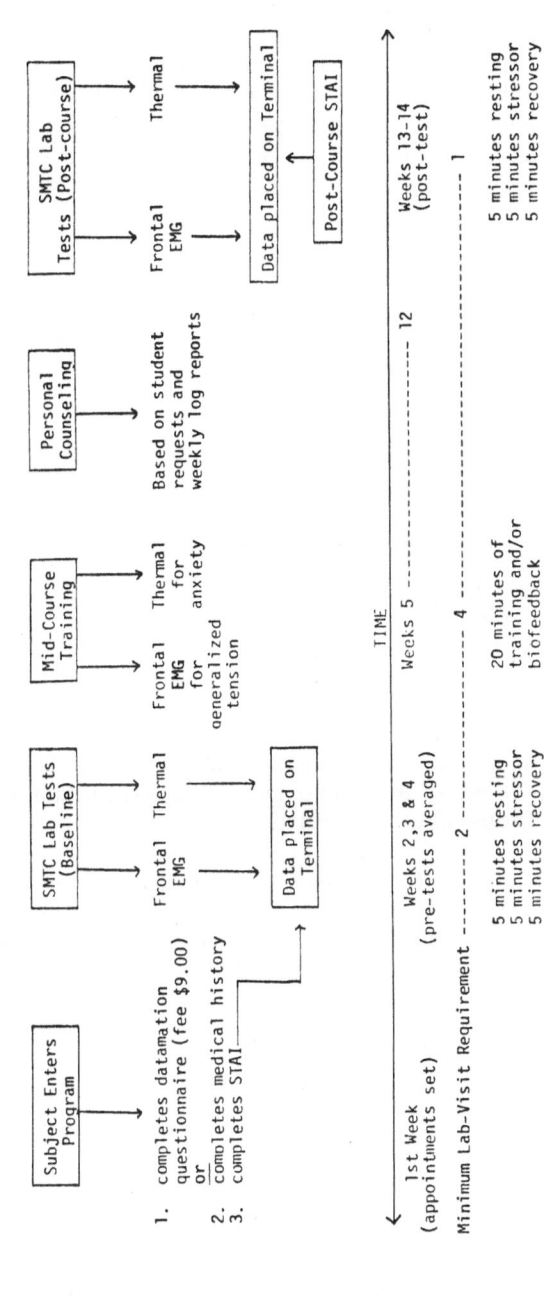

Figure 3. Center for Improved Stress Management & Tension Control objectives and operation protocol.

almost without exception, I am able to demonstrate to the serious
student evidence of reduced tension which coincides with his or her
auto-sensory perception of being relaxed. This reinforcement is usually supportive. Figure 3 illustrates the SMTC laboratory process.

Somewhat less dramatic feedback, but equally important, are the
individual experiences of conferences. Students understand that this
program provides them with one hour per week for either SMTC lab work
or conference, depending on their personal needs or desires. I believe a great deal of support is given to students during these meetings. Often, very close and rewarding relationships grow out of our
exchanges.

Other feedback includes a daily diary completed and returned each
week, a personal evaluation of progress, and telephone conversations.
No more powerful educational device exists than the accessibility of
students to their instructor, an individual who is hopefully a credi-
ble model for their own goals and needs.

BRIEF SUMMARY OF PROGRAM STUDY RESULTS

Since 1979, information has been collected on program participants to provide a data base for research designed to address the
question of long- and short-term impact. Specifically, this study was
designed to describe the participants who enroll, and measure, by way
of frontal EMG and combined hand temperature, if changes occur due to
course involvement.

We found that tense participants among the 172 subjects were
typically older, female, and reported stress-related symptoms of hypertension and sleep-onset insomnia.

At the conclusion of the program, tense subjects were able to
demonstrate significantly reduced frontal EMG values and higher combined hand temperatures. We are encouraged by our findings and will
continue to refine and expand our services to the community we serve.

REFERENCES

American Cancer Society, "1974 Cancer Facts and Figures," 72-375M-8/72-
     No.5008-LE.
Bernstein, D., and Borkovec, T. D., 1973, "Progressive Relaxation
     Training," Research Press, Champaign, Illinois.
Cannon, W. B., 1939, "Wisdom of the Body (Revised)", W. W. Norton
     Company, New York.
CRM Random House, "Essentials of Life and Health," 2nd edition, 1977,
     New York.

Edwards, M. A., 1971, The role of tension control in education, <u>presented at</u>: First Meeting of the American Association for the Advancement of Tension Control, Chicago.

Gottschalk, E. C., 1983, Stress is more severe for collegians today: counselor's keep busy, <u>The Wall Stress Journal</u>, June 1, 1-20.

Greenberg, J. S., 1983, "Comprehensive Stress Management," W. C. Brown, Dubuque, Iowa.

Holmes, T. H., and Holmes, T. S., 1974, How change can make us ill, <u>Stress</u>, 25(1).

Jacobson, E., 1976, "You Must Relax (Revised)," McGraw-Hill Book Company, New York.

Johnson, J., 1938, "Change of Air or the Pursuit of Health and Recreation," London.

Newspaper Associates Enterprise, 1983, "World Alamnac," New York.

Scopp, A., 1976, A model of holistic health, <u>Yoga Journal</u>, Nov.-Dec., pp.4-5:56.

Selye, H., 1939, "The Stress of Life," (Revised) McGraw-Hill Book Company, New York.

Stoyva, J., 1975, A Psychophysiological model of stress disorders as a rational for biofeedback training, <u>presented at</u>: Second Meeting of the American Association for the Advancement of Tension Control, Chicago.

Switzer, J. R., 1977, The stress epidemic: A challenge to health education, <u>Journal of Mississippi Association of Health, Physical Education and Recreation</u>, pp.9-10.

WORKSITE STRESSORS AND UNIVERSITY FACULTY

N. Blithe Runsdorf

Assistant Professor
Department of Administration & Management
Chatham College, Pittsburgh, Pennsylvania

Universities today are faced with mounting pressures. Account-ability, declining applicant pools of traditional college age cohorts, and resource allocation conflicts exacerbate competition between and among departments and between traditional disciplines and newer pro-grams. Interinstitutional conflict about goals has the schools and colleges of universities clashing for pieces of a diminishing resource pie when they need to work together.

University colleges of arts and sciences are complex professional organizations whose goals are research (sustaining the creating of knowledge), teaching (spreading knowledge), and service (applying knowledge). With an apparently dispersed authority structure, final authority for meeting institutional goals is in the hands of indi-vidual professionals — the faculty members who function with a high degree of self-determination. While undergraduate teaching is an institutional activity, on which basis faculty are paid, a professor's professional career depends on activities done in the time not teaching, i.e., research.

Wolk (1975) observed the growing conflict between research and teaching which had "become the absurd paradox that only scholars engaged in research can be good teachers and they have too little time for teaching." Knapp (1962) pointed to the current academic norm of a research oriented faculty which makes it difficult to maintain the intrinsic traditional relationships between teaching and research. Role activities are fractionalized, Knapp argued, in that they do not relate to or mutually enrich each other; research conducted is not introduced back into the undergraduate classroom. On the other hand conflicts among functions seem to be divisions of time and energy resolved within the individual faculty member who teaches, conducts research, consults, and has administrative responsibilities, so that:

> ... what a professor does is not a job; what a
> professor does is to lead a distintive way of life
> within a particular community of professional
> people — a college or university faculty ...
> Academic work is a total activity, a way of life
> ... faculty are frustrated (in organizations) by
> not having enough time to accomplish all they
> want. The professor is always an academic.
> Partitioning professorial work into parts for
> analysis distorts the nature of the activity
>
> (Fulton and Trow, 1974, p.77).

The nature of a person's work interacts with personality. Tannenbaum (1962) found that persistent conditions in the work environment caused personality and behavioral changes which were tied significantly to adjustments and satisfactions of members and organizational performance. Charns, Lawrence, and Weisbord (1977) found significant blurring, by individual professionals, of the distinctions among their several functions, which seems to be what many do, might result in acitivities not appropriate for organizational performance. Bernard and Blackburn (1972) found that faculty experienced the greatest role conflict with respect to self-set performance standards in their multiple roles: researcher; teacher; community servant.

Organizational and self expectations are strong, positive, independent predictors of role behavior. A high correspondence was found by deVries (1975) between the value placed on and amount of time spent on a task, although it appears to be more difficult for faculty to match behaviors with expectations for teaching and research, those functions are highly valued. This may reflect role blurring attributes since conflict may be an expected price for success in academic professions. Shull (1972) considered this issue in terms of role stress and suggested that conflict may result from administrative failures to provide different organizational strategies to reflect a professional role that is pluralistic in terms of multiple behaviors and fractionated in terms of task behaviors.

It would seem that different functions require both different structural arrangements and individual cognitive and emotional responses. When internal perception and understanding about the significant variables for a function match the reality of that function, it is more likely that high performance will be achieved.

Differentiation is important in terms of the organizational characteristic of formality of structure and the behavioral characteristics of orientations to goal, time and interpersonal relations. When people in a particular department share attitudes and interests that focus clearly on departmental goals and have time horizons consistent with their tasks, they will be more effective. When they have developed interpersonal modes that are appropriate to the nature of their work, their performance should be effective. Organizational practices that provide appropriate latitude or control consistent with the nature of the task will also facilitate the department's dealing with its part of the environment. It is clear from Lawrence and Lorsch's work that different orientations are required for different tasks and that these orientations are measureable.

While universities may now be open systems that was not always so. Universities originated as closed systems in a certain world. As conditions increased the necessity for openness, and certainty yielded to uncertainty, universities continued to maintain traditions that failed to take change into account.

In the university, the structure of the organization is based on disciplines. That is, disciplines are represented by different units, or groups of individuals in the organization. This type of differentiation is similar to that found among clinical specialities and primary departments in the academic medical center but rather different from the structural differentiation found among the functional groups of production and sales, for example, in industrial organizations.

The grouping of faculty in the university does not reflect a separation of functions of that institution. Rather, the key professionals, the faculty, perform a combination of the functions of research, education, and service. An individual performing more than a single function may provide a means of coordination among functions, for this individual has information relevant to all of the tasks, hence producing appropriate integration. On the other hand, if the functions have differentiated task characteristics, they may need to be differentiated structurally for effective performance by the individual within the organization.

The University of Pittsburgh was selected as the primary site for this research and was chosen for the following reasons:

(1) It's representative status as a member of the 52 Association of American Universities;

(2) the representativeness of its faculty who are dispersed among the almost three dozen disciplines and areas within the three divisions of the College of Arts and Sciences, Humanities, Social Sciences, and Natural Sciences and

(3) its accessibility.

The literature considers the College of Arts and Sciences as the
traditional and historical location of the multiple goals of a university. As such it seemed a reasonable beginning point when considering
the university as an organization to which to apply the differentiation theory of Lawrence and Lorsch. A large sample was necessary to
achieve enough faculty respondents, making follow-up accessibility a
requirement. In addition, accessibility was a factor in order to
determine if final nonresponders differed in any systematic way,
especially on the basis of functions performed, from those faculty
members who participated.

Faculty for this study (n=546) were defined by the <u>Bylaws of the
Faculty of Arts and Sciences as amended, October 11, 1975</u> as:

> . . . all persons who hold <u>full-time academic
> appointments in the University of Pittsburg with the
> titles, Instructor, Assistant Professor, Associate
> Professor, and Professor</u>, and with a primary
> appointment in the Faculty of Arts and Sciences;
> those Research Assistant Professors, Research
> Associate Professors, Research Professors, and
> part-time Faculty who are in the tenure stream; the
> Chancellor; the Provost; the Deans and Associate
> Deans of the Faculty of Arts and Sciences, the
> College of Arts and Sciences, and the School of
> General Studies are members of the Faculty of Arts
> and Sciences . . . . (paragraph 11, section 2).

Two major questions provided the framework for this study:

1) Can faculty members be considered multiple function
professionals? That is, do faculty in a university perform
some combination of the functions of a university, e.g.
teaching, research, service?

2) Do faculty perceive differences of the work characteristics
among the functions performed? That is, on the basis of
several differentiating task characteristic measures
identified in industry and academic medical center studies,
do faculty perceive differences between or among their
teaching, research, service or administrative activities?

FACULTY PERCEPTION OF FUNCTIONS

College of Arts and Sciences <u>faculty</u> members of the University of
Pittsburgh <u>identified themselves as multiple function professionals</u> by
indicating that they perform more than one of the functions associated
with the role of faculty member at a university. Research and teaching (undergraduate or graduate) were the most frequently combined

functions performed as well as the most frequently named individual
functions performed. Between three fourths and four fifths of the 164
responding faculty members reported some combination of research and
teaching and almost three fourths reported doing research and both
undergraduate and graduate teaching. Approximately two-thirds of the
faculty respondents considered it important to work on a combination
of functions. Teaching, research, and service were the most frequently identified functions of a university as was expected from previous
research (Rotem and Glasman, 1977; Parsons and Platt, 1968; Etzioni,
1959), although service was reported as an activity being done by
fewer than half of all respondents and by only a third when combined
with research and teaching activities.

The emphasis on research necessary for an institution to be a
member of the 52 Association of American Universities and the research
reward emphasis reported at the University of Pittsburgh and others,
may explain the low reporting of service activities. It may also be
that some research is conducted through the delivery of service to the
community.

FACULTY PERCEPTION OF DIFFERENTIAL TASK CHARACTERISTICS

College of Arts and Sciences faculty members at the University of
Pittsburgh perceived that the different functions they perform (teaching, research, service) have different requirements for the task
characteristic measures of time, relationships, goals, and structural
formality associated with them. That is, in the performance of their
roles as teacher or researcher, faculty perceive differences in the
amount of time it takes to know results of their work, in the amount
of necessary interaction with others; in the clarity of goals, and
with the amount of procedural control exerted on them.

FORMALITY OF STRUCTURE

Consistently low or moderate requirements for organizational
constraints on faculty activity were reported, which seems to reflect
the perceived uniqueness and non-anticipatory nature of the activities
associated with the functions of teaching and research and with the
high degree of professional autonomy of faculty members for their own
work. This is consistent with autonomous attributes of professionals
identified by Greenwood (1957), Gouldner (1957), Strauss and Rainwater
(1962), Freidson and Rhea (1965), Hall (1968), and Weisbord, Lawrence,
and Charns (1978).

The need for structural formality, e.g., formal rules, procedures, and external control, increases as the tasks performed become
more certain or repetitive. The lower ratings reported for all of the
functions performed by faculty indicate the uncertain nature of these

activities. Research, which within the functionally divided industrial organization was expected to be the most uncertain function,
and, therefore less formally structured, was reported by faculty as
being the most certain of the functions performed, and having the
greatest amount of repetition, rules, procedures, and constraints
associated with it. This result perhaps reflects the rigor and con-
straint associated with the procedures for the conduct of research.

Research, undergraduate teaching, graduate teaching, and service
were reported as successively uncertain functions that required successively less structural formality. In other words, the characteristic of structural formality did differentiate among the functions
performed by faculty.

INTERPERSONAL ORIENTATIONS

The task characteristic measure about the requirement for interpersonal orientation versus concern for technical matters appears to
consistently differentiate among the functions of teaching, research,
and service performed by faculty. In all of the combinations of
functions where each occurs, results indicated that research required
the most concern for tasks or technical matters. This result is
consistent with the relationship between task-orientation and both
highly certain and uncertain functions identified by Lawrence and
Lorsch (1969) and also reflects the greater autonomy associated with
research activities. It is on research that a faculty member is
judged by colleagues in the discipline.

Successively increasing concern with interpersonal relationships
is found in teaching (both graduate and undergraduate), service, and
administration. As tasks are identified as moderately certain or
uncertain, that is, containing an approximately equal mix of certain
and uncertain activities, styles develop requiring greater interpersonal interactions. All of the functions could be considered to
have moderate uncertainty based on obtained results. Research, perceived as the most certain (relatively) was expected to, and was
identified as being most task oriented. Teaching activities appear to
represent the greatest balance between technical matters and interpersonal relationships. This seemed to reflect the differences involved in the preparation for and actual teaching, the contact with
the students being taught and the collegial discussions regarding
courses as curricular matters.

GOAL ORIENTATIONS

In all of the combinations of function in which they occured,
each of the functions was reported as having moderately to mostly
clear goals, with which at least half of all faculty agree. In the

University, which is not a functionally differentiated organization, the goals and functions of the organization may be found in each department and in each faculty member. This appears to be reflected in the results of perceived goal clarity and agreement obtained in this study.

TIME ORIENTATIONS

Time orientations were concerned with differentiating among functions on the basis of how long it takes to know the results of activities associated with the functions of reseach, teaching, service and administration. Results obtained reflect faculty members' perceptions of the actual percentages of feedback about functional activities within five time frames. In general, for the shorter time frames, relatively small percentages of feedback were reported for all functions. As the time frames lengthened, faculty members' reported knowing more about the results of their work. The time orientation task characteristic measure seemed to discriminate best between the functions of research and undergraduate teaching. Faculty reported knowing significantly more about undergraduate teaching results in the shortest time frames, within a day and within a week, while reporting small percentages about their research accomplishments. For the longest time spans opposite results were obtained. Faculty do seem to regard the pieces of their work as different.

Differences in the functions do not preclude working toward a larger organizational goal nor the need for collaboration and integration includes the effective management of such conflict. Lawrence and Lorsch (1969) found that the most effective mode of a conflict management is confronting, followed by forcing. A confronting mode of conflict resolution implies the existence of a super-ordinate goal held by both parties in conflict. In the industrial setting, the profit of the firm is typically a commonly agreed upon measure. In an extension that bargaining was the preferred conflict resolution mode, followed by smoothing, a mode associated with lower quality integration in industry. Confronting and forcing, the modes of choice for conflict resolution within industry, were the least used in academic medical centers.

No more than moderate conflict was observed in the comparison between any pair of functions. However, the conflict reported for each pair 'in general' was consistently higher than for the conflict 'in a faculty member's own work.' Student's t tests were performed on each of the pairs of functions. For those pairs of functions that contained either service or administration, the function conflict reported from 'in general' to 'in a faculty member's own work' was significant at the .05 level.

For issues among faculty in a department, unilateral action --
individuals acting on their own decisions -- was the perferred method
of conflict management. Bargaining, a method of conflict management
in which individuals try to maximize their own interests by negotiat-
ing differences, was most preferred for managing conflicting issues in
all other settings (college/university; faculty/college administration, between departments within colleges). They do not confront,
rarely force and often avoid, and frequently prefer smoothing.

Organizational performance is related to the quality of two
strategic decisions. The first is a decision about the organization's
mission as it relates to the external environment. From the literature it can be seen that universities have traditionally defined three
missions: teaching, research, and service. The second strategic decision is the design of the organization to support the work that
derives from those mission decisions. Organization theory suggests a
separate organizational design for each function, major piece of
direct work, or product. Among several design issues, decisions about
structure, and reward systems have relevence within the context of
this study.

In universities, functional differentiation is not reflected in
structural differentiation; there are not departments of teaching,
research, or service. Instead, the structure of the organization is
based on academic diciplines in which faculty potentially perform work
contributing to each and all of the missions of the university and
presumably provide whatever integration or resolution of the
differences occurs.

It seems unlikely that a strictly functional organization would
be satisfactory in universities. There is nothing inherently wrong
with structuring by disciplines. But other structural elements may
need to be designed to support achievement of the three missions. For
example, a matrix organizational structure would provide coordination
between one or more disciplines, the organizational unit, and one or
more of the functions, the work tasks of the organization. In addition, it may be important to examine other aspects of organizational
design to better understand productivity, or the lack of it.

An organization should be designed to fulfill its own mission.
However, the university seems unusually influenced by goals external
to the organization. The extent that others -- professional organizations, research funding sources -- are impacting on the institution's achievement of its goals may be a reflection of the organization's failure or inability to equally reward the functions of teaching, research, and service. In other words, both the external forces
and the internal failure or inability are driving a single goal to the
detriment of the others.

A number of issues, including the development of performance evaluation systems and congruent rewards, related to the optimal functioning of a university deserve further research attention, with design issues probably the most promising for consideration.

Beyond the organizational design issues just discussed, there is one other of importance. That is the fit of the individual and the work of the organization. For the faculty member, membership in the professional organization, reflected in the organizational design of the discipline-based department, the issue is also one of rewards. In this study, faculty report considerable professional recognition and academic advancement associated with research activities, but little or none for their teaching and service functions. The university's support for the performance of any of the missions — teaching, research, service — in terms of financial remuneration is perceived as low or moderate. This suggests that goals of the organization are being performed for reasons that may have less to do with the organization and more to do with external factors. Interesting research questions at the micro level, the individual faculty member, appear to be concerned with the reward system and possible conflicts with the goals of the supra-organization.

At this point some psychosocial stress factors become especially significant for the faculty member. For example: extreme ambiguity or rigidity in relation to one's tasks; too much or too little conflict; too much or too little responsibility, especially for people; daily task variability or routinized stability of activities. Often for the organization to survive, the employees are encouraged to define their egos in terms of the organization, to depend on the organization, and to restrain emotional reactions.

The expectation for everyone to do everything might seem to fly in the face of individual differences. Implications are for how one could design the individual faculty position. Consider the possibilities of some doing teaching, some doing research, as well as combinations of the various functions. A combination or focus on a single function could be allowed to vary in several ways, for example, from term to term, at phases in career.

The most promising research directions suggested by this study seem to be in the areas of the two strategic choices to be made when managing an organization for high performance. For both the faculty member and the organization, a number of questions are suggested about the lack of congruence between the way the university is designed to support the missions of teaching, research, and service. What fertile territories for the application and development of the strategies of our groups.

REFERENCES

Bernard, W. W., and Blackburn, R. T., 1972, Faculty Role Conflict in a Rapidly Changing Environment, ERIC Document, ED 065093.

Charns, M. P., Lawrence, P. L., and Weisbord, M. R., 1977, Organizing Multiple Function Professionals: An Analysis of the Organizaiton of Academic Medical Centers, North-Holland/TIMS Studies in the Management Sciences, 5:71-88.

deVries, D. L., 1975, Role Expectations Related to Faculty Behavior, Research in Higher Education, 3,2:111-129.

Etzioni, A., 1959, Authority Structure and Organizational Effectiveness, Administrative Science Quarterly, 4,1:43-67.

Freidson, E., and Rea, B., 1965, Knowledge and Judgement in Professional Evaluations, Administrative Science Quarterly, 10:107-124.

Fulton, O., and Trow, M., 1974, Research Activity in American Higher Education, Sociology of Education, 47:29-73.

Gouldner, A. W., 1957, Cosmopolitans and Locals — Toward an Analysis of Latent Social Roles I, Administrative Science Quarterly, 2:281-306.

Greenwood, E., 1957, Attributes of a Profession, Social Work, 2,3:45-55.

Hall, R. H., Professionalization and Bureacratization, American Sociological Review, 168, 33:92-104.

Knapp, R. H., 1962, Changing Functions of the College Professor, in: "The American College," Sanford, ed., Wiley, New York.

Lawrence, P. R., Lorsch, J. W., 1969, "Organization and Environment: Managing Differentiation and Integration," R. D. Irwin, Inc., Homewood, Illinois.

Parsons, T., and Platt, G. M., 1968, "American Academic Program: A Pilot Study," Report for the National Science Foundation.

Rotem, A., and Glasman, N. S., 1977, Evaluation of University Instructors in the U.S.: The Context, Higher Education, 6,1:75-92.

Shull, F. A., Jr., 1972, Professorial Stress as a Variable in Structuring Faculty Roles, Educational Administrative Quarterly, 8,3:49-66.

Strauss, A. L., and Rainwater, L., 1962, "The Professional Scientist," Aldine, Chicago.

Tannenbaum, A. S., 1962, Control in Organizations: Individual Adjustment and Organizational Performance, Administrative Science Quarterly, 7:236-257.

University of Pittsburgh, 1975 (amended), "Bylaws of the Faculty of Arts and Sciences," University of Pittsburgh, Pittsburgh.

Weisbord, M. R., Lawrence, P. R., and Charns, M. P., 1978, Three Dilemmas of Academic Medical Centers, Journal of Applied Behavioral Science, 14,3:284-304.

Wolk, R. A., 1965, Today's Multiversity: An Institution Torn By Conflicting Pressures, Atlanta Economic Review, 15,8:3-6.

TEACHING DYNAMIC RELAXATION

AT THE UNIVERSITY OF THE THIRD AGE - PARIS VI

Suzanne Masson*

Neuropsychiatrist
Hopitaux de Paris
Paris, France

The University of the Third Age was founded nine years ago for the continuing education of elderly people. Over the past two years, it has been modified to become a general city university, open to all, in order to bring the different generations together.

The experience recounted in this paper began right from the start of the university, and is primarily concerned with the training of elderly persons. In practice, these elderly students can choose, if they wish, to undertake a course of physical activity in parallel with their theoretical and scientific courses. We are concerned with the physical and mental education of men and women whose ages generally range from 60 to 80 years.

A variety of activities were proposed including mime and bodily expression, swimming, basic gymnastics and several sports. For material reasons, it was feasible to undertake only gymnastics in the group of directed and controlled activities. We wanted the exercise to take the form of dynamic relaxation, inspired by the work of Edmund Jacobson, of Feldenkrais and of Tai-Chi Chinese gymnastics. This included some elements of rhythmical gymnastics, which seemed to us to be closer to the needs of people of this age group.

An initial compulsory medical examination was provided for all participants. A medical advisor supervised all the physical activities for six years. Blood pressures and pulse rates were taken before, during and after each training sessions. In most subjects, blood pressure tended to diminish and the pulse to slow down with practice. So there was no counter-indication that this kind of exercise was unsuitable for elderly or sick people.

* ‾‾‾‾‾‾‾‾‾‾
Translated by J. Macdonald Wallace.

From the 90 elderly students wishing to take part, three groups
were formed according to level of ability (non-random): Group 1 was
made up of all those able to take part in all sports, and the Chinese
gymnastics exercises. Group 2 was made up of those able to take part
in selected activities and the Chinese gymnastics. Group 3 was made
up of those able to undertake only the Chinese gymnastics.

Therefore, the only activity found to be suitable for all participants was the Chinese gymnastics, "Tai-Chi", save only those suffering from a recent infarct or a serious progressive illness. Over the
first six years of investigation, no accident occured with any student; therefore, three years ago, the presence of the medical advisor
at each exercise session was no longer considered necessary.

METHOD

The exercise program involved minimal demand upon the cardiovascular system, an absence of static effort and heavy workload as
well as sudden movement, and an emphasis on breathing. Thus it was
possible to have fairly long sessions focused specifically upon development of psychomotor education. In order to achieve adequate effect
on static and dynamic equilibrium, coordination, concentration and
muscular tone (which are so important at this age) long hours of
training are required.

The voluntary capacity for relaxation diminishes progressively as
people grow older, therefore it is important to counter this regression, which is not inexorable, when taught properly. The reduction of
these tonic difficulties increases articular mobility and freedom of
movement by linking it to kinetic harmony. It facilitates elegance
and effective movements as well as correction of posture. An improvement is always possible, whatever the age (even after the age of 80)
if one uses indirect means to attain tonic regulation. These can be
taken from the work of Feldenkrais (1977), of Gerda Alexander (1970)
or of Edmund Jacobson (1929).

We prefer to have the students exercise in a standing or a sitting position. To have participants exercise while lying down on the
floor is not suitable for elderly people. We keep them standing still
only for short periods, dynamic activity being by far more suitable.
We worked with groups of 40-50 people for periods of 1 1/2-2 hours.

To obtain a reduction of tension in the students, many of whom
were rather stiff, we used:

1) concentration in breathing, which deepens progressively,
   and at the same time reduces muscle tension;

2) paying attention to the cutaneous sensations from the soles of the feet;

3) paying attention to proprioceptive signals;

4) relaxation of the face with head up, looking down and mouth slightly open;

5) the use of postural reflexes by pushing down with the hand which is placed on top of the head;

6) relaxation of one side of the body, with the head turned to the opposite side;

7) a sequence of contraction and relaxation, working particularly with the muscles of the face and of the eyes (Jacobson, 1929).

When the student is moving, we asked for:

1) concentration on the movement to be carried out, with sustained attention, never allowing the action to become automatic;

2) oscillations of the body around an axis (Feldenkrais, Alexander, 1970);

3) some slow to-and-fro movements, making alternate demands on agonists and antagonists;

4) some auto-massage, and passive movements of joints, principally those of the feet;

5) moving the weight of the body from one foot to the other (the Yin and Yang of Tai-Chi);

6) paying attention to the demonstrator, and indentifying with his movement, that is, a mirror activity without voluntary elaboration of a motor image and its control, thus leading to a state of reduced tension favourable to relaxation (Tai-Chi);

7) continuity of movement: the end of one movement is the beginning of the next (Tai-Chi);

8) slow, deliberate movement in rhythm with breathing (this adaptation of action to respiratory rhythm is done unconsciously, thus increasing relaxation);

9) the poetic atmosphere of Tai–Chi, suggested by the symbols illustrating "the formulae" (Masson 1983b), which represent gestures and not meaningless movements;

10) silence and concentration during the session; the tutor does not speak, but demonstrates without pause;

11) very little criticism and correction from the tutor, at least, in the early stages;

12) no pursuit of immediate success.

If these points are followed, the physiological and psychological effects of the activities are achieved even before the student knows how to carry out the whole movement.

In conclusion, the results seem reasonably valid to us given the fact that following the program there was a dramatic:

1) decrease in pulse rate and in blood pressure, probably due partly to reduction in muscle tone and in anxiety (especially important because elderly students are usually slightly hypertense);

2) increase in thoracic capacity and improvement in abdominal breathing;

3) improvement in muscle tone (which aids in walking) and in psychomotor training with the development of greater balance, coordination and flexibility.

4) reduction in anxiety and improvement in sleeping habits, concentration and memory;

5) a feeling of calmness and well–being is noted by all at the end of the session, in general there appeared to be a re–birth of self-confidence and a renewed awareness of body image. There seemed to be an acceptance of self bringing about serenity, optimism, and a better ability to get on with other people. All of these outcomes are important as the subject grows older in order to foster happiness and healthy living practices.

REFERENCES

Alexander, G., 1970, L'eutonie, EMC Psychiatrie, 3–1970,3780,B30, pp. 1–6.
Feldenkrais, M., Conscience du corps, in "Réponses," Robert Laffont, ed.

Feldenkrais, M., "La santé en douze leçons," Marabout Service, Ms. 211.
Feldenkrais, M., 1977, "Awareness Through Movement," Harper & Row,
        New York.
Jacobson, E., 1929, "Progressive Relaxation," University of Chicago
        Press.
Jaxquin, J., 1977, Les activités sportives du 3e age de l'Universite
        Pierre et Marie Curie, (unpublished M.D. thesis)
Liang, T. T., 1977, Tai Chi Ch-uan for Health and Self-Defence,
        Random House.
Masson, S., 1983a, "Généralités sur la rééducation psychomotrice et
        l'examen psychomoteur," Presses Universitaires de France, Paris.
Masson, S., 1983b, "Les relaxations," Presses Universitaires de France,
        Paris.

METHODS OF COPING WITH STRESS

COGNITIVE STRATEGIES TO REDUCE STRESS IN

COMPETITIVE ATHLETIC PERFORMANCE

Dorothy V. Harris

Professor
The Pennsylvania State University
University Park, Pennsylvania

INTRODUCTION

One of the most common problems each of us has to deal with day in and day out is stress. Some of us live with an ever-present, rather subdued, level all the time. We cannot associate it with any particular event so we think this is just the way we are! Other stress responses are much more overt, reflected in such signals as a racing heart, disrupted breathing patterns, sweaty palms, feelings of weakness, and so on. Many times we can associate this with the specific situation we anticipate or in which we find ourselves. Those who experience situationally-specific stress and tension are much more aware of the cause and are in a better position to do something about controlling it. Regardless of the cause of the stress, the more years we experience it at any level, the greater the cost. It is impossible to go through life without stress, nor would we wish to. Without it, we would not learn to cope or to control it.

Bernard (1968) distinguished two types of stress: "eustress", the pleasant stress, and "dystress", the unpleasant stress. In other words, stress is not all bad! The "dystress" is actually stress out of control to the point that our system begins to short-circuit and breakdown under the constant pressure to adapt without any relief. Stress, as the term is used today, has the reputation of being undesirable and generally bad for us. However, stress and tension do not have to be detrimental:* many of us thrive on the state and discover that many stress-producing events can produce a pleasurable "high".

---

* Editor's note: We would go even further, viz., tension (i.e., muscular contraction) is necessary for living: overtension constitutes the problem and, by definition, it is always undesireable.

In fact, without stress and tension to prod us, some of us would never learn to adapt and to use the energy generated to our advantage; life would become tedious and boring. Bernard (1968) said this good stress, "eustress," is associated with excitement, adventure, thrilling experiences; it enhances vital sensations, releases energy, "turns people on," and is fun. This stress is the thing that activates our "juices" and stimulates our creativity, our culture, our motivation, and our survival. Without the challenges of stress and tension, not much would get done!

However, the real challenge may be to help each individual find out what his or her optimal level of arousal or response is for maximizing potential performance and productivity. We need to learn at what point we should begin to control it so the level does not reach that which causes dysfunction. Handled well, stress can add to our personal growth and development and enhance our ability to maximize our potential. Handled poorly, or allowed to run out of control, stress can become our enemy.

THE MIND/BODY INTEGRATION

If we were told that the body and mind did not interact, we would most likely disagree. We could cite examples of cognitively perceiving a threat to our well-being which was immediately accompanied by sweating, a racing heart, altered breathing, and feeling of muscular weakness. We literally think with our muscles! In fact, under normal conditions, we cannot think without physiological responses. Jacobson (1930) demonstrated that in a series of studies over 50 years ago.

The body is a highly complex entity composed of a multitude of different, yet highly integrated, biological systems which promote effective interaction between our internal and external environments. These highly differentiated systems are integrated and monitored by the nervous system. The nervous system is anatomically divided into the central and the peripheral nervous systems. The brain and the spinal cord compose the central nervous system while the network of nerves connecting the various organs and systems of the body to the central nervous system makes up the peripheral nervous system. Approximately one-half of the nerves are categorized as afferent and lead to the central nervous system; the other half are efferent and lead from the central nervous system. While thought and memory are the responsibility of the central nervous system, the entire nervous system allows the body to interpret consciously or unconsciously our external and internal environments.

All relaxation techniques and stress management skills are designed to interrupt the cycle of stimulation and response at one point or another. The majority of the techniques focus on the mental control or the efferent nerve control which controls the stimulation from the central nervous system to the muscle. This, in turn, leads to the relaxation of the muscles. Progressive relaxation, on the other hand, focuses on the afferent portion of the nerves, reducing the stimulation to the brain which reduces the mental stimulation to the muscles, resulting in relaxation. The end result of any approach to relaxation is the same; reducing stimulation in either the efferent or afferent half will interupt the circuit of stimulation that is needed to maintain muscular tension.

People respond to stress in two basic ways: Those who manifest their stress bodily or somatically may suffer from headaches, backaches, muscular tics, gastrointestinal disorders, or the like, while those who manifest stress mentally or cognitively become anxious, depressed, or generally irritable and angry. One might argue that since the mind/body is so integrated, these cannot be separated into two different response patterns. In fact, some argue that there has to be a cognitive acknowledgement before a somatic response can occur. If we accept the fact that we do "think with our muscles", this would be true. Further support of this integrated response is provided by the generally accepted belief that it is not the situation per se that causes the stress response, it is our cognitive perception of the situation.

This round-about introduction leads me to the position I wish to take for the remainder of this paper. Accepting the fact that the mind/body is so integrated that we literally think with our muscles, then it is not the situation per se but our perception of the situation that gets us into trouble, and that the stress that is most easily identified, therefore controlled is situationally-specific brings me to the major point of my argument. Exercise and particularly competitive sport provide the best laboratory for teaching youngers how to manage and regulate their arousal or response to situations they perceive as threatening to their self-esteem.

RATIONALE FOR TEACHING STRESS MANAGEMENT THROUGH COMPETITIVE SPORT

There are several theories that provide support for why exercise and competitive sport may serve as the best laboratory for teaching an awareness of mind/body responses so that we can become "tuned in" to our body. This is a necessary prerequisite to acquiring self-regulatory skills that will allow us to control our arousal so that it does not develop into worry and anxiety leading to dysfunction.

Sport situations are among the first encountered by youth where their performance requires an integrated, finely-tuned response. Most youngsters start out enjoying their participation in sport. They are highly motivated to learn and to practice skills and strategies that will help them get better. In sport and exercise situations, they can learn early on how to control and regulate their worries and anxieties so they can avoid making mistakes. The coping skills and strategies that work in sport are the same ones that can be applied to all other situations throughout life which cause worry, anxiety and stress. In fact, competitive sport may provide the first and best opportunity to teach youngsters how to cope with stress. They can readily see the dysfunction that occurs when they do not learn to regulate their arousal. In sport, we call it "choking". However, the principles are the same whether we blank out on an exam, blow our lines on stage, or become dysfunctional in any situation where we are trying too hard because we are anxious and worried about our performance.

Bandura's (1977) theory of self-efficacy model for psychotherapeutic change lends some support as well. According to Bandura, "... cognitive processes mediate change but ... cognitive events are induced and altered most readily by experience of mastery arising from effective performance". Exercise and sport provide good opportunity for feedback through the experience of mastery arising from effective performance. If we persist in exercise or sport, we become more fit and better at it. We adapt to the stress of exercise in such a way that the body's response to stress-producing stimuli becomes more tolerant and efficient. This, in turn, leads to another rationale for exercise and sport's contribution to countering stress. The more physically fit we are, the less dramative the physiological response to stress. Further, the more fit we are, the lower our resting heart rate. This means that our heart has a much greater range for adjusting to stress. That is, if our resting heart rate is 55 and we become quite anxious and stressed, it may increase considerably and reach 100. On the other hand, if we are unfit and have a resting heart rate of 85 and encounter the same stress as a fit person, our heart rate might be 150 or more. The more fit we are, the better prepared to adjust to stress when we encounter it; we also return to the pre-stressed state much more rapidly.

According to Selye (1974), stress is stress whether the stressor is perceived psychologically or is a real physical one such as exercise. In either case, the body reacts in a predictable manner. Any stress, whether it be exercise, fear, worry, a deep emotional reaction such as love, or whatever, produces similiar physiological responses. The body can only respond in a limited number of ways. During an anxiety attack, we experience the same cardiorespiratory response as when exercising: pounding of the heart, breathlessness, sweating, subjective feeling of distress. When this is associated with anxiety, we feel out of control. This perpetuates an anticipatory anxiety state of wondering when panic is going to strike again. Exercise, on the other hand, produces similar cardiorespiratory responses but provides a feeling of being in control because we can generate and reduce them at will.

Further support for exercise and sport countering stress and tension comes from the evidence that suggests that "exercise is nature's best tranquilizer," or to say it another way, "action absorbs anxiety." Increasing evidence suggests that muscular tension and anxiety are reduced through exercise and that moods become more positive. In addition, the limited amount of research relating exercise and sleep patterns leads to the conclusion that sleep comes sooner and is more restorative with regular exercise. Inasmuch as insomnia is one of the problems of the chronically anxious and stressed person, this offers a positive, potential solution to disturbed sleeping patterns and resulting fatigue.

One of the most difficult solutions to recommend to individuals who are chronically fatigued and who have no medical explanation for their condition is regular exercise. It is now accepted that one has to use energy to generate further energy. This explains why those who exercise regularly consistently demonstrate more energy and vigor that those who are sedentary.

More recently research endeavors have focused on biochemical changes associated with exercise. These changes reportedly influence moods in a positive direction and contribute to the "feeling better" syndrome associated with regular exercise. Exercise increases epinephrine and norepinephrine levels; vigorous exercise produces high levels of these hormones. This connection had led many to speculate and to look for a biochemical rationale for explaining why exercise can counter chronic anxiety and depression. The most popular biochemical theory involves the role endorphine may play. More and more neurotransmitters are being discovered with each new research study and increasing knowledge is being generated about brain blood barriers. We may find that stress which is perceived as pleasurable produces an entirely different response at the neurotransmitter levels than stress which is distressful. Much more research is needed in this area.

Exercise is a selfish adventure, no one stands to gain but the participant. It is something that we must do for ourselves; no one else can do it for us under normal conditions. In the process of increasing physical power, stamina, endurance and energy, our mental power, stamina, endurance and energy will increase as well. As a result, we develop a sense of mastery and control over our body which generalizes to other situations. Many who suffer from chronic tension feel helpless and that their behavior has no effect on improving the situation in which they find themselves. Exercise and sport help to reduce this feeling because regular exercise produces changes that are readily observable.

There are numerous research studies supporting the increased self-efficacy and self-esteem in youngsters who participate in exercise and sport programs. These programs provide one of the first means of positive feedback for accomplishment and achievement. After all, in sport you can be the best in the world as a teenager. This

does not occur in other pursuits. Youngsters can be taught how to
evaluate their own progress in relation to where they began rather
than in relation to "being best" or "being first." They can learn to
monitor several indicators which will assist them in this evalutation.

TEACHING SELF-REGULATORY SKILLS THROUGH EXERCISE AND SPORT

With all the concern about the undesirable effects of chronic,
long-term stress it becomes apparent that the sooner one can become
aware of the detrimental effects of stress, the better. Teaching
youngsters skills and strategies to control stress through exercise
and sport situations that they find enjoyable, and where the relevance
of what is being taught is apparent, may be one of the best lessons
they can learn. Children can be taught to readily observe the difference in their performance in relation to how they feel and what
they are thinking. They learn that there is an optimal level of read-
iness or arousal, that their performance suffers when they are not
aroused enough as well as when they are too aroused and worried about
how they will perform. They learn that the mental aspects of performance are just as important as the physical aspects and that they
have to develop skills in both of these to maximize their performance.
They soon learn that they have to become aware of the body's signals.

Worrying about how we are going to perform leads to becoming
anxious about our performance which increases the stress and tension.
This sequence occurs when we worry about anything whether it be taking
an exam, making a speech, meeting important people, or playing in the
most important sports event of our life. We do not worry just in our
heads; our whole body worries. As a result, anytime we worry, we
experience some reaction in our body as well as in our mental state.
This fact is a critical one regarding performance in exercise and
sport situations and children can learn it quickly.

Whenever we worry and become stressed we experience disruption
and dysfunction to some degree. The more worried we become the more
tension and stress we experience; the more tension and stress we
detect, the more worried we become! Each time we perform we have
certain expectations and hopes about how that performance will be.
Expectations are generated by past performances: past performances
shape the expectations about future performance. For worry to take
place, we have to perceive a difference between what we hope for and
what is actually happening or what we expect to happen. Fortunately,
worry does not begin immediately at the point we perceive the difference between our expectancies and our perception of the situation.
As a result, we have time to begin to regulate and control our response to the situation. We have two possible response patterns to
the initial worry. We can either direct it toward a productive end
and cope with the situation or we can worry in circles in an unpro-
ductive manner and generate ever greater stress.

Identifying what we are trying to accomplish, that is, what our expectancies are, within exercise and sport is critical to the adherence and success of participation. This is accomplished through teaching youngsters how to set realistic and specific goals that can be evaluated. Goals should be set for short-term, intermediate, and long-term stages and related to all aspects of training, conditioning and fitness. Setting goals which can be identified and evaluated teach youngsters how to monitor their own progress. They learn mastery and control and how to assume responsibility for directing outcomes. What we set is what we get!

It has been said many times that the only difference between our best performance and our worst performance is the variation in our self-thoughts and self-talk. Teaching an awareness of the type of cognitive processing that all of us do all the time is a big step toward changing thoughts that are negative to more positive and constructive ones. The expectancies we have about our performance serve as self-fulfilling prophecies. What we expect is generally what we get!

The mental aspects of exercise and performance cannot be separated from the physical. There are no marked changes in our physiological capacity, skill level or biomechanical efficiency during a competition or between two competitions close together. What does change is the psychological control. Fluctuations in psychological regulation can be prevented to avoid performance decrements by developing coping skills and strategies to manage worry and anxiety about performance. Mental barriers that are self-limiting and mindsets that do not allow for normal progress can also be eliminated with goal setting and changing negative thoughts to positive ones.

Teaching relaxation skills is the best antidote for stress and tension resulting from worry about how we will perform. The proper form in movement and sport involves using the right amount of muscular tension necessary to move the body in the most efficient manner with the least amount of effort. Too much muscular tension interferes with the execution of any skill or movement. Just think how being afraid of the water elevates the tension levels and affects the performance of the beginning swimmer!

Relaxation techniques and skills are best learned after a physical workout since exercise tends to be one of nature's best tranquilizers. The training sessions for athletes learning to relax should follow practice. However, once they have learned the skill, the sports practice can be started with a relaxation session to ease the accumulation of the day's tensions and stresses so that concentration can be improved. You will find that practices will become much more productive and require shorter periods of time when this procedure is followed.

Athletes discover that sometimes during practice and competition, instead of trying harder when they are having difficulty, they need to relax more and "just let it happen". Putting too much effort into a skill or movement is nearly always counter productive. Further, during competition they can learn to utilize differential relaxation techniques to improve their form by using only those muscles necessary to execute the skill. This will reduce and possibly eliminate mental errors, turnovers, overhitting, using too much force, losing their "touch" or "choking." In fact, reducing tension to the point that they have optimal arousal improves their alertness and awareness in such a way that their performance will be maximized and they can play their best with the least amount of effort. In short, teaching athletes relaxation skills and strategies teaches them how to play at consistently high levels at or near their potential performance during practice and competitions. When they see this relationship, they begin to use relaxation skills in all situations where they wish to avoid disruption and maximize their potential.

One of the biggest hurdles to learning to relax is lack of concentration. Inability to maintain and to regain concentration when it lapses is also one of the biggest problems for athletes. In fact, losing concentration is the most used attribution for failure to meet one's expectancies in sport performance. Concentration or paying attention to what we are doing and what is going on is a skill that can be learned. It must be practiced on a regular basis just as relaxation to maintain a high level of efficiency. This is an easy concept to teach through sport as lapses in concentration are obviously there for everyone to see.

Directing one's attention to the breathing rate and concentrating on the rhythm in a passive manner is useful in teaching relaxation as well as concentration strategies. This procedure also helps the athlete develop an awareness of bodily responses. If the mind wanders between breaths, the athlete can be taught to let those thoughts pass on through and to come back to directing his or her attention to the breathing rate. Another advantage to using breathing is that it is always there; if it ins't, it really doesn't make much difference whether one trys to relax and concentrate! In short, when athletes can develop relaxation and concentration skills that are prepotent over worry and stress about performance, they can develop consistency in performance of their skills. The association of relaxation and focused concentration with maximal performance is generalized to other pursuits the athlete has outside of sport.

After athletes have been taught to keep self-thoughts and self-talk positive and constructive, to relax away unnecessary tensions, and to maintain concentration without having thoughts wander all over the place, they are ready to develop the skills of using their imagination. They can be taught how to mentally rehearse, to visualize everything that they are trying to accomplish. According to Brown (1980), the use of imagination is the "germ substance" of the super-mind. Youngsters can be taught that imagining that they are executing a skill or movement just as they would like makes a significant contribution to their accomplishment of that task. Having them see/feel that they are doing a particular skill in their mind's eye demonstrates clearly to them that they "think with their muscles." This teaches the mind/body link in a very relevant fashion. They learn quickly that negative and unsuccessful images tend to produce unsuccessful performances. They also learn that positive images reduce the worry and stress of not performing as they would like and result more often in better performances. They learn that they have to assume the responsibility for regulating their own arousal in such a way that they can optimize their potential in whatever they pursue. Conversely, they learn that the penalty in sport performance for not assuming responsibility for self-regulatory skills is great. If they cannot cope sufficiently to maintain some consistency in their performance, they get eliminated from the system!

EVALUATION OF THE EFFECTIVENESS OF TEACHING THESE COGNITIVE SKILLS AND STRATEGIES

Setting goals that are individualized, specific and measurable provides one of the best means of evaluating whether they are attained or not. When someone embarking on an exercise program or training for sport can actually see progress occurring as they attain one goal after another, they know that something has to contribute to that progress! Teaching individuals how to keep good daily logs of what they set out to accomplish, what they did, how they did it, and all the other aspects of their life that might have influenced their success or lack of success, they readily observe the relationship between what they can control and regulate and what happens. This provides the best individualized means of evaluating their effectiveness of applying cognitive skills and strategies.

Obviously, the evaluation of significant others is also an important factor in determining the success of learning and practicing these skills. When others note our improvement and compliment us, or when we get positively reinforced in any way by others for our efforts, we are motivated to continue to apply these cognitive skills and strategies to everything that we do.

In our Sports Psychology Laboratory at Penn State, in addition to observing improvement in athletic performance we are assessing the effectiveness of teaching cognitive skills and strategies to athletes by measuring their ability to regulate cognitive and somatic responses prior to training and comparing it to how well they do after training. We have been able to demonstrate significant changes in the self-regulation of both cognitive and somatic manifestations of worry. We are now testing the effectiveness of and application of those skills in the field under stressful situations. Our initial efforts have been rewarding to date. We are using equipment designed for biofeedback for our assessments and have just obtained telemetry equipment which will aid in the research in the field.

IMPLICATIONS

Just think of the magnitude of the contribution that could be made to society through teaching all youths self-regulatory skills through physical education, exercise and sport programs! Think how much more helpful it would be if all teachers and coaches in these programs understood the integrated mechanisms that underlie arousal, worry and stress and could apply that knowledge in situations that are relevant and meaningful to the learner. The emphasis would be on the participant's own capacity and ability to feel and to experience what is going on in his or her body, to experience cognitively and physically in a fine-tuned, integrated manner what involvement in exercise and sports is all about. The focus would be on the participant's own internal state as a function of performance in every endeavor, sport or otherwise. Each participant would become conscious of learning how to tune in to his or her body, to read the signals, and to regulate them to maximize the potential that each has. This approach places heavy emphasis on self-awareness and on the individual assuming the responsibility for his or her own arousal regulation under all conditions. Creative insight into how to extend the limits of our abilities could be gained.

Only in physical education, exercise and sport programs within formal education is the opportunity provided to teach each individual how to integrate in an holistic manner both physical and cognitive awareness in such a way that each student can become all he or she is capable of becoming. This places physical education, exercise and sport in a central position within the educational arena.

REFERENCES

Bandura, A., 1977, Self-efficacy: Toward a unifying theory of
    behavioral change, Psychol. Review, 84:191.
Bernard, J., 1968, The eudaemonists, in: "Why Man Takes Chances,"
    S. Z. Klausner, ed., Doubleday & Company, Inc., Garden City
Brown, B. B., 1980, "Supermind," Harper & Row, New York.
Harris, D. V., 1973, "Involvement in Sport: A Somatopsychic Rationale
    for Physical Activity," Lea & Febiger, Philadelphia.

Harris, D. V., 1982, Maximizing athletic potential: Integrating mind
        and body, JOPERD., 53:3-31.
Harris, D. V., In press, "Mental Skills for Physical People: An
        Athlete's Guide to Sport Psychology," Leisure Press, West
        Point.
Jacobson, E., 1930, Electrical measurements of neuromuscular states
        during mental activites, Am. J. Physio., 91:567.
Ledwidge, B., 1980, Run for your mind: Aerobic exercise as a means of
        alleviating anxiety and depression, Canad J. Behav. Sci./Rev.
        Canad. Sci. Comp., 12,2:126.
Selye, H., 1974, "Stress Without Distress," J. P. Lippincott Co., New
        York.

PHYSICAL ACTIVITY AND STRESS REDUCTION:

A REVIEW OF SELECTED LITERATURE

Marigold A. Edwards

Associate Professor
University of Pittsburg
Pittsburg, Pennsylvania

According to Lazarus (1977) stress management can be directed
to three levels. First is the external environment which one prepares
to master, confront, avoid or flee from by using formal strategies
such as time management, assertiveness and decision making as well
as informal, essential strategies such as vacations and scheduling
'R & R' activities to reduce the internal costs of daily living.
The second level is that of changing perception. As Selye (1974)
said, "it's not what happens to a man that counts, it's how he takes
it." Perception determines whether the stress response is triggered
or not. The third level is that of changing the sress response itself. The resultant response management directly reduces the sequelae
of the stress response. This level provides the last possible opportunity for intervention.

Logically, exercise could contribute at levels two and three.
The role of exercise in changing perception could be preventive, so
that the stress response is less easily triggered or subdued if not
avoided. Lazarus (1977) puts it into context: "coping behavior pre-
cedes emotion and helps regulate its form and intensity." Exercise,
in changing the stress response itself, reflects a cultural analog of
the biologically inherited 'fight or flight' survival response followed naturally by relaxation and rest. At this level, exercise
capatalises on the mind-body connection in reverse — a somatopsychic
effect that also changes perception. Exercise thus fits Lazarus'
(1977) 'ameliorative palliative procedure.' Exercise is treatment.

Davidson and Schwartz (1976) among others, have referred to the
exercise paradox. Exercise is a stimulant, therefore, how can it
relax at the same time? How can strategies from opposite ends of the

effort continuum have comparable outcomes? Yet everyone knows that exercise reduces stress! We all know how 'good', virtuous, self-righteous, 'high', relieved-that-it-is-over-for-another-day, pleasantly fatigued, in control of our behavior and so on, we feel after exercise. Byrd (1963) reported that 92% of 438 San Francisco physicians prescribed one or more of four physical activities: walking, swimming, golf and bowling in that order: presumably, since exercise improves physical health, it also improves affect via the mind-body connection. In a booklet prepared by the National Heart, Lung and Blood Institute (1981) the feeling better benefits often experienced by people who exercise appropriately include these statements: "helps in coping with stress", "helps counter anxiety and depression," "helps you to relax and feel less tense" and "improves the ability to fall asleep quickly and sleep well". What then is the definitive evidence? It is not overwhelming. The early research was plagued by design problems (Morgan, 1982). Cross-sectional studies explained psychological differences between active and sedentary groups on the basis of exercise or fitness. Nevertheless, Morgan (1982) states that even with their inherent weaknesses these correlational and similarly afflicted longitudinal studies do provide important support for the role of exercise in stress reduction. Out of this confusion two researchers, Herbert deVries and William P. Morgan have been involved in a series of pertinent investigations.

deVries, interested in the effect of exercise on tension states in skeletal muscles, as early as 1968 spoke of a possible 'tranquilizer effect' of exercise already accepted by physical educators and coaches on the basis of subjective experience. To his surprise there was no experimental evidence in the literature. His hypothesis for these early studies was that vigorous activity would reduce the neuromuscular level measured by electromyography (EMG). A pilot study (de Vries, 1968) with eight subjects cycling for 30 minutes showed non-significant reductions due, no doubt, to the small number.

Experiment I (deVries, 1968) therefore, compared 29 subjects (mean age, 22) after five minutes of bench stepping with 15 minutes rest on consecutive days. Results showed very small pre-post changes on control (rest) days and substantial changes on exercise days. Specifically, elbow and thigh EMG's decreased 58% and 32% respectively within one hour post-exercise compared to a 1.5% decrease and 7% increase on non-exercise days. Four subjects achieved electrical silence in the arm muscle post-exercise and 10 likewise for the leg muscle enhancing the results. This apparent 'floor effect' the author interprets as the uncompromised natural ability of fit youth to relax at will. From acute effects, deVries turned to longer term exercise.

Experiment II (deVries, 1972) engaged 11 male faculty (mean age, 40) in one hour of strength and aerobic exercise two to three times weekly for 17 sessions. Combined change EMG's for elbow and thigh showed a 25% mean decrease compared to a corresponding increase of 24%

for seven control subjects. The six exercisers with neuromuscular complaints benefitted most accounting for over 80% of the EMG change, hence deVries concluded that vigorous physical activity may provide significant relief from hyperactive states. It may be too, that longer term exercise, while not dramatically reducing tension for the normal person, may offset its accumulation. With muscle tension effects of acute and long term exercise for young and middle-aged in hand, deVries focused on the elderly.

Experiment III (deVries and Adams, 1981) compared exercise to meprobamate, a popular tranquilizer at the time. Ten subjects, aged 52-70 years, selected for high EMG, high blood pressure and high anxiety scores participated in five different treatments repeated three times. There were no significant differences among meprobamate (400 mg), placebo (pill) and control (sitting reading). The 15 minute treadmill walk at heart rate (HR) 100 bts/min showed elbow EMG reductions of 20%, 23%, and 20% at 30, 60 and 90 minutes post-exercise. Changes at exercise HR 120 though similar were more variable and did not reach significance. As deVries points out, given side-effects and potential dependence on pharmacologic agents, particularly at higher doses, the efficacy of 15 minutes of light exercise as a tranquilizer for these older persons should not be overlooked. The physiological mechanism for this phenomenon is not yet well understood. Yet, physical activity is probably the most important sensory stimulation for maintaining homeostasis (Haugen, 1960) and decreasing depression (Oster, 1979).

Experiment IV (deVries, Wiswell, Bulfalian and Moritani, 1981) measured spinal relfex rather than EMG to confirm his earlier findings since the tranquilizer effect still remained controversial (deVries, 1981). There is much conflicting evidence regarding the use of frontalis EMG to reflect general bodily tension (Stoyva, 1976; deVries, 1981; deVries et al 1981). Therefore, an alternative method, electrical stimulation of the tribial nerve and the resultant H/M ratio (method of Angel and Hoffman), was used with ten subjects on three exercise days. A mean reduction of 18.2% in H/M ratio is consistent with the 20-23% reduction previously reported for EMG (deVries, 1981). Five subjects in a similar paradigm showed comparable mean reductions in the H/M ratio; i.e., 16.15% post-exercise with a 7.85% increase on non-exercise days (deVries, Simard, Wiswell, Heckathorne and Carabetta, 1982). In addition, mechanical stimulation of the Achilles tendon showed post-exercise tendon reponse reductions at six of seven hammer heights at a mean of 15.3% and the usual consistent increases on control days. In summary of his work, deVries (1981) cautiously claimed that both acute and chronic exercise of the appropriate type, namely walking, jogging, cycling and bench stepping, of moderate intensity i.e., 30-60% maximum and 5-30 minutes duration provides a significant tranquilizer effect for young, middle-aged and old persons.

The psychological effect of exercise is reported universally as "a feeling of well being". Exercise can apparently lower muscle tension measured by EMG. What is the comparable psychometric evidence? Since non-clinical depression appears unchanged by exercise, i.e., healthy subjects do not become less depressed (Morgan et al, 1970) the specific and now voluminous literature relating exercise and depression is not within the scope of this paper.

Morgan, has followed a line of research involving exercise with state anxiety as the most frequently used dependent variable. It is defined by Spielberger, Gorsuch and Lusheme (1970) as a complex, emotional reaction or state evoked by situations interpreted as personally threatening and characterised by feelings of apprehension, nervousness, heightened activation of the autonomic nervous system, tension, etc. It could be viewed as an emotional manifestation of stress response which has physiological, cognitive and behavioral consequences. Morgan (1979) noted that 10 million Americans suffer from anxiety neurosis; that 10-30% of the patients seeing physicians are anxious neurotics and 30-70% seeing internists are there for conditions arising from unrelieved stress. Concerned about this problem, Morgan and co-workers explored the role of exercise in anxiety reduction with objective psychometric evidence.

The first study (Morgan, Roberts and Feinerman, 1971) with 120 male faculty exercising at workloads eliciting a HR 150-180 bts/min did not differentiate psychological state. The second experiment was designed in light of the Pitts-McClure hypothesis (Morgan et al, 1971) which cast doubt on the efficacy of physical activity because they had found that lactate infusion precipitated anxiety. Yet increased blood lactate is a normal response to vigorous exercise. The implication was extended to normal persons under stress as well as anxiety neurotics. Therefore, Morgan et al (1971) reduced the workload heart rates to much less than 150 with 36 male and female students and found no effect on anxiety; not even state anxiety (STAI) changed yet subjects reported exhiliration and feeling better. The authors, while stating that the tests were not sensitive to these feeling changes, concluded that light to moderate activity does not improve psychological state, albeit a popular notion. Since the high work loads in experiment one did not increase anxiety and skeptical of the Pitts-McClure hypothesis applied to normals, Morgan et al (1976; 1979) increased the workloads in subsequent studies to produce post-exercise lactates up to 120 mg%. Forty males showed an immediate post-exercise increase in anxiety, however, 20-30 minutes following a vigorous 45-minute workout anxiety level fell well below (p < .001) baseline. These results were replicated with 15 males on a 15 minute run leading the authors to conclude that vigorous activity reduces anxiety. A series of experiments were designed, therefore, to evaluate the validity of the Pitts-McClure hypothesis of exercise-induced anxiety (Morgan et al, 1976, 1979). Twenty eight males taken to maximum oxygen consumption (VO_2max) via walking and running on two separate days showed an immediate post-exercise increase on the modified four-item STAI supporting the hypothesis. To study the anxiety response in recovery, these same subjects repeated the workloads and conditions with the four-item STAI administered at 3-5 minutes post-exercise. That the reduced state anxiety reported earlier at 20-30 minutes post-exercise begins within 3-5 minutes of quitting was clearly demonstrated. To study the anxiety response during exercise, the four-item STAI was administered to these same subjects during exercise to exhaustion (approximately 23-24 minutes) in both modes. State anxiety increased throughout the first half of each exercise mode, reached an asymptote and remained stable for the second half of the exercise, falling to baseline within five minutes. According to Morgan et al, (1979) this immediate drop in self-reported anxiety is what makes high intensity exercise eustress, and the resultant lowered levels in the next 20-30 minutes perhaps part of the runner's 'high'. The study was replicated with 28 additional males. Conceivably, since anxiety neurotics might produce higher exercise lactates giving some basis for the Pitts-McClure hypothesis, Morgan (1979) ran six anxiety neurotic and six normal males to exhaustion. Finding identical pre- and post-exercise lactates and no evidence of an anxiety attack he replicated the study and findings with 17 females. Moreover, state anxiety scores were also reduced post-exercise for the high anxious. To evaluate the effect of perception on the anxiety response to exercise Morgan et al (1979) ran 18 previous subjects to exhaustion with two of three groups using a placebo pill or Benson's non-cultic meditation to dampen the fatigue and discomfort. The resultant blunting of the exercise-induced anxiety but not the biochemical responses in both groups supports perceptual differences which, in addition to the dramatic drop in anxiety post-exercise may, they theorized, partly determine the eustress/distress label. In summary, this set of experiments by Morgan, Horstman and colleagues demonstrated that state anxiety increases with vigorous exercise but falls within a few minutes to below baseline levels which are maintained at 20-30 minutes post-exercise in both normals and clinically anxious, thereby refuting the Pitts-McClure hypothesis.

At this point Morgan et al (1979) viewed this function of exercise as a means to self-regulate anxiety on a daily basis by preventing a cumulative effect. Comparing exercise with non-cultic mediation and quiet rest as control he was suprised to find equal and significant reductions in anxiety for 75 normal and highly anxious males (Bahrke and Morgan, 1978). To explain this discrepant but apparently common role of active and passive interventions in anxiety-reduction, Bahrke and Morgan (1978) now suggest exercise as 'time out' or diversion from the concerns of the day. Although their view of the role of exercise in managing anxiety had to change (Bahrke and Morgan, 1978), Morgan (1978; 1979; 1982) still recommends it as the method of choice because of the many physical and physiological benefits that accrue with exercise alone. Nor does this view preclude the notion of fitness-related brain chemistry to account for 'feeling better' but of

which measurement is still elusive. In Morgan's most recent research
(Farrell, Gates, Maksud and Morgan, 1982), that exercise stimulates
increased endorphins was clear; the psychological and physiological
meaning was not. The authors suggest cautious interpretation of
exercise-induced endorphin research which is currently burgeoning
since its physiological significance is not known. As well, the
relationship of peripheral to central levels and assessment techniques
are currently limited to peripheral measures and extremely high in-
fusions have failed to change behavior. The findings of Baekelund
support a mild exercise addiction in man. It is attractive to hy-
pothesise that since exercise stimulates increased endorphins, they
may provide the biochemical base for this behavior (Farrell et al
1982). Lately, the need of exercise-adapted persons has been compared
to that of drug-adapted persons. The popular vocabulary (psychological dependence, tolerance and addiction) has a negative connotation; it is inappropriate, and in the opinion of the author, it is
generally harmful. Regardless, the exercise-induced endorphins may
well be the explanation for adherence to a lifestyle that includes
appropriate exercise. The analogy provides an easy next step; if the
addict cannot get his 'fix' he/she is known to become hyperkinetic,
edgy, nervous, anxious and so on. Exercise for the chronic exerciser
may well be the 'shot in the arm' that reduces 'this' anxiety. This
empirical case example suports Bahrke and Morgan's (1978) earlier view
of the psychological role of exercise. At this point, opposing ends
of the arousal continium, exercise on the one hand and meditation or
quiet rest on the other, both produce significant and comparable
reductions in anxiety. Herein lies a paradox, to which I will return.

Given that exercise does reduce stress, for whatever reason,
research agreement on the optimum level of exercise intensity is
lacking. Morgan et al's (1976; 1979) research dictates high intensity
exercise and deVries calls for moderate levels. Is it reasonable to
think that persons who cannot or choose not to exercise vigorously
never experience this potential benefit? As Dienstbier et al (1981)
point out, how can 'time-out' value be exclusive of the moderate
exerciser? Several studies suggest that exhaustive exercise interferes with performance of mental and motor tasks, although where
fitness is a factor, high and medium fit subjects are less affected
(Gutin and Di Gennaro, 1968). Dienstbier et al (1981) found that
moderate exercise improves stress tolerance more than exhaustive exercise, but not for all subjects. Twenty-three students including five
females from a semester class preparing to run the marathon participated. Following a marathon run, a six mile pace run and a no-run
day, a variety of psychologial tests and physiological responses were
monitored, specifically: capillary constriction, GSR and subjective
rating of a "stress" sound tape; the Mood Adjective Checklist and Buss
and Plomin temperament inventory - which were also administered at the
beginning of the course; and subjective rating of a cold environment.
The across-the-board reductions in negativity except cold rating, from
no run to short run suggests "a remarkable improvement in well being

for our running subjects after a moderate run" (Dienstbier et al, 1981). Every positive factor increased. In both cases, the marathon conditions were less clear but often parallel. Very significant differences in responses were demonstrated when subjects were divided into two groups on the importance of running to self-concept. In general, the high running self-concept group maintained and even increased the benefits of a short run following a competitive marathon whereas the smaller and more variable gains in stress tolerance for the low running self-concept group with moderate exercise were lost and even reversed at exhaustive levels. Dienstbier et al (1981), while providing clear support for moderate rather than exhaustive exercise to improve stress tolerance, have introduced the unstudied factor of commitment into the study of psychological outcomes of exercise. Moreover, the results contradict a 'time-out' (Bahrke and Morgan, 1978) role for the positive psychological changes of exercise in several ways:

1) the unequal effects based on running self-concept,

2) the reduced effects of the ultimate 'time out' marathon and

3) the longer term effects of up to five hours (Dienstbier et al, 1981).

Coben and Robertson (1982) reported significantly lower heart rate and blood pressure responses (p < .05) to a mental arithmetic test at 30 minutes following maximum exercise. However, these benefits were not sustained at 45 minutes post exercise. It is possible that benefits of exercise are limited to the immediate 30 minutes post-exercise? Empirical evidence would suggest that these are spurious results.

Several studies (Mitchum, 1976; Wood, 1977) have reported changes in anxiety with light to moderate exercise. A jogging program compared with stress inoculation training (Long, 1982) over a ten-week period with men and women from the community reported that reductions in anxiety were maintained at three-month follow-up. The joggers, however, had additional benefits of improved fitness.

Davidson and Schwartz (1976) hypothesized that since anxiety is multidimensional, a matching of treatment, condition and outcome measure would reflect the impact of an intervention most sensitively. Sime (1977), like Morgan et al (1971), failed to find significant reductions in STAI scores with light to moderate exercise compared to a resting placebo group. The significant reductions in EMG, HR and electrodermal response (EDR) reported for the exercise group are to be expected with a somatic intervention. Sime's subjects and setting (high test anxious students pre-final exam) reflect overriding cognitive anxiety so that STAI may not identify all that is going on

somatically at these low levels of exercise i.e., HR 100–110 for ten
minutes. Failure to find STAI responsive at moderate activity levels
implies a limited usefulness of exercise for stress reduction.

What Sime (1977) didn't plan for was the resultant dilution of
the effectiveness of both exercise and meditation by the placebo/rest
procedure. Similarly, Bahrke, Morgan (1978) surprised and discouraged
by the equal and significant redutions in STAI scores of the sitting-
at-rest controls were moved to re-state the 'feeling better' role for
exercise as a 'time-out'. Yet Michaels, Huber, McCann (1976) had
already reported that quiet rest was equally capable of dampening
arousal as Transcendental Meditation (TM). As an informal strategy,
while not guaranteeing elicitation of the relaxation response, quiet
sitting will often work for many people.

Having proposed multimodal therapy to fit the multidimensional
nature of anxiety, Schwartz, Davidson, Goldman (1978) in a retrospective study of 77 long-term exercisers and meditators found that exercisers reported less somatic and more cognitive anxiety while the
reverse was true for the meditators. These results lend credibility
to the comprehensive theoretical conception pertaining to the differential effects of relaxation developed by Davidson and Schwartz (1976).

The effect of emotional state on exercise, given societal conditions, the prevalence of hypertension and the apparent increase in
Type A (Chesney and Rosenman, 1980) warrants attention. Tharp and
Carson (1975) reported increases in emotionality scores for an emotional strain of rats following forced swimming which represented
exhaustive exercise because of the emotional stress of survival.
In a unique approach to this question in humans, Schwartz et al (1981)
reported the differential effects of six emotional states on blood
pressure and heart rate at rest and during exercise. It was clearly
demonstrated that emotional stress elicits a very clear pattern of
response. It is particularly pertinent to examine the responses to
the anger condition. At rest, anger-induced Diastolic Blood Pressure
(DBP) and mean arterial pressure were higher than any other emotional
state suggesting a harmful pressor effect. Exercise in each emotional
state showed that anger increased heart rate (HR) by more than 33
beats per minute. The increase was significantly greater than for
fear, more than twice that of normal exercise and six times that of
relaxed exercise. Systolic Blood Pressure (SBP) in anger, was especially delayed in recovery to baseline. The anger-induced DBP response
at rest was normalized during exercise which provides support for the
role of large muscle activity in dissipating the potentially harmful
pressor effect of suppressed anger. Schwartz, Weinberger and Singer
(1981) concluded, "Exercising in an angry state produced over-whelm-
ingly different cardiovascular demands than exercising in a normal or
relaxed state." The implication and caution is that it is not necessarily 'what' you do, but 'how' you do it, which is only a slight
variation of Selye's original theme.

Although Schwartz et al (1981) did not quantify the exercise, the data suggests a unique cardiovascular efficiency in the relaxed exercise condition. Benson, Hartly and Howard (1978) have demonstrated that the physiological cost, measured by oxygen consumption, was decreased (p < .05) with simultaneous elicitation of the relaxation response, although HR did not change. The implication for the judicious use of exercise by those with any coronary heart disease risk factor is clear but one wonders if the recommended 30-minute workout squeezed into a 25-minute slot in an already overpacked, hectic day might not only be counterproductive to the benefits of exercise, but also hazardous to the health.

Blumenthal, Saunders, Williams and Wallace (1980) reported that 21 subjects classified A on the Jenkin's Activity Scale showed a significant reduction in magnitude of A following a ten-week moderate exercise program. It has been demonstrated in animals (Ul'ianinski et al, 1981) that experimentally induced emotional stress produces various cardiac arrythmias which normalise with moderate activity but exacerbate with heavier exercise. Both trained and untrained animals exposed to exhaustive exercise while emotionally stressed show increased blood lipids and decreased adrenal function that favor atherogenesis. However, stress below complete exhaustion was accompanied by normal, stable function in trained animals (Khomula et al, 1980). The foregoing research evidence pertains to the acute effects of exercise on stress with dose, emotion and behavioral style as factors for appropriate use. It is not clear yet whether there are chronic effects.

It appears that significant stress adaptation accrues from chronic exercise. Several studies (Michael, 1957) support the view that chronic exercise sensitizes and enlarges the adrenals thus enhancing stress adaptation: i.e., exercise of the appropriate dosage over time conditions the stress response mechanisms so that the response to any stressor is more efficient and with greater reserves. Frenkl, Salay and Csakvary (1969) in an earlier series of experiments reported that trained rats exhibited a reduced steroid response to the training stimulus compared to untrained controls and a more rapid blood clearance of exogenous steroid for both trained animals and man. They replicated these plasma steroid differences in trained and untrained rats following 60 minutes of physiologically equated nonexhaustive swimming. The trained rats again showed significantly smaller post-exercise steroids returning to baseline in about 45 minutes; the untrained, at three hours post-exercise, remained at double their resting value. Similarly in male University students, peak values were lower and recovery to baseline faster in exercise adapted subjects. The evidence suggests that the stress response with training is reduced in intensity and duration.

If a concept such as cross-tolerance exists via a 'training effect' of emotionality following chronic exercise (Dienstbier, 1981) then exercise adaptation results in a smaller stress response and a quicker return to homeostasis.

Frenkl (1971), had demonstrated ulcer-resistance in trained rats, and noted a lower incidence of ulcers in trained male humans. In his series of studies, Frenkl (1971) demonstrated that ulcer-resistance was transmissable from trained humans and rats to untrained rats and guinea pigs. Comparing animals injected with serum from trained or untrained subjects, the gastric acid response to histamine, the rectal hypothermia and recovery to histamine and serotonin and the plasma steroid response to restraint all showed blunted stress responses to these ulcerative agents in favor of the exercise-adapted donors. Frenkl concluded that physical activity is an antiulcerogenic agent. Similarly favorable results have been reported for treadmill running by male rats (Johnson and Tharp, 1974). Also in support of cross-tolerance, Bartlett (1956) showed that daily exercise inhibited the hypothermic response of restrained rats exposed to cold ($p < .01$). Restraint hypothermia is a stress response for which exercise can provide the necessary adaptation.

Support for the concept of stress adaptation comes from Selye's (1956) theory of stress and his three phase General Adaptation Syndrome (GAS). In this case exercise elicits a non-specific alarm response which, when repeated over time promotes general adaptation, rather than exhaustion, to a new higher level. As Michael (1957) said, "The GAS of Selye concerns the adrenal cortex and a 'learning' process of defense against future exposures of the same stress." Given Selye's (1956) non-specific stress and the concept of and support for cross-tolerance implicit in the work of Frenkl (1971), Bartlett (1956) and Johnson and Tharp (1974), for example, chronic exercise can be said to promote a 'learned' process of defense against future exposures to a wide variety -- indeed, to any, -- stress. Therefore, exercise of an appropriate dosage over time stimulates or conditions the stress response mechanisms in an adaptive way so that the response to any stress is more efficient.

Birrell and Roscoe (1978) found that animals on an intensive aerobic program showed no adverse changes in ventricular tissue or microcirculation in response to a five-day stressing procedure but sedentary weight-controlled and sedentary free-feeding animals showed dilated valves, platelet agglutination and fatty infiltration in the myocardium. All animals initially lost body weight, an accepted index of experimental stress, but the exercise rats lost least and on day four gained weight while the other two groups continued to lose, suggesting an earlier adaptation to stress. Thus, conditioned hearts withstand stress better and exercised animals respond to stress better!

Weber and Lee (1966) studied the effect of prepubertal exercise on the emotionality of rats at puberty and found that forced exercise animals exhibited significantly less emotionality, as measured by exploratory behavior, compared to voluntary exercised and inactive controls. Later, these results were replicated in an almost identical study (Tharp and Carson, 1975) with additional controls. Caution and experimental constraints bar extrapolation to human subjects but there is wide support for the concept and reality of autonomic conditioning to account for stress-related disorders and their recovery (Schwartz, 1979). The role of exercise as an autonomic conditioner has been thoroughly studied in physiologic terms (Tharp and Buck, 1974; Dienstbier, 1981; Gantt, 1964; Von Euler, 1974), but is elusive in the psychologic realm though its presence is widely accepted.

Exercise both energises and elicits the physiologic and psychologic responses of relaxation. Various authors (Gantt, 1964; Frenkl et al, 1969; Frenkl, 1971; Tharp and Buck, 1974) concluded that exercise-induced adaptation produced a more efficient stress response with greater reserves due to the increased sensitivity and hypertrophy of the adrenal gland accompanied by adaptive changes in the hypothalamic-hypophyseal axis to explain chronic exercise effects for stress reduction.

Gellhorn's (1969) elegant analysis of the processes involved in the absence of movement in stress interprets our logic and experience for acute exercise counteracting stress. Based on the connectedness of mind and body, the case is made for the neurotic potential of continuing sympathetic discharge in the absence of movement. The role of exercise is direct. Skeletal muscles used in exercise provide proprioceptive reinforcement enhancing sympathetic activation; on stopping exercise, the proprioceptive impulses cease and the exercise-primed natural parasympathetic rebound both terminates sympathetic activation and returns the system to homeostasis and muscle relaxation. This reciprocity between ergotropic and trophotropic systems explains the physiologic and psychologic concomitants of relaxation regardless of their means of achievement. Unfortunately, we have learned to override most natural responses or relationships. Too many of us live self-destructive lifestyles as reflected in the statistics for hypertension, coronary heart disease and prescriptions for tranquilizers. We need more balance in our eating, drinking, exercise and stress behaviors. In our mechanized, urbanized society stimulation is prolonged, thus reducing the chance for the natural parasympathetic rebound, and underactivity is the norm, allowing uninhibited erogotropic discharges to accumulate and produce a variety of dis-stressing symptoms. Gal and Lazarus (1975) note that activity, threat-related or not, is useful in modifying the associated emotional state.

Exercise is a very simple and practical means for release of bodily arousal. Some researchers have interpreted this release as 'feeling better' or 'calming down' so often reported. As with fitness, the effects are transient though persistent, not permanent, when repeated over time. Moreover, we know that the human organism must move or it will deteriorate. We would do well to take Sheehan's (1977) advice:

"We must go back to being an animal; first be an animal."

REFERENCES

Bahrke, M. S., and Morgan, W. P., 1978, Anxiety reduction following
    exercise and meditation, Cogn. Ther. Res., 2:323.
Benson, H. D., Hartley, T., and Howard, L., 1978, Decreased VO_2
    consumption during exercise with elicitation of the relaxation
    response, J. Hum. Stress, 4:38.
Bartlett, R. G., 1956, Stress adaptation and inhibition of restraint-
    induced (emotional) hypothermia, J. Appl. Physiol., 8:661.
Birrell, P., and Roscoe, C., 1978, Effects of intensive aerobic
    exercise on stress reactivity and myocardial morphology in
    rats, Physiol. and Behav., 20:687.
Blumenthal, J. A., Sanders, W., Williams, R., and Wallace, A., 1980,
    Effects of exercise on the Type A (coronary prone) behavior
    pattern, Psychosom. Med., 42:289.
Byrd, O. E., 1963, A Survey of beliefs and practices on the relief of
    tension by moderate exercise, J. Sch. Health, 33:426.
Chesney, M. A., and Rosenman, R. H., 1980, Type A Behavior and the
    work setting, in: "Current Concerns in Occupational Stress,"
    C. L. Cooper and R. Payne, eds., Wiley, New York.
Coben, J. H., and Robertson, R. J., 1982, Effect of maximal voluntary
    exercise upon physiological response to stress, Med. Sci.
    Sports, 14:166.
Davidson, R. J., and Schwartz, G. E., 1976, The psychobiology of
    relaxation and related states: a multi-process theory, in:
    "Behavior Control and Modification of Physiological Activity,"
    D. I. Mostofsky, ed., Prentice-Hall, Englewood Cliffs.
deVries, H. A., 1968, Immediate and long-term effects of exercise upon
    resting muscle action potential level, J. Sports Med. Phys.
    Fitness, 8:1,
deVries, H. A., and Adams, G. M., 1972, Electromyographic comparison
    of single doses of exercise and meprobamate as to effects on
    muscular relaxation, Am. J. Phys. Med., 51:130.
deVries, H. A., 1981, Tranquilizer Effect on Exercise: A critical
    Review, Physic. Sports Med., 9:47.
deVries, H. A., Wiswell, R. A., Bulbulian, R., and Moritani, T., 1981,
    Tranquilizer effect of exercise: acute effects of moderate
    aerobic exercise on spinal reflex activation level,
    Am. J. Phys. Med., 60:57.
deVries, H. A., Simard, C. P., Wiswell, R. A., Heckathorne, E., and
    Carabetta, V., 1982, Fusimotor System involvement in the
    tranquilizer effect of exercise, Am. J. Phys. Med., 61:111.
Dienstbier, R. A., Crabbe, J., Johnson, G. O., Thorland, W.,
    Jorgensen, J. A., Sader, M. M., and Lavelle, D. C., 1981,
    Exercise and stress tolerance, in: "Psychology of Running,"
    M. H. Sacks, and M. L. Sachs, eds., Human Kinetics, Champaign.
Farrell, P. A., Gates, W. K., Maksud, M. G., and Morgan, W. P., 1982,
    Increases in plasma B-endorphin/B-lipotropin immunoreactivity
    after treadmill running in humans, J. Appl. Physio., 52:1245.

Frenkl, R., Csalay, L., and Csakvary, G., 1969, A study of the stress reaction elicited by muscular exertion in trained and untrained man and rats, Acta Physiol. Acad. Sci. Hung., 36:365.

Frenkl, R., 1971, Humoral mechanisms of ulcer-resistance of the organism adapted to physical exercise, Acta. Med. Acad. Sci. Hung., 28:66.

Gal, R., and Lazarus, R. S., 1975, The role of activity in anticipating and confronting stressful situations, J. Hum. Stress, 1:4.

Gantt, W. H., 1964, Autonomic conditioning, Ann. N. Y. Acad. Sci., 117:132.

Gellhorn, E., 1969, The consequences of the suppression of overt movements in emotional stress: a neurophysiological interpretation, Confin. Neurol., 31:289.

Gutin, B., and Di Gennaro, J., 1968, Effect of a treadmill run to exhaustion on performance of long addition, Res. Quart., 39:958.

Haugen, G. B., Dixon, H. H., and Dickel, H. A., 1960, "A Therapy for Anxiety Tension Reactions," MacMillan, New York.

Johnston, I. H., and Tharp, G. D., 1974, The effect of chronic exercise on reserpine induced gastric ulceration in rats, Med. Sci. Sports, 6:188.

Karvonen, M. J., 1959, Effects of vigorous exercise on the heart in: "Work and the Heart," Rosenhaum and Belknap, eds., Hoeber Medical Books, New York.

Khomulo, P. S., Kadushkina, N. N., and Zharova, I. P., 1980, Effect of physical activity on blood lipids and adrenal function during emotional stress, Biull Eksp. Biol. Med., 89:428.

Lazarus, R. S., 1977, Cognitive and coping processes in emotion, in: "Stress and Coping," A. Monat, and R. S. Lazarus, eds., Columbia University, New York.

Long, B. C., "A comparison of aerobic conditioning and stress inoculation as stress management interventions," (prepublication), University of British Columbia, Vancouver.

Michael, E. D., 1957, Stress adaptation through exercise, Res. Quart., 28:50.

Michaels, R. R., Huber, M. J., and McCann, D. S., 1976, Evaluation of Transcendental Meditation as a method of reducing stress, Science, 192:1242.

Mitchum, M. L., 1976, "The effect of participation in a physically exerting leisure activity on state anxiety level," Master's Thesis, Florida State University.

Morgan, W. P., Roberts, J. A., Brand, F. R., and Feinerman, A. D., 1970, Psychological effect of chronic physical activity, Med. Sci. Sports, 2:213.

Morgan, W. P., Roberts, J. A., and Feinerman, A. D., 1971, Psychologic effect of acute physical activity, Arch. Phys. Med. Rehab., 52:422.

Morgan, W. P., and Horstman, D. J., 1976, Anxiety reduction following acute physical activity, Med. Sci. Sports, 8:62.

Morgan, W. P., Horstman, D. H., Cymerman, A., and Stokes, J., 1979,

Use of exercise as a relaxation technique, <u>J. S. Carolina M.A.</u>, 75:596.

Morgan, W, P., 1979, Anxiety reduction following acute physical activity, <u>Psychiatr. Ann.</u>, 9:141.

Morgan, W. P., 1982, Psychological effects of exercise, <u>Behav. Med. Update</u>, 4:25.

National Heart, Lung and Blood Institute, 1981, "Exercise and Your Heart," Public Health Service, Washington.

Oster, C., 1979, Sensory Deprivation and Homeostasis, <u>J. Am. Geriatr. Soc.</u>, 27:364.

Pitts, F. N., and McClure, J. N., 1967, Lactate Metabolism in Anxiety Neurosis, <u>New England Journal of Medicine</u>, 277:1329-1337.

Schwartz, G. E., Davidson, R. J., and Goleman, D. J., 1978, Patterning of cognitive and somatic processes in the self-regulation of anxiety: effects of meditation versus exercise, <u>Psychosom. Med.</u> 40:321.

Schwartz, G. E., 1979, Disregulation and systems theory: a biobehavioral framework for biofeedback and behavioral medicine <u>in</u>: "Biofeedback and Self-Regulation," N. Birbaumer and Kimmel, eds., Erlbaum, Hillsdale.

Schwartz, G. E., Weinberger, D. A., and Singer, J. A., 1981, Cardiovascular differentiation of happiness, sadness, anger, and fear following imagery and exercise, <u>Psychosom. Med.</u>, 43:343.

Selye, H., 1956, "The Stress of Life," McGraw Hill, New York.

Selye, H., 1974, "Stress without Distress," J. B. Lippincott, Philadelphia.

Sheehan, G., 1977, <u>in</u>: "Coping With Life on the Run," Sports Productions, Shiller.

Sime, W. E., 1977, A comparison of exercise and meditation in reducing psychological response to stress, <u>Med. Sci. Sports</u>, 9:55.

Spielberger, C. D., Gorsuch, R. L., and Lushene, R. E., 1970, "Manual for the State Trait Anxiety Inventory," Consulting Psychologists Press, Palo Alto.

Stoyva, J., 1976, Self-regulation and the stress-related disorders: A perspective on biofeedback, <u>in</u>: "Behavior Control," Mostofsky, ed., Prentice-Hall, Englewood Cliffs.

Tharp, G. D., Buck, R. J., 1974, Adrenal adaptation to chronic exercise, <u>J. Appl. Physiol.</u>, 7:720.

Tharp, G. D., Carson, W. H., 1975, Emotionality changes in rats following chronic exercise, <u>Med. Sci. Sports</u>, 7:123.

Ulianinski, L. S., Urmancheava, T. G., Stepanian, E. L., Fufacheva, A. A., and Gritsak, A. V., 1981, Effect of motor activity on the development of arrthythmia in experimental emotional stress, <u>Kardiologiia</u>, 21:64.

Von Euler, U. S., 1974, Sympatho-adrenal activity in physical exercise, <u>Med. Sci. Sports</u>, 6:165.

Wood, D. T., 1977, The relationship between state anxiety and acute physical activity, <u>Am. Corr. Ther. J.</u>, 31:67.

Weber, J. C., and Lee, R. A., 1966, Effects of differing prepuberty exercise programs on the emotionality of male albino rats, <u>Res. Quart.</u>, 39:748.

THE PHYSICAL MIRROR OF STRESS

Kathryn Curtis Lake

Osteopath, Private Practice, Sussex, England
and
The British School of Osteopathy, London, England

The aim of this paper is to present a hypothesis to explain why stress, when prolonged for extended periods of time, or in a situation of sudden stress overload, causes a relatively local patho-physiological disease state which varies from individual to individual.

INTRODUCTION

Stress is defined as any stimulus or succession of stimuli of such magnitude as to disrupt the homeostasis of the individual; the adaptive mechanisms of the organism fail, become disproportionate or incoordinate, resulting in disease, disability or death. This stimulus, the stressor, has been defined by Spielberger (1979) as a situation, physical or mental, which in some way threatens the individual. The individual's perception of the stressor is totally dependent on his compensation and adaptation to his environment, in relation to all his past experiences and inherited characteristics. Hence the stressor stimulus varies considerably from one individual to another.

The physiological responses to the stressor are widespread through out the body. They take the form of changes in neural function, hormone release, and cardio-respiratory responses. Their combined effects alert the body's total defensive and adaptive mechanisms to its external environment. Our concern is focused upon what causes the final breakdown of homeostasis and the manifestation of disease in a particular area or system of the body in any one individual?

As an osteopath I am primarily interested in the physiology and mechanics of the whole individual which is reflected in the 'Physical Mirror'. If we look at any mechanical system, if there is a weakness

within its structure, when it is subjected to prolonged stress or
sudden stress overload, it is that weakest part which will show the
signs of strain first, the magnitude of the strain being directly
proportional to the severity of the stress. Therefore, in any one
individual, the final expression of 'Dis-ease' following prolonged
stress or sudden stress overload will manifest itself through its
weakest homeostatic link.

Structure and function are totally interdependent in the maintenance of homeostasis within the internal environment (Hoag, 1979).
It is the structure and function as expressed through the musculoskeletal system that I am going to concentrate on now, and demonstrate
its key role in the final localization of patho-physiological disease
states.

THE GROSS EFFECTS OF THE MUSCULOSKELETAL SYSTEM
ON THE HOMEOSTASIS OF THE INTERNAL ENVIRONMENT

To demonstrate these gross effects, I have chosen one of the
fundamental functions of the musculoskeletal system as a whole, which
is, to adapt and compensate for the force of gravity in the erect
posture.

The ideal standing posture is one where there is least work load
on the musculoskeletal system, and therefore one which is most economic on the resources of the internal environment (Gray, 1973).
This is achieved when:

(a) The center of gravity line of the body in the anterior/post-
    erior plane falls from the mastoid process to the apex of the
    longitudinal arch of the foot just anterior to the medial
    malleollus, the line of force tending to exaggerate the
    spinal curves, as in Figure 1 (A).

(b) In the lateral plane, the center of gravity line falling from
    the external occipital protuberance, through the line of the
    spine, to midway between the two feet, as in Figure 1 (B).

As soon as the center of gravity line moves away from the ideal,
there will be an increased work load on the musculoskeletal system.
For example, if the center of gravity line of the body moves too far
posteriorly, there will be an increased work load on the anterior
musculature to maintain the erect posture. It would be totally inappropriate if the body in this posterior weight-bearing position was
just in a posterior tilted position, and mechanically, because of the
configuration of the spine, this is not possible. The alteration in
the center of gravity line further exaggerates the spinal curves in
the anterior/posterior plane, and as the body will always try to keep
the eyes pointing forward in the horizontal plane, the net result is

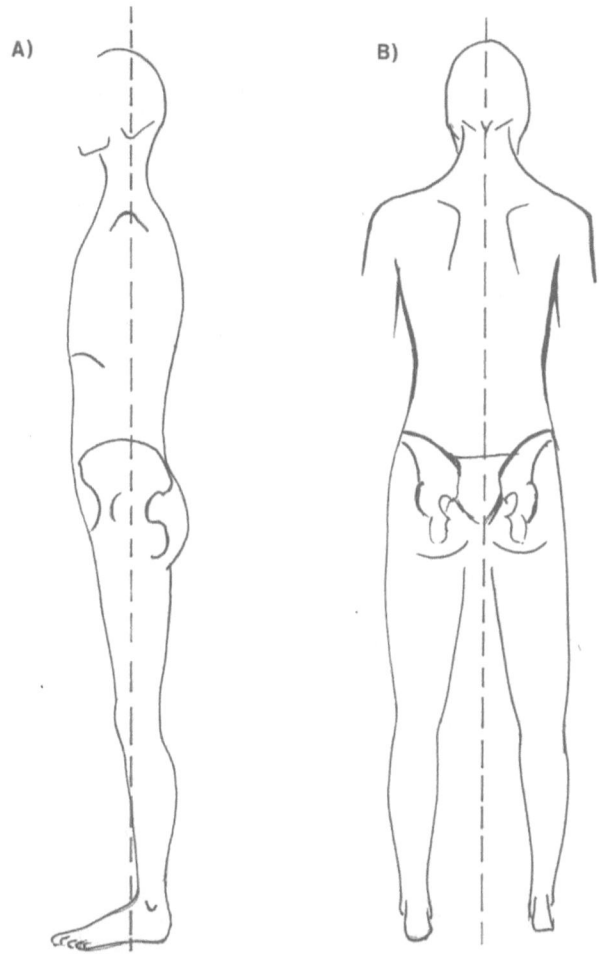

Figure 1. Diagram to illustrate the ideal posture in:
(A) the anterior-posterior plane (B) the lateral plane
Each vertical dotted line is a center of gravity line

an increase in the thoraric and lumbar curves, and a decrease in the
cervical curves (Figure 2). Consequently, there are local areas of
maximum workload on the musculoskeletal systems so that these areas
are under a degree of physical stress, and in this situation they have
a reduced potential for adaptation and compensation.

As already stated, structure and function are totally interdependent. The alterations in the mechanics of the standing posture
also reflect on the internal environment both directly and indirectly.
There is an alteration in the morphology of the rib cage and therefore
in the mechanics of respiration. There is a change in the tension of

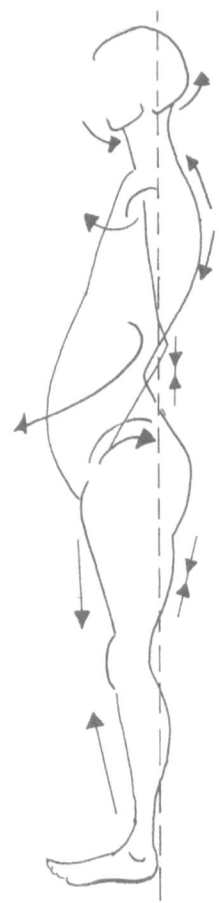

Figure 2. Postural compensation when the center of gravity falls
aft in the anterior posterior plane. Arrows indicate force direction.

the anterior abdominal wall, which will affect the development of the
optimum pressure differentials between the abdomen and thorax. It
effects the positional relationships of the visceral organs in the
abdomino-pelvic cavity, the suspension of which will reflex in mechanical tensions within the organs themselves, which will affect local
fluid fluctuations, capillary beds etc.

These effects can be clearly demonstrated if one compares two
examples of the extremes of compensation in the anterior/posterior
plane. Figure 3 illustrates these two postures, highlighting the
work load on the musculoskeletal system in each case.

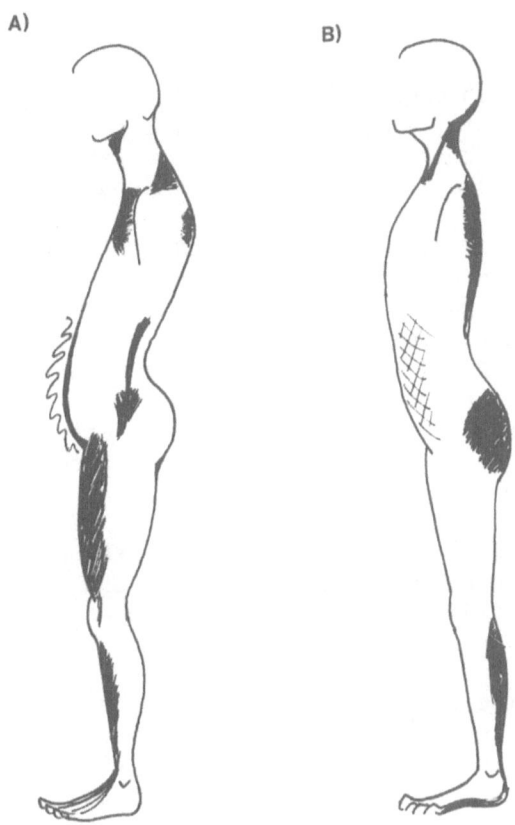

Figure 3. The work load on the musculoskeletal system in:
(A) anterior weight bearing - (B) posterior weight bearing.
Shaded areas represent areas of maximum stress.

     Let us now consider the gross effects of these mechanical stress
patterns on the homeostasis of the internal environment, in relation
to the basic requirements of every constituent cell in the body,
i.e.,: every cell requires a good supply of oxygen, nutrients and
trophins, and a means of removal of waste products. This necessitates
a good arterial blood supply and adequate drainage via the venous and
lymphatic systems. Therefore I am considering the effects of these
mechanical stress patterns on the body's cardio-respiratory mechanics.

   The chief areas which I am going to discuss are:

     The thorax and carriage of the shoulder girdle.

     The thoraco-abdominal cavity, the diaphragm.

     The abdomen and pelvis.

THE THORAX AND CARRIAGE OF THE SHOULDER GIRDLE

In the posterior weight bearing posture there is an overall
depression of the rib cage anteriorly. This is induced by an increase
in the dorsal kyphosis, exaggerated in the upper thorax by the pro-
traction and medial rotation of the shoulder girdle, holding the ribs
of the upper thorax in a relatively expirational position.

The center of gravity line falls through the thorax from a su-
perior/anterior to a posterior/inferior position, increasing the com-
pressive force through the thorax in an anterior/posterior plane.
Therefore, in respiration the least line of resistance is in the lower
lateral expansion of the rib cage, any anterior expansion pivoting
from the mid to lower dorsal spine. The maximum tension in the musculature is seen in the upper thorax. The net effect is an embarass-
ment to the usage of the upper thorax causing a relative lack of
exploitation of the upper lobes of the lungs.

In the anterior weight bearing posture, there is a tendency for
the whole thorax to be pulled upward and forward into an inspirational
position. The shoulders are retracted and externally rotated putting
maximum work load on the posterior musculature of the upper thorax.

The mass of the thorax lies anterior to the center of gravity
line, further forcing the ribs upwards and forwards and reducing the
dorsal kyphosis. This posture increases the work load of the total
extensor apparatus of the dorsal spine, which binds in the lower ribs,
decreasing their lateral expansive capacity. Therefore on inspiration
the whole of the thorax tends to move forward en masse, anteriorly and
superiorly.

THE THORACO-ABDOMINAL JUNCTION, THE DIAPHRAGM

Both these postures affect the tension suspension of the diaphragm, and therefore its potential superior/inferior excursion during
respiration.

In the posterior weight bearing posture, the mechanical stress
pattern of the thorax cause the diaphragm to be held in a relatively
expirational position at rest, i.e.: the domes of the diaphragm are
high. The crowding in around the thoraco-abdominal junction, greatly
reduces the potential descent and flattening of the domes of the
diaphragm during inspiration. This will have a direct effect on the
development of pressure differentials between the abdoment and thorax,
reducing the efficiency of the diaphragm, not only during respiration
but also in venous return.

In the anterior weight bearing posture the diaphragm is held under a constant degree of tension, therefore the domes are already slighty descended and flattened at rest. However, in contrast to above, there is no crowding in around the thoraco-abdominal junction, which is opening out posteriorly to anteriorly. Therefore utilisation of the diaphragm in this case in its respiratory and venous drainage function is more efficient than above.

THE ABDOMEN AND PELVIS

In the posterior weight bearing posture there is an increase in the lumbar lordosis, which hinges backwards into extension at approximately the level of lumbar vertebra 3 & 4. The pelvic bowl is relatively tilted backwards. This configuration puts the muscles of the anterior abdominal wall at a mechanical disadvantage, reducing their ability to maintain an optimum resting tone.

There is a marked shortening in the posterior intersegmental muscles of the lumbar spine, with the greatest work load thrown on the ilio-psoas muscle to prevent the total collapse of the lumbar vertebra into extension.

The net result is to place tension onto the more posteriorly positioned visceral organs of the abdomino-pelvic cavity. There is a potential limitation in the development of the optimum intra-abdominal pressure, causing a relative pooling of blood in the posterior/inferior visceral structures, i.e., the rectum and uterus.

In the anterior weight bearing, the center of gravity line falls relatively anterior to the 5th lumbar vertebra and sacrum, increasing the lumbo-sacral angle. The combined effect of this plus, the superior/anterior attitude of the thorax causes the pelvic bowl to tilt forwards. In this posture there is a relative increase in tension of both the anterior abdominal wall and the spinal extensors. Together with the diagphragmatic tension, this would tend to increase the intra-abdominal pressure, i.e., the maximum tension within the visceral organs falling on those in the most anterior/inferior position (the bladder).

CONCLUSION

In general terms there is a tendency for an overall 'sluggishness' in the cardio-respiratory mechanics in the posterior weight bearing posture, whereas, in the anterior weight bearing there is a tendency for an overall tension in the cardio-respiratory mechanics.

These factors in themselves would tend to predispose the individual with the posterior weight bearing posture to manifest first with congestive problems below the diaphragm. In contrast, the individual with the anterior weight bearing posture would tend to manifest first with tension and secondary congestive problems above the diaphragm.

*** NOTE ***

To illustrate a point, I chose the mechanical stress patterns in two classical compensations for alteration in the center of gravity in the anterior/posterior plane (See Figure 3). There are many variations, and furthermore, one must superimpose on these patterns any asymmetries in the lateral plane, and these will further localize the effects of these mechanical stress patterns on the cardio-respiratory mechanics unilaterally.

Ironically, the body is totally capable of absorbing these mechanical stresses. We adapt and compensate for them under the direction of the central nervous system, which, in response to the total afferent input from the external and internal environments, integrates the complex reflex mechanism of the spinal cord to meet the demands of the body as a whole. It is the sympathetic division of the autonomic nervous system that mediates between the visceral and somatic systems, directing the resources of the internal environment to the areas of the musculoskeletal system in proportion to demand at any given time (Korr, 1975).

Disease will not become manifest until the adaptive mechanisms of the organism fail, or they become inappropriate, disorganised or aberrant. However, when there is an area of increased muscle tone, hypertonia, that is maintained irrespective of physical activity, and which does not represent any actual tissue damage, the adaptive mechanisms of the musculoskeletal system are failing and becoming inappropriate.

THE LOCAL EFFECTS OF THE MUSCULOSKELETAL SYSTEM
ON THE HOMEOSTASIS OF THE INTERNAL ENVIRONMENT,
MEDIATED THROUGH THE SOMATIC AND SYMPATHETIC NERVOUS SYSTEMS

Somatic and sympathetic efferent pathways are multisegmental and under the control of the higher centers, but their activity is under the continual influence of the afferent input, which produces the reflex responses that are essential to our adaptation to the moment to moment changes that are occuring in a given system at any one time.

The maintenance of muscle tone is a function of the spinal reflex mechanisms of the muscle spindle (Becker, 1975). When there is an area of sustained muscle hypertonia which is inappropriate, the afferent volley has dominated and taken precedence over the vertically organized patterns they originally served. Figure 4 illustrates the neurological pathways that may be involved. This can be further complicated by the mechanical effect of the sustained hypertonia on the nerve itself. Nerve trunks are susceptable to stretch, angulation, constriction, torsion and ischemia. Anatomically they are most

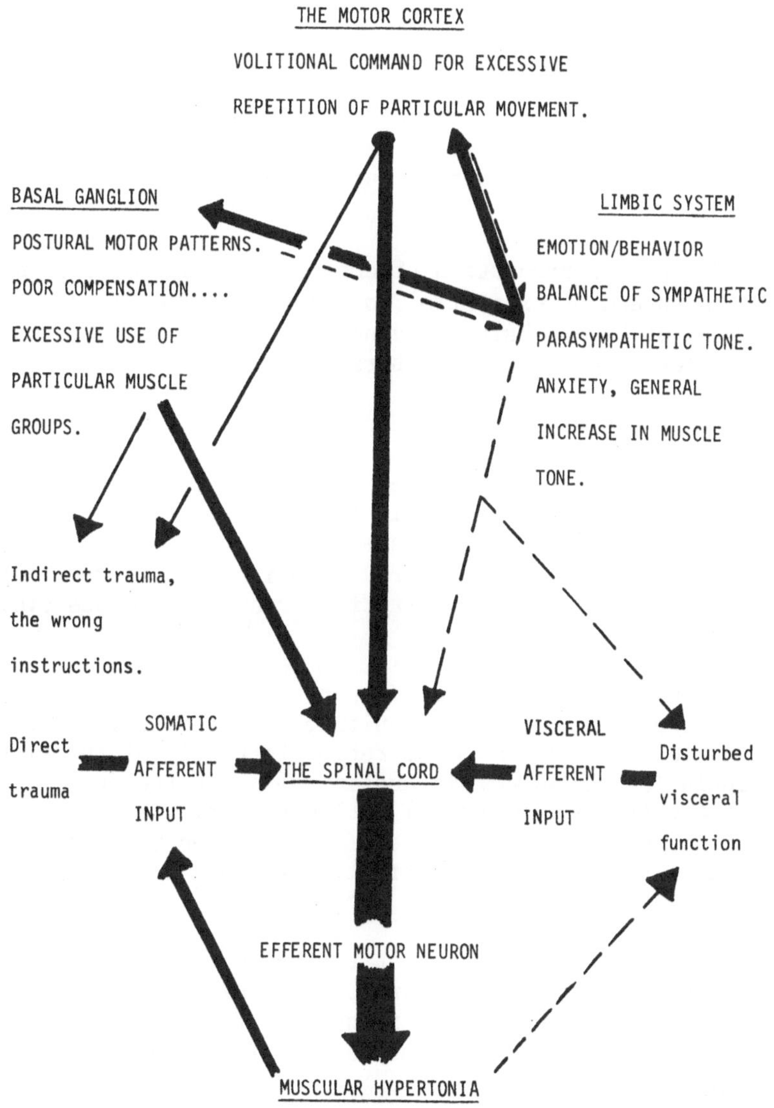

THE MOTOR CORTEX

VOLITIONAL COMMAND FOR EXCESSIVE

REPETITION OF PARTICULAR MOVEMENT.

BASAL GANGLION
POSTURAL MOTOR PATTERNS.
POOR COMPENSATION....
EXCESSIVE USE OF
PARTICULAR MUSCLE
GROUPS.

LIMBIC SYSTEM
EMOTION/BEHAVIOR
BALANCE OF SYMPATHETIC
PARASYMPATHETIC TONE.
ANXIETY, GENERAL
INCREASE IN MUSCLE
TONE.

Indirect trauma,
the wrong
instructions.

Direct
trauma

SOMATIC
AFFERENT
INPUT

THE SPINAL CORD

VISCERAL
AFFERENT
INPUT

Disturbed
visceral
function

EFFERENT MOTOR NEURON

MUSCULAR HYPERTONIA

Figure 4. Neurological Pathways that may be involved in producing an area of maintained hypertonia of a particular muscle or muscle group.

vulnerable to these phenomena at their receptor origin, when travel-
ling through bony canals, when transversing tissue interfaces, when
crossing bone and during their passage through skeletal muscle. The
initial effect of the mechanical strain, before it is sufficient in
intensity to cause actual partial or complete conduction block, is
disturbed neurophysiology, manifesting as an irritable focus of hy-
perexcitability (Sutherland, 1978). This hyper excitability expresses
itself in the following ways:

(a) The nerve becomes spontaneously active at the point of irri-
    tation, with the production of impulses that can travel in
    either direction; those travelling in the wrong direction
    known as antidromic impulses.

(b) There is amplification of impulse traffic along the nerve at
    the site of irritation, i.e., an impulse travelling down the
    nerve will set off a train of supernumerary impulses at this
    point.

(c) There can be the production of emphaptic transmission, where
    there is spread of excitation to adjacent nerve fibers, the
    phenomena of "Cross Talk" tending to occur from the larger to
    the smaller diameter nerve fibers.

The net result is a total disturbance of impulse transmission along
the course of the nerve. There is an increase in the afferent input
which does not truly represent the activity in the effector organs or
any external environmental stimuli; the activity in the efferent
neurons causing an effector response that does not truly represent the
central orders, the latter further disrupting the afferent input.
This confusion of nerve impulse traffic creates a vicious cycle
(Korr, 1976).

The increase in the afferent input has a summation effect on the
interneurones of that segment, resulting in sensitization of the
segment and lowering of the motor reflex threshold of the efferent
neuron population, both somatic and autonomic (Patterson, 1976). This
segment is facilitated and normal subliminal stimuli will cause an
effector response. When the subject is subjected to mechanical or
emotional stimuli, they are the first to respond and the last to
recover (Korr, Wright and Thomas, 1962).

The autonomic function is that expressed through the sympathetic
nervous system. These segments are showing all the signs of being in
a state of alarm; their response is not related to any rational demand, therefore the intrisic organization of the spinal cord has
become segmentally non-adaptive.

Every one of us has our own physiological alarm pattern which
manifests through these segmental reflex aberations. This was demon-

strated by the topographical representation of electrical skin resistance of the trunk. It was shown that every subject had areas of
lowered skin resistance; within this pattern of lowered skin resistance there were areas that showed a high incidence of reproducibility
over an extended period of time (Korr, 1958). These areas are correlated with the areas of the individual under greatest mechanical
strain.

THE ETIOLOGY OF INDIVIDUAL PHYSIOLOGICAL STRESS PATTERNS

When an individual is subjected to prolonged stress or sudden
stress overload, the neuronal responses are exaggerated at those
segments that are already in a state of alarm (Korr, 1978).

These patterns are the expression of the individual's total
adaptation and compensation to their internal and external environments, directed by all their past experiences, inherited characteristics and present circumstances.

Structure and function are totally interdependent. The mechanical stress patterns that develop as a result of the individual's
final compensation to the force of gravity in the erect posture are
reflections of the additive effects of the above, affecting the individual's physiological adaptive capacity through the cardio-respiratory and neurological pathways.

At the moment I am carrying out a Clinical Research Program the
aim of which is to show the significance of the individual's past
medical history to their present health status, emotional and physical. This program is still in its pilot stage, but already certain
characteristics are developing:

Subjects often exhibit somatic and visceral breakdown in the
local adaptive capacities, which appear to be precursors to
subsequent manifestations of disease of that area.

The onset of disease is usually associated with periods that
are stressful (mentally, emotionally or physically).

It is becoming apparent that the predisposition to the development of stress-related diseases occurs in response to emotional and
physical expression of compensation and adaptation of the individual
to musculoskeletal strain. This has a profound affect on the homeostasis of the internal environment through the cardio-respiratory mechanics and the inherent final integration of the body's adaptive mechanisms within the spinal cord with a subsequent influence upon health
and disease.

REFERENCES

Becker, R. F., 1975, The gamma system and its relationship to the
        development and maintenance of muscle tone, Journal of American
        Osteopathic Association, 75:170-187.
Gray, H., 1973, Locomotion and Posture, in: "Gray's Anatomy," R.
        Warwick and P.L. Williams, eds., Longman, Edinburgh.
Hoag, J. M., 1979, The Muscoluskeletal System: A major factor in
        maintaining homeostasis, Journal of American Osteopathic
        Association, 78:562-75.
Korr, I. M., 1975, The Sympathetic Nervous System as Mediator between
        the Somatic and Supportive Processes, in: "The Physiological
        Basis of Osteopathic Medicine," I. N. Kugelmass, ed.,
        Insight Publications.
Korr, I. M., 1976, The spinal cord as organizer of disease processes:
        Some preliminary perspectives, Journal of American Osteopathic
        Association, 76:35-89.
Korr, I. M., 1978, Sustained Sympathicotonia as a Factor in Disease,
        in: "The Nuerobiological Mechanisms in Manipulative Therapy,"
        I. M. Korr, ed., Plenum Press, New York.
Korr, I. M., Thomas, P. E., and Wright, H. M., 1958, Patterns of
        Electrical Skin Resistance in Man, Sonderbruck aus Band XVII,
        Heft 1-2.
Korr, I. M., Wright, H. M., and Thomas, P. E., 1962, Effects of
        Experimental Myofacial Insults on Cutaneous Patterns Of
        Sympathetic Activity in Man. Sonderbruck aus Band XVII, Heft 3.
Patterson, M. M., 1976, A Model Mechanism for Spinal Segmental
        Facilitation, Journal of American Osteopathic Association,
        76:62-72.
Spielberger, C., 1979, Stress and Strain, in: "Understanding Stress
        and Anxiety," L. Kristal, ed., Harper and Row, London.
Sutherland, S., 1978, Traumatized nerves, roots and ganglion:
        Musculoskeletal factors and neuropathic consequences, in: "The
        Nuerobiologic Mechanisms of Manipulation Therapy,"
        Plenum Press, New York.

THE MITCHELL METHOD OF PHYSIOLOGICAL RELAXATION

Laura Mitchell

Physiotherapist (Retired)
London, England

A stressed person is one in whom the normal physiology is displaced by a disturbance of homeostasis which may lead to pathology. When the body reacts to threat, the anterior pituitary gland secretes adrenocorticotrophic hormones, which in turn stimulate the thyroid gland to release thyroine and the medullae of the adrenal glands to secrete norepinephrine and epinephrine into the blood stream. The resulting sympathetic nervous system responses include:

<u>Increased</u>

1. Heart rate and blood pressure

2. Metabolism and oxygen consumption

3. Muscle work of selected muscle groups

4. Carbon dioxide

5. Breathing rate and depth

6. Dilation of bronchioles

7. Blood glucose and fatty acids

8. Sweating

<u>Decreased</u>

1. Peristalsis

2. Gastro-intestinal glandular activity

3. Blood supply to the skin

4. Kidney function

The approach to dealing with these stress responses is to focus upon the pattern of muscle work in order to break through this vicious circle, because sustained muscle work produces excess lactic acid which of itself can perpetuate a state of anxiety (Benson, 1976). Typical faulty muscle-use patterns include the following: face frowning, teeth clenched (possibly grinding), tongue on roof of mouth, shoulder girdles raised, arms adducted, elbows flexed, hands clenched. The breathing may be restricted with frequent deep gasps, or quick and shallow breaths. When sitting, the legs are often crossed or one leg may be wound round other. When standing, there may be much nervous movement. Every stressed person does not show all these signs; but the more stressed the individual is, the more signs are displayed. If this behavior is adopted for short periods of time accompanied by the other fight or flight changes, and the stimulus is discharged into positive action (rigorous exercise) then no harm may be done (Carruthers, 1974). In fact in some people stress acts as a welcome occasional stimulus. However, if the stress response is maintained chronically then there is the danger of pathology. It is the response to the stress situation, not the situation itself, which determines the result. "Health involves what is coming to be called 'The Whole Man'" (Wright, 1975).

If we can find a safe physiological path to change the stress muscular pattern into a pattern of muscular ease, then we will have a choice whether we wish to maintain the stress stimulus or to replace it with one of ease. All the other fight or flight physiological reactions then fade out if muscle relaxation is achieved.

PHYSIOLOGICAL LAWS GOVERNING MUSCLE PATTERNS

If we examine the muscle pattern laws closely we will find help and a clear way through to our objective. These are:

1. All muscular activity is developed as habitual patterns of movement (Knott and Voss, 1963).

2. All patterns are learned by repetition and stored in the sensory area of the brain (Guyton, 1972; O'Connell and Gardner, 1972).

3. Servo-motor area of the brain initiates the stimuli for activity and muscle work (Basmajian, 1967; Russell, 1975; Wells, 1971).

4. Sherrington's Law of Reciprocal Innervation states that when one group of muscles is working, its antagonist is inhibited (O'Connell and Gardner, 1972).

5. Position of joints and skin pressures register the body posture by afferent impulses to the brain rather than from muscles (Buchwald, 1967; Gardner and Osburn, 1973; Gowitbe and Milner, 1980).

The Mitchell method of physiological relaxation is the application of all the above laws so that any person in the stressed pattern of movement can change it, at will, to one of total relaxation by giving selected orders to himself and carrying them out. It took me from 1957 to 1963 to work out and test the method on hundreds of people of varying types so that it would be:

1. Simple to understand, perform and achieve success.

2. Easy to demonstrate using lay language that would apply in any position: lying, sitting, side lying, leaning forwards or backwards, standing.

3. Useful for full relaxation for rest or sleep, or to be used in parts of the body while working with other parts.

4. Suitable for men, women or children and for teaching by lay persons as well as professional therapists.

5. The instructions must be unambiguous e.g. the word 'Relax' could not be used as its meaning is not clear to some people.

TEACHING TO AN INDIVIDUAL OR A CLASS

The person may choose a position where he/she :

1. sits leaning back on a high backed chair with arm support,

2. sits leaning forward onto a table with pillows for arms and head to rest upon,

3. lying on the back on the floor with one pillow under the head, or

4. lying on the side with a pillow under head and atop knee.

The lighting in the room should be normal and no music is ever
used. The class discusses positions of strain, and general body awareness. The instructor demonstrates reciprocal innervation of flexors
and extensors of fingers. The following instructions are to be made:

1. To MOVE exactly as ordered.

2. To STOP the part moved and hold it.

3. To FEEL skin pressures and exact new joint position.

Upper Extremity Self Orders

Shoulders 'Down'.
"Pull your shoulders towards your feet" (The depressors of the
shoulder girdle are activated. Therefore the elevators are reciprocally relaxed.)
"Stop" (The depressors of the shoulder gridle relax).
"Feel" (Your neck is longer and your shoulders further away from
your ears.)

Elbows 'Out and Open'.
"Keep your arms touching their support and push your elbows
gently away from your sides. Open the angle at your elbow by pushing
your lower arm slightly away from your upper arm." (The abductors of
the shoulders and extensors of the elbows are active therefore the
adductors of the shoulders and flexors of the elbows are reciprocally
relaxed.)
"Stop" (The abductors of the shoulders and extensors of the
elbows relax).
"Feel" (The weight of the arm pressing the skin on the support
and that the elbows are away from the sides and open.)

The Hands 'Long and Supported'.
"Make your fingers and thumbs long. Stretch out and separate
your fingers and thumbs extending your wrists." (The extensors and
abductors of the fingers, thumbs, and wrists are active therefore the
flexors and adductors are reciprocally relaxed.)
"Stop." (The extensors and abductors relax - and the hands fall
onto the support.
"Feel." (The pads of the fingers are touching the support. The
thumbs are separated and heavy, and supported, fingers and thumbs are
long, not clenched.)

The sequence around the body is arms, legs, body, head, breathing, face, mind. This sequence may be altered once the technique has
been thoroughly practised. The orders for each area must never be
changed as they fulfill all the laws of body change and body awareness
for the purpose of obtaining relaxation at will.

List of self-orders for whole body

Shoulders:
"Pull your shoulders towards your feet."
"Stop."
"Feel your shoulders are further away from your ears."

Elbows:
"Elbows out and open."
"Stop."
"Feel your upper arms touching the support and away from your
  sides. Feel the open angle at the elbows.

Fingers and Thumbs:
"Fingers long and supported." The fingers and thumbs are stretched
  out 'long' with wrists extended.
"Stop." The fingers fall back on to the support.
"Feel the fingertips touching the support and the thumbs heavy."

Legs:
"Roll your thighs outwards."
"Stop."
"Feel your turned out legs."

Knees:
"Move your knees very gently if not comfortable."
"Stop."
"Feel your comfortable knees."

Feet:
"Push your feet away from your face, bending at the ankle."
"Stop."
"Feel your dangling feet."

Body:
"Push your body into the support."
"Stop."
"Feel your body lying on the support."

Head:
"Push your head into the support."
"Stop."
"Feel your head lying in the support (pillow)."

Breathing:
"Breathe in gently lifting your lower ribs upwards and
  outwards towards your armpits, and a slight bulging above
  your waist in front. Breathe out easily and feel the ribs
  fall back." Repeat once only (Last, 1978).

<u>Face</u>:
   "Keep your mouth closed and drag your jaw down."
   "Stop."
   "Feel your separated teeth. Place your tongue low in your
     mouth."
   "Close the eyes lowering the top eyelids only."
   "Stop."
   "Enjoy the darkness."
   "Smooth the forehead up into the hair, continue over the
     top of the head and down backwards."
   "Stop."
   "Feel the hair move."

(A full description of these sequences with pictures may be found in
Mitchell (1977).)

REFERENCES

Basmajian, J. V., 1967, "Muscles Alive," Williams and Watkins,
        Baltimore.
Basmajian, J. V., 1976, "Primary Anatomy," Williams and Wilkins,
        Baltimore.
Benson, H., 1975, "The Relaxation Response," Morrow, New York.
Buchwald, J. S., 1967, Proprioceptive Reflexes and posture, <u>American
        Journal of Physical Medicine</u>, Vol.46, pp.104-113.
Carruthers, W., 1974, "The Western Way of Death," Davis-Poyntor,
        London.
Gardner, W. D., and Osburn, W. A., 1973, "Structure of the Human
        Body," Saunders, Philadelphia.
Gowitbe, B. A., and Milner, M. 1980, "Understanding the Scientific
        Bases of Human Movement," 2nd edition, Williams and Wilkins,
        Baltimore.
Guyton, A. C., 1972, "Structure and Function of the Nervous System,"
        Saunders, Philadelphia.
Knott, M., and Voss, D., 1963, "Proprioceptive Neuromuscular
        Facilitation," Hoeber-Harper, New York.
Last, R. J., 1978, "Anatomy Regional and Applied," Churchill
        Livingstone, London and New York.
Mitchell, L., 1977, "Simple Relaxation," John Murray, London.
Mitchell, and Dale, B., 1980, "Simple Movement: The Why and How of
        Exercise," John Murray, London.
O'Connell, and Gardner, 1972, "Understanding Scientific Bases of Human
        Movement," Williams and Wilkins, Baltimore.
Russell, R., 1975, "Explaining the Brain," Oxford University Press,
        Oxford.
Wells, K. F., 1971, "Kinesiology," Saunders, Philadelphia.
Wright, H. B., 1975, "Executive Ease and Dis-ease," Gower Press,
        Epping, England.

INDEX

Acetylcholine, 117, 120
Acupuncture, 118, 119
Adrenalin, 117, 120
Agoraphobia, 10, 201
Alexithymic, 92
Alzheimers Disease, 120
Amitriptylene, 134
Amnesic, 92
Anger, 380
Angina, 165, 333
Angina Pectoris, 54
Angiography, 155, 156
Anorexic, 92
Anoxia, 153
Anti-depressants, 8
Antidromic, 396
Anxiety, 3, 18, 60, 62, 131
Arousal, 319
Arrythmias, 381
Arthritis, 7, 250
  rheumatoid, 125
Arthrosis, 53
Assertiveness, 162
Asthma, 250
Atherosclerosis, 225, 248
Autistic, 64
Auto-massage, 355
Autogenic Training, 11, 20, 160,
      170, 259-285, 299
Autonomic Response Specificity,
      93
Autonomy, 348

Backache, 8, 61
Beta-blocker, 150

Biofeedback, 20, 60, 75, 76,
      157, 165, 320, 333
  EMG, 9, 31
  thermal, 9
Bloodflow, cerebral, 22
Blood pressure, 93, 100-118,
      150, 158, 169, 228, 231,
      314, 353, 356, 379, 380
Brachialgia, 56
Buerger's disease, 165
Bulemic, 92
Bureaucratization, 245
Burnout, 288
Bursitis, 3

Cancer, 250, 332
  breast, 37
  cervical, 38
  lung, 36
Capsalsin, 119, 121
Carcinoma, bronchial, 37
Cardiometer, 310
Cartilage, 127
Catastrophizing, 92, 93
Catecholamines, 133, 157, 227,
      228, 235, 247, 249, 251,
      252, 315, 316
Causalgia, 154
Claudication, 165
Cognitive restructuring, 254
Cold pressor test, 93
Colitis, 333
Colon, spastic, 3
Congruence, 351
Cortex, 118